Finite Elements-based Optimization

Electromagnetic Product Design and Nondestructive Evaluation

Finite Elements-based Optimization

Electromagnetic Product Design and Nondestructive Evaluation

S. Ratnajeevan H. Hoole
Yovahn Yesuraiyan R. Hoole

CRC Press is an imprint of the
Taylor & Francis Group, an **informa** business

CRC Press
Taylor & Francis Group
6000 Broken Sound Parkway NW, Suite 300
Boca Raton, FL 33487-2742

Printed on acid-free paper

International Standard Book Number-13: 978-1-4987-5946-5 (Hardback)

Visit the Taylor & Francis Web site at
http://www.taylorandfrancis.com

and the CRC Press Web site at
http://www.crcpress.com

Printed and bound in Great Britain by
TJ International Ltd, Padstow, Cornwall

Contents

Preface

The finite element method is now a well-developed art. Research funding for developing methodology is therefore scarce. Funding for work that keeps our research engines oiled and ready to go, has necessarily shifted to applications of the method rather than to developing the method itself – in our case to investigating methods of shape optimization and nondestructive evaluation. Responsibility for developing methodology grows in small steps in companies developing professional finite element software and tends to include algorithmic development besides methodological.

This text was therefore developed for a graduate course combining students focused on

- Learning the finite element method *per se* together with shape optimization, and developing it to cover coupled field optimization;
- Their theses in other areas (such as nondestructive testing) where the method would be used, these being the largest group in the class;
- Computer Science students interested in the algorithmic aspects associated with the finite element method, such as GPU computation and mesh generation. Indeed, the finite element method which once began in engineering is now often classified under computer science in major research universities.

This then describes the coverage of this book that was designed for a semester-long graduate course at Rensselaer Polytechnic Institute and Michigan State University that was taught by S. Ratnajeevan H. Hoole. The book is light on homework assignments because it is rich in repeating demonstrative examples, which have been used in place of homework effectively.

This book covers the work done over the course of S. Ratnajeevan H. Hoole's career with hundreds of graduate students. The descriptions here owe much to the contributions of his doctoral students Srisivane Subramaniam, Konrad Weeber, T. Pham, Sabapathy Krishnakumar, Victor Karthik, and Sivamayam Sivasuthan, all Doctors of Philosophy now. Rather surprisingly, the work of undergraduates at Harvey Mudd College and University of Peradeniya as homework assignments and undergraduate projects was also turned into ISI-indexed papers, proving the point of the late Peter P. Silvester of McGill University that the work underlying this subject is just a little beyond high school mathematics, and can be mastered easily. Although many people think, mistakenly, that it is heavily mathematical, it is indeed mathematical but not heavily so, and any engineering graduate and most senior undergraduates can master the subject easily.

Finite element field computation is a subject we have enjoyed immensely, and we wish you the reader the same pleasures that we derived from its study, programming, and teaching.

S. Ratnajeevan H. Hoole
Jaffna, Sri Lanka

Yovahn Yesuraiyan R. Hoole
Urbana Champaign, Illinois

Acknowledgments

A book like this is developed over a lifetime of work. Much of it is from class notes. However, naturally, many of the results are the outcome of the doctoral work of students under my supervision. It is appropriate to acknowledge the following doctoral students who developed the work appearing in this book under my supervision:

1. Dr Srisivane Subramaniam
2. Dr Konrad Weeber
3. Dr T. Pham
4. Dr Sabapathy Krishnakumar
5. Dr Sivamayam Sivasuthan
6. Dr Victor Karthik

Other students also contributed by testing some of the ideas contained here through assigned homework – particularly some of the brilliant students I was privileged to teach at Harvey Mudd College and University of Peradeniya.

A big thank you to all of my students.

S. Ratnajeevan H. Hoole

Authors

S. Ratnajeevan H. Hoole, BSc (Eng), MSc, PhD (Carnegie Mellon), retired as a professor of electrical and computer engineering from Michigan State University in the United States. For his accomplishments in electromagnetic product synthesis the University of London awarded him its higher doctorate, the DSc (Eng) degree in 1993, and the IEEE elevated him to the grade of fellow in 1995 with the citation "For contributions to computational methods for design optimization of electrical devices." He is presently a life fellow.

His paper on using his inverse problem methods from design for nondestructive evaluation is widely cited, as is his paper on neural networks for the same purpose. These appear in *The IEEE Transactions on Magnetics* (1991 and 1993, respectively). He has authored six engineering texts published by Elsevier, another by Elsevier (now carried by Prentice Hall), Oxford, Cambridge (India), WIT Press, and CRC Press.

His research has been funded by the US Army's Army Tank Automotive Research Development Center (TARDEC), Northrop's B-2 Bomber Division, the National Science Foundation, NASA (Dryden and Pasadena), IBM, and other industry research groups.

Prof Hoole has been the vice chancellor of the University of Jaffna, Sri Lanka, and as member of the university grants commission, was responsible, along with six others, for the regulation of the administration and academic standards of all 15 Sri Lankan universities and their admissions and funding. He has contributed widely to the learned literature on Tamil studies and been a regular columnist in newspapers. Prof Hoole has been trained in human rights research and teaching at the René Cassin International Institute of Human Rights, Strasbourg, France, and has pioneered teaching human rights in the engineering curriculum. He received the coveted Gowri Gold Medal from India's IETE in 2015 for his work on professional ethics. His latest book *Ethics for Professionals: An Internationalist Human Rights Perspective* was published in 2018 by Cognella Press, San Diego, CA.

He was appointed a member of the three-member election commission in November 2015 for a five-year term by the President of Sri Lanka on the recommendation of the constitutional council and the federation of university teachers' associations known as FUTA under the Nineteenth Amendment to the Constitution of Sri Lanka.

Yovahn Yesuraiyan R. Hoole is a graduate student at the University of Illinois at Urbana Champaign. He holds a BS in Computer Science and a BA in Electrical Engineering from Rice University. He is currently working toward a doctorate in Electrical Engineering. His research interests are in Optimization, Machine Learning, and their applications to real world engineering problems.

1

Analysis versus Design through Synthesis

1.1 From Make-and-Test to Analysis and Now Synthesis

When the field of electromagnetism began and culminated in the landmark work of James Clark Maxwell, design problems were traditionally tackled in closed form – using symbols for variables like dimensions, permeability, permittivity, etc. – and electromagnetic fields were expressed in terms of these symbols by solving the governing equations. This is analysis and is to be distinguished from true design, namely synthesis. Nearly everything we teach in an undergraduate class is analysis.

In analysis, given the description of a device, we compute how it would perform (see Figure 1.1); for example, given a coil, the task of analysis would be computing its inductance (which is performance in this design sense). Real design usually starts with performance and poses the task of synthesizing the device that would yield the performance we want. That is starting from the right and going to the left of Figure 1.1.

Except when working with closed form (explicit expressions) of the solution in terms of algebraic parameters describing a system, there is no known one-step process of synthesis. Instead, as shown in Figure 1.2, we cyclically (i.e., iteratively) synthesize the design, getting ever closer to the final synthesis. We repeat our steps of analysis which we are comfortable with, but are limited by the closed form analysis tools we have, which in turn are limited to simple geometries like circles and rectangles for which closed form analysis works.

For realistic geometries, that are not confined to circles and rectangles, which enable closed form solutions, up to the 1960s, synthesis involved repeatedly undergoing "make-and-test," a long and costly process which is

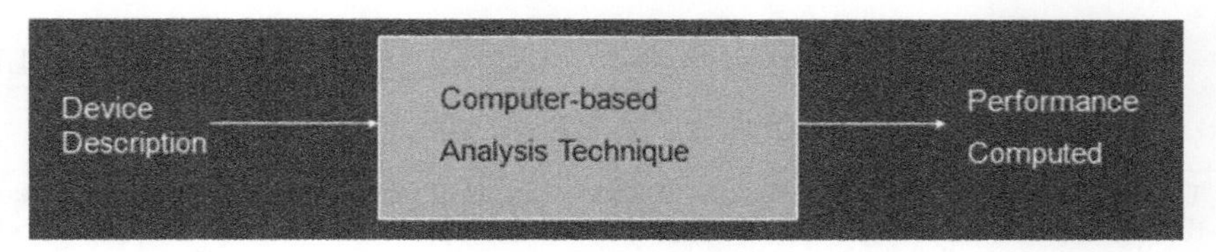

FIGURE 1.1
Analysis – the reverse of synthesis.

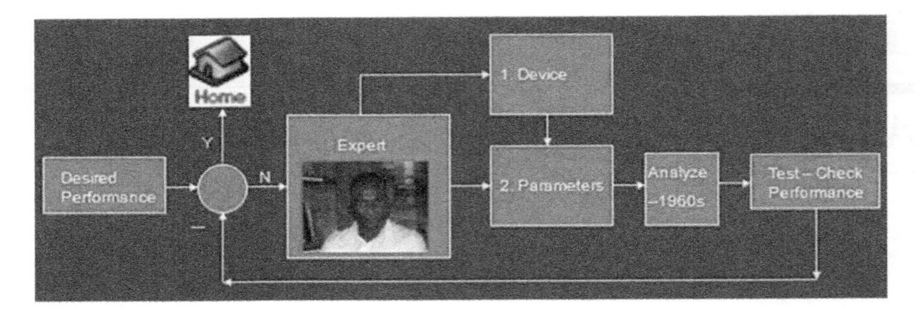

FIGURE 1.2
The field expert cyclically adjusts parameters of design as performance is measured by make-and-test or analysis from the 1960s.

still practiced in many industries. In testing (by measurements and later by analysis – see Figure 1.2) when the performance did not match the desired results, an expert altered the parameters of device description such as dimensions, materials, and sources like current and charge knowing as an expert how to alter them so as to nudge the performance of the current design toward that which is desired. The device was made, now for the new dimensions, and tested again. Real experts were few and costly. By the late 1960s when digital computers became more widely available, instead of making and testing by measurement of the resulting performance, it was through digital analysis that this same process was carried out with great savings in time. Numerical methods like the finite difference analysis and the later finite element analysis from around 1967 finally permitted real geometries that were not confined to circles and rectangles.

The most recent techniques (see Figure 1.3) employ software in place of the expert. Instead of the expert determining how to change parameters, artificial intelligence with rules and mathematical optimization methods is brought to bear.

The software uses a mathematical function defined to be at a minimum when the design object* is realized. Thus, if we want a certain flux density B in a region, we would define n measuring points i, shape the device according to the current values of the parameters of description $p_1, p_2, \ldots, p_n$, compute by analysis the B values at the n measuring points, and determine F with, for example, a least-square definition according to:

$$F(p_1, p_2, \ldots, p_n) = \sum_{\text{Measuring Point } i=1}^{n} \left(B_{i\text{-computed}} - B_{i\text{-desired}}\right)^2 \tag{1.1}$$

* The original word was object; and an objective was a military object like a hill to be taken through battle, according to G.H. Fowler. However, as the field of management developed and military terms like campaign, aggressive marketing, etc., were brought into everyday English, engineers followed suit. Here the word object is used because two words, each serving a different meaning, always makes a language more powerful.

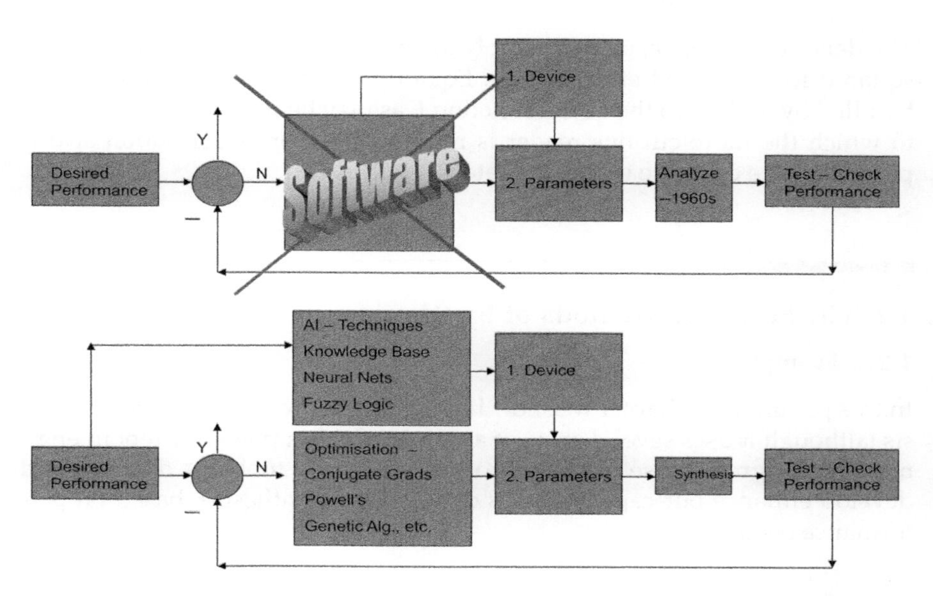

FIGURE 1.3
Software taking the place of the expert in modern design.

Here clearly F is a numerical function of $p_1, p_2, \ldots,$ and p_n in that we can only compute F as a number for a given set of ps and the value of F depends on the value of the ps. Thereafter, mathematical optimization techniques are employed to find the set of ps for which F is zero when the desired and computed values will match. Of course, if the object is impossible to realize, then F would never be zero but get as close to it as possible.

This book is about the optimization methods implemented through the process involving the computational directive:

$$\text{Minimize } F = F(p_1, p_2, \ldots, p_n),$$

subject to the inequality constraints $g_i(p_1, p_2, \ldots, p_n) \leq 0$ $i = 1, \ldots,$ the number of inequality constraints m and the equality constraints $h_i(p_1, p_2, \ldots, p_n) \leq 0$ $i = 1, \ldots,$ the number of equality constraints

When a function F has to be maximized, we simply replace it by $-F$ which has to be minimized. This keeps the formulation simple.

An inequality constraint for a geometric length p_2 can be as simple as $p_2 > 0$ since a physical length has to be positive. In the required form it is $-p_2 \leq 0$. Similarly, a length p_2 subject to constraints on both sides, $a \geq p_2 \geq b$ may be cast in the required form:

$$\frac{(p_2 - a)(p_2 - b)}{a - b} \leq 0 \tag{1.2}$$

The denominator $(a-b)$ is there simply to normalize values and may even be squared for improved effectiveness. Equality constraints may similarly be handled by adding to the object function F as may be found in the literature to which the more curious reader is referred for more information at this point until we examine it in detail later on in this text.

1.2 The Power of Methods of Synthesis

1.2.1 Examples

In this preambular chapter we show how the switch to synthesis from analysis (although it uses several steps of analysis) yields a powerful tool in engineering design. We use examples to show how the tools we describe and develop enhance our capabilities as engineers to synthesize, based on performance needs

1.2.2 Pole Shaping

A simple problem of pole shaping and its optimized result are shown in Figure 1.4. It is now a classic problem because it is easily understood and concerns the optimization of the shape of a pole face so as to produce a constant flux above it as in an electrical machine. The symmetric left half is shown. The initial geometry is on the top left and the initial design is on the top right. As to be expected, the expected performance is not achieved because of fringing (top right), the flux above the left corner of the pole face falls as shown in the lower graph on the lower-left corner.

As a result of optimization, we want the reluctance between the horizontal piece of iron at the top and the left corner of the pole to decrease so that the reduced reluctance might cause the flux there to increase and off-set the effect of fringing. The object function would be the sum of the square of the difference between the computed and expected B values at the measuring points:

$$F = \frac{1}{2}\sum_{i=1}^{m}(B_i - B_0)^2 \tag{1.3}$$

where m is the number of measuring points and the half is there for simple elegance so that it would vanish upon differentiation of F with respect to each B_i as required in optimization using derivatives.

This pole face–shaping problem is a problem with multiple solutions as a result of which early attempts at optimization in the 1960s (when the theory of mathematical optimizations was far from well developed) yielded jagged pole contours, the geometry rising and falling in shape, thereby making such design results mathematically valid but practically useless. With constraints

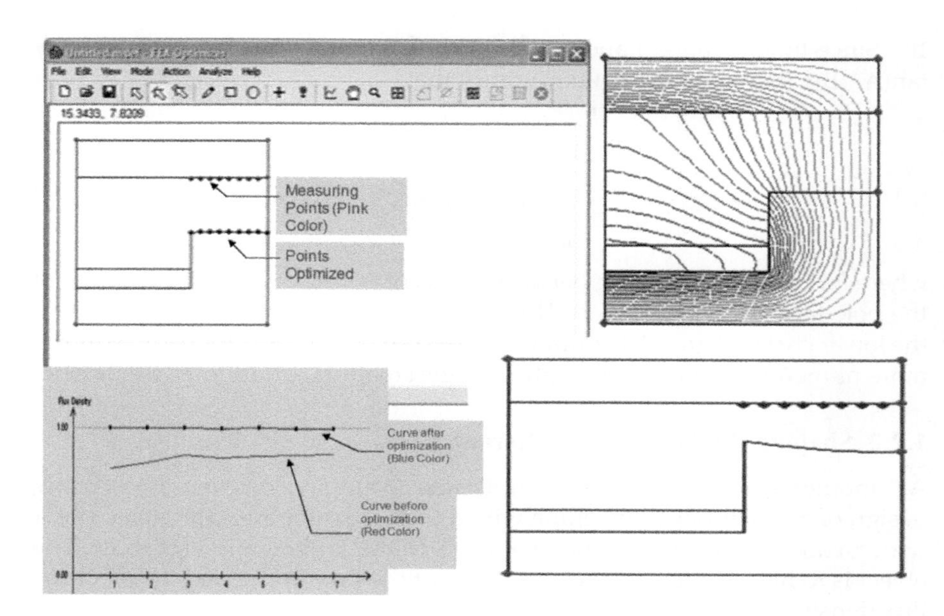

FIGURE 1.4
Pole face and its optimization for constant air-gap.

however, the situation is different. To have a smooth contour, we may say, as shown in Figure 1.5, that for adjacent points say 1, 2, and 3 whose heights p_1, p_2, and p_3 are being optimized,

$$p_1 \geq p_2 \geq p_3 \tag{1.4}$$

so as to ensure the absence of jagged contours going up and down. However, we have assumed *a priori* (i.e., from theoretical deductions rather than practical observations) knowledge from what we know of magnetic circuits that the final design would match this condition, as opposed to

$$p_1 \leq p_2 \leq p_3 \tag{1.5}$$

because we know from our understanding of the physics of magnetic reluctance that the left edge needs to rise.

To make the process general, we may impose something like this: that the difference between θ_1 and θ_2 of Figure 1.5 be less than some tolerance, say

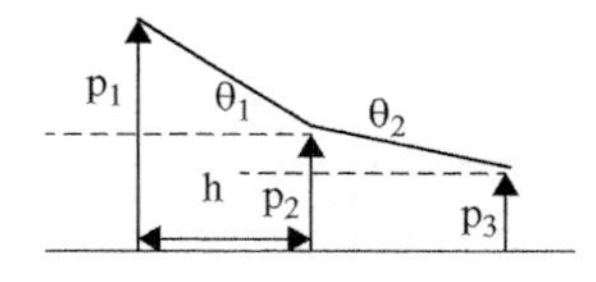

FIGURE 1.5
Smoothing the pole face contour through constraints.

18°. Since tan(θ) = tan(−θ) and it falls with θ, we may say, from the identity tan(A − B) = (tanA − tanB)/(1 + tanA.tanB) that

$$\frac{\frac{p_1 - p_2}{h} - \frac{p_2 - p_3}{h}}{1 + \frac{(p_1 - p_2)}{h}\frac{(p_2 - p_3)}{h}} \leq \tan 18^{\circ} \tag{1.6}$$

where h is the equal spacing between adjacent points on which the height of the pole-face p was optimized. This is what was used to obtain the result in the lower part of Figure 1.4 as our optimized solution. For smoother solutions, more numerous subdivisions with an angle smaller than 18° may be chosen.

1.2.3 Shaping the Rotor of an Alternator

As another similar but more complicated example, consider the starting design of a synchronous machine with a rotating rotor and stationary stator with an air-gap in between the two. The left-half is shown in Figure 1.6. The object is to have a sinusoidal distribution of the flux from E to F (peak of the flux density at E and zero at F). Constraints are that:

a) The geometric parameters be positive and less than a constant in the air-gap so that during machine vibrations the rotor does not hit the stator;

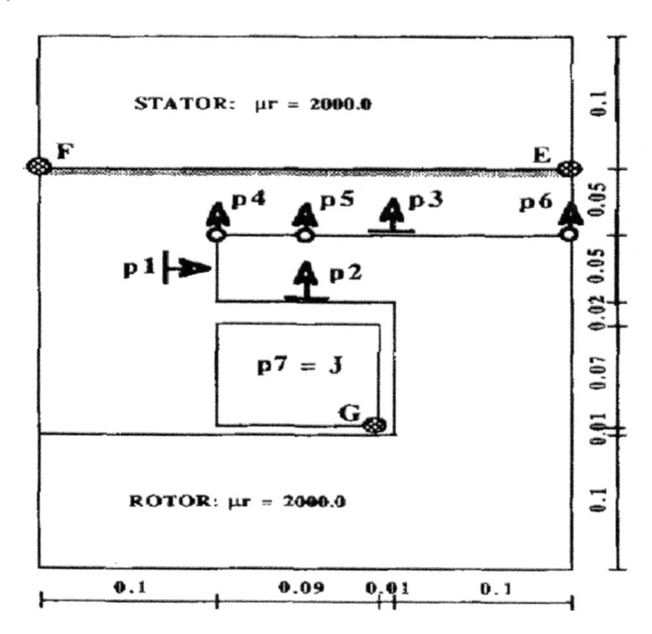

FIGURE 1.6
Starting design of a parameterized synchronous machine.

b) The current density p_7 in the copper windings be less than what copper can withstand, say 2 to 4 A/mm^2; and

c) The leakage flux, the flux going up through the rotor and bending back to the rotor without permeating the stator to be useful by crossing the windings, is less than a desired percentage of the total flux.

Figure 1.7 shows the final optimized shape – that is the design – corresponding to the constraints Copper Current Density $J < 4$ A/mm^2, Air-gap < 2 mm, and Leakage Flux $< 20\%$ of Main Flux.

1.2.4 Shaping a Shield

Another application needing a slightly different formulation – showing that the formulation of the optimization problem is critical to the validity of the result – is that of determining the shape of a copper shield so as to keep flux below a minimum value B_0 in areas to be shielded. The study arose because the R&D Labs of Southern California Edison had power cables running around them, permanently magnetizing the reinforcement. As a result, computer screens were showing distorted images.

Now, if we formulated the optimization problems as having $B = B_0$ at selected points as the object, the process would lead to that result. However, it must be realized that without the shield the flux density would have been below B_0 at many places and above B_0 at others. The shield is meant to correct B only in places where it would be above B_0, whereas the course of formulation suggested would make $B = B_0$ at all measuring points. The way out is to reformulate the problem as one of minimizing the volume of the shield. In two-dimensions this would be the sum of the areas of all the finite element triangles (described later in Chapter 2, but for now to be understood as the area of numerical analysis where the geometry is divided into triangular

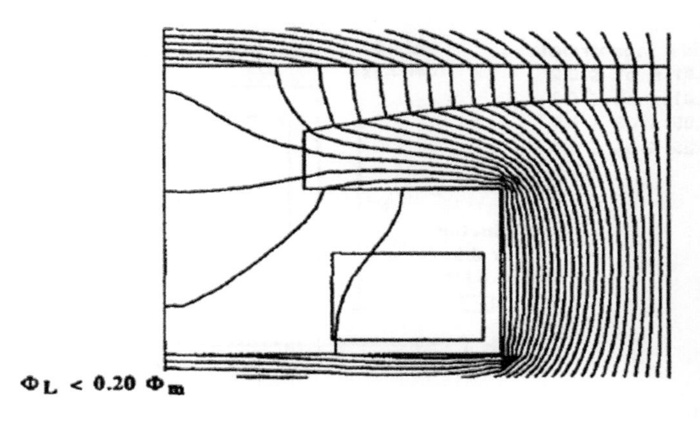

FIGURE 1.7
The final optimized geometry.

shapes). And $B < B_0$ would become a constraint at all measuring points. That is, the optimization directive is:

Minimize F = Volume of Shield $(p_1, p_2, \ldots, p_n) = \sum_i$ Area of Element i on the shield,
subject to the inequality constraints $g_i(p_1, p_2, \ldots, p_n) = B_i - B_0 \leq 0$ for $i = 1, \ldots,$ the number of measuring points m and the additional constraints $g_i(p_1, p_2, \ldots, p_n) \leq 0$ $i = m+1, \ldots,$ the number of additional constraints l

These additional constraints would impose the requirement that all lengths be positive to avoid the mathematically valid but physically invalid negative values for physical dimensions.

This is now an eddy current problem because in shielding structures we are dealing with eddy currents governed by

$$-\frac{1}{\mu}\nabla^2 A = J - j\omega\sigma A \tag{1.7}$$

where μ is the magnetic permeability, σ is the conductivity, ω is the angular frequency, A is the magnetic vector potential, j is the imaginary number $\sqrt{-1}$, and J is the current density. Solving these equations is dealt with at the end of the next chapter, and the equations are derived from Hoole (1989a, 1989b).

Figure 1.8 shows the starting configuration prior to optimization and shows the measuring points, the parameters for optimization, and the constraints on them in the symmetric upper-half of the system. This problem can be analyzed for, one, when the shield is ungrounded and, two, when it is grounded. When it is ungrounded the currents have little freedom to flow since they must sum to zero – that is current flowing into one part of the shield must come out in another since there is no path to ground.

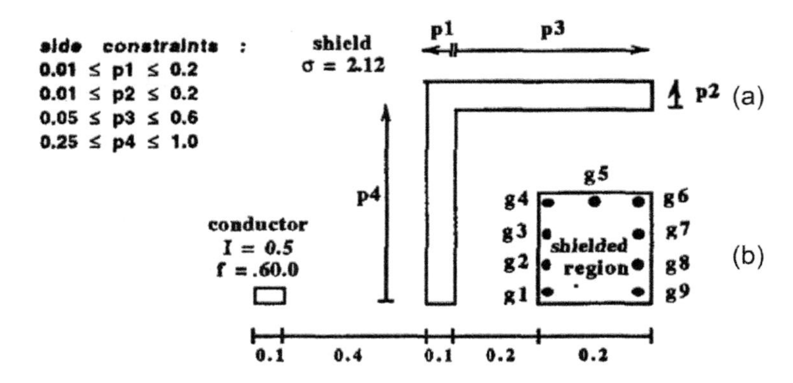

FIGURE 1.8
Starting configuration of shield: a) p_1 to p_4: Variables to be optimized b) g_1 to g_9: Measuring points.

This requirement is accomplished by an additional constraint on the current. The second grounded situation allows currents to flow as necessary to accomplish our design objects. The results are shown in Figure 1.9 for the grounded shield, Figure 1.10 for the ungrounded shield, and Figure 1.11 shows the fluxes at the nine measuring points for the starting configuration and the two final designs.

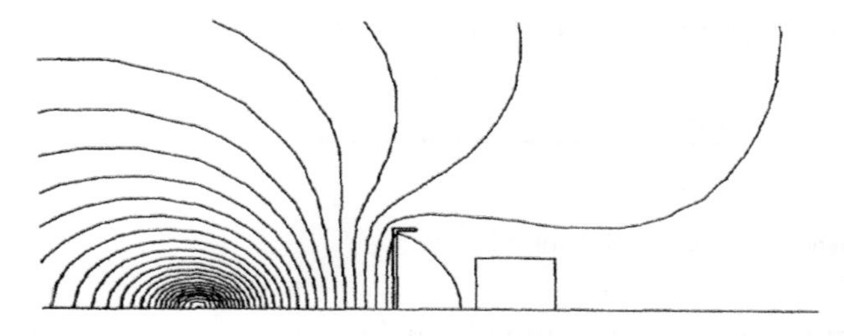

FIGURE 1.9
The field lines (where fluxes flow) for the optimized grounded shield.

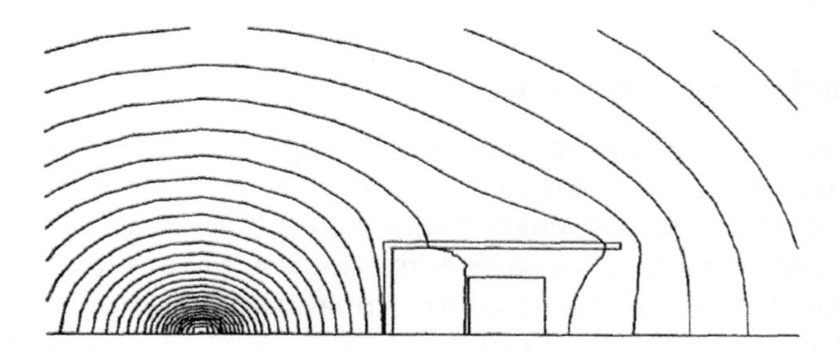

FIGURE 1.10
The field lines for the optimized ungrounded shield.

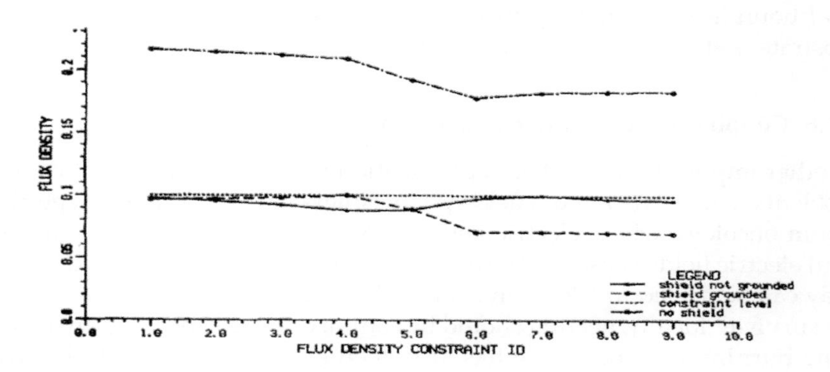

FIGURE 1.11
The initial flux densities at measuring points for the starting design and two final designs

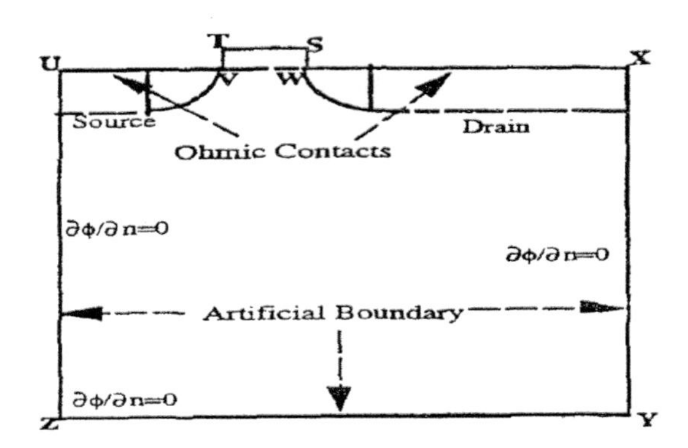

FIGURE 1.12
An npn transistor miniaturized by optimization.

The effectiveness of a grounded shield, something well known to old-fashioned practitioners, is readily apparent here. This experience shows how young engineers without experience can be greatly aided by these new software tools, which can take over the role of experts.

1.2.5 Miniaturizing a Transistor

The same concepts as for the shield may be applied to the miniaturization of an npn transistor. We want the transistor to be as small as possible while keeping the electric-field strength below a threshold so as to avoid electric breakdown of the material as parts of the transistor at different voltages are brought close together in the process of minaturization.

Figure 1.12 shows the final design of such an npn transistor optimized for minimum volume subject to all electric fields being below a specified breakdown strength of the material used so as to avoid breakdown. (The governing equation is nonlinear and is not discussed here). The three artificial boundaries mimic far away conditions at zero potential because the substrate is effectively an open boundary.

1.2.6 Coupled Field Problems: Electroheat

Another important area where optimization is necessary is that of coupled problems where two or more field systems interact. An example is hyperthermia in oncology where electric and thermal fields interact as Joule heating from electric fields is used to burn off cancerous tissue. It is known that if the body can be heated to 117°F cancerous cells tend to die off while healthy tissue survives. Ideally, heating should be confined to the cancerous areas since some burning of good tissue has also been experienced. The design trick in hyperthermia is to ensure that only the diseased tissue is heated while areas of healthy tissue are kept at safe temperatures through mathematical constraints. Other applications are in electric-machine design where we do not want overheating,

In hyperthermia treatment, electrodes carrying AC currents outside the human body induce E-fields in the torso or limbs, for example, where the cancer is. This E-field must be computed by solving the diffusion equation, which we have already encountered. This alternating E-field, because of the conductive nature of human tissue induces Joule heating at power density $q = 0.5\sigma_e E^2$, where σ_e is the electric conductivity in the body (different for different body parts such as meat, fat, bones, blood, etc.). This sources a second field problem, a Poissonian problem for temperature T:

$$-\sigma_t \nabla^2 T = q \tag{1.8}$$

where σ_t is the thermal conductivity of the various materials (see Figure 1.13).

The complications are that the parameters of description are in the electrical system and the object function in the thermal system. Further, as heating occurs, the electric and thermal conductivities will change, making each stage of solving the coupled electrothermal problem iterative because it is nonlinear. And then, at each stage of the optimization iterations, each electrothermal iterative process will need to be repeated. This has implications for the method of optimization we need to adopt but we will not detain ourselves here for such an esoteric subject beyond the scope of this book. We note that for more accurate modeling, allowing for blood flow also must be worked in to reflect the heat carried off by the blood.

To demonstrate that it can be done, we took a rectangular conductor. As eddy currents flow in it and heat it, the equi-temperature contours outside it,

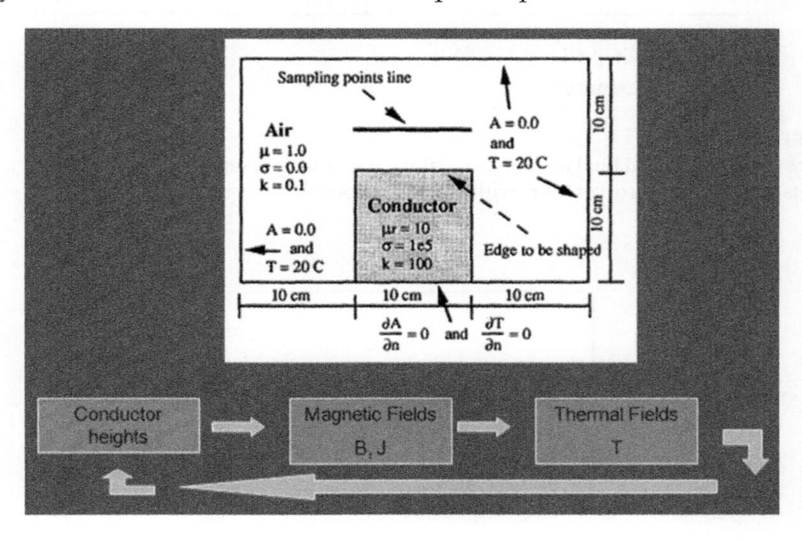

FIGURE 1.13
Optimizing temperature in a coupled electro-heating problem.

we expect, will be circles slightly distorted near the corners of the conductor. Thus, at a line just above the conductor and along a line parallel to an edge of the conductor the temperature will not be constant. We will see how it can be made constant by shaping the starting rectangular configuration of the conductor. The initial design is shown in Figure 1.14 where the conductor is a rectangle. The final design and the temperatures along the measuring points for the initial and final designs are presented in Figure 1.15.

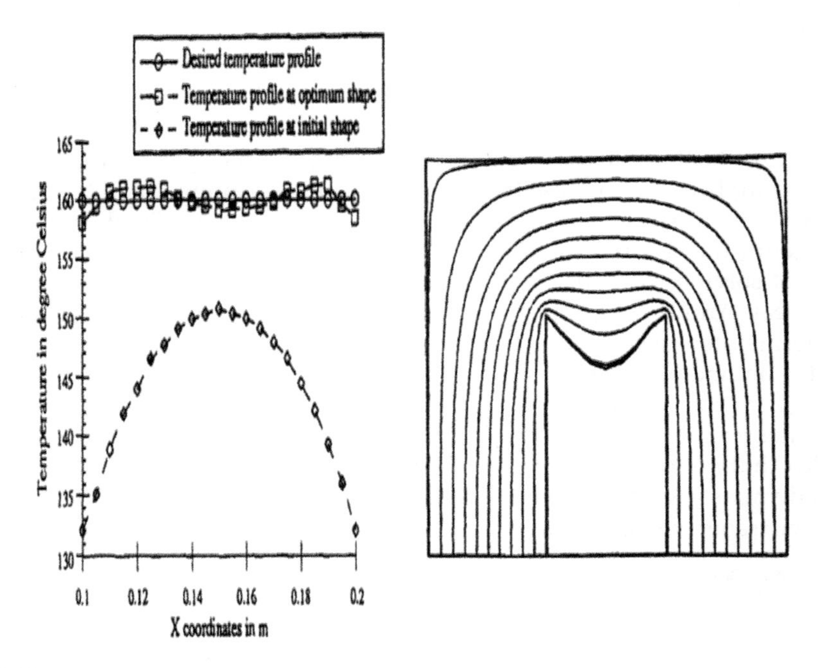

FIGURE 1.14
Graphs of the initial and final temperatures along the measurement points and the final design with its equi-temperature contours with a flat line along sampling points.

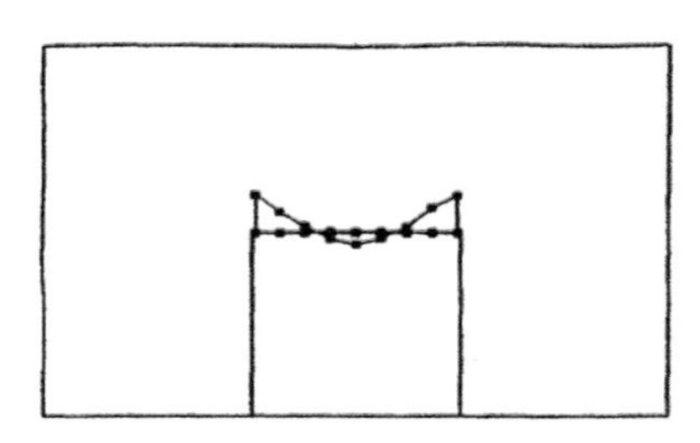

FIGURE 1.15
Initial and final designed shapes of the conductor.

1.2.7 Nondestructive Evaluation

As a final note on this section on design, it behooves us to point out that the way the design problem is set up, it is ideally suited for nondestructive testing as well. The design problem we have posed is: To yield such and such a performance, what is the shape of the system? Here if we replace performance with measurements, the slightly altered question becomes: To give such and such measured field values, what should the shape of the system be? If we measure the fields outside a system, the synthesized shape to yield those measurements would show any defects (if any) in the system around which measurements were made; and this would be with only exterior measurements to determine defects in inaccessible regions!

As a demonstration, Figure 1.16* shows a sensor with wheels and motors which is rapidly moved through a gas pipeline being tested for defects. The sensor measures the fields caused by currents flowing within it. The inset in Figure 1.16 shows the magnetic fields flowing from the sensor being altered by a defect in the pipe.

Figure 1.17 shows the computational process. We note that in place of "Desired Performance" we have "Measured Field." What matches this is the "Predicted Defect Profile" in place of the "Desired Design." Figure 1.18 shows the predicted profiles matching closely the measured profiles. Computationally our actual input is not the measured shape of defect, but the output fields from the defect. The actual shapes were measured to demonstrate the power of the method.

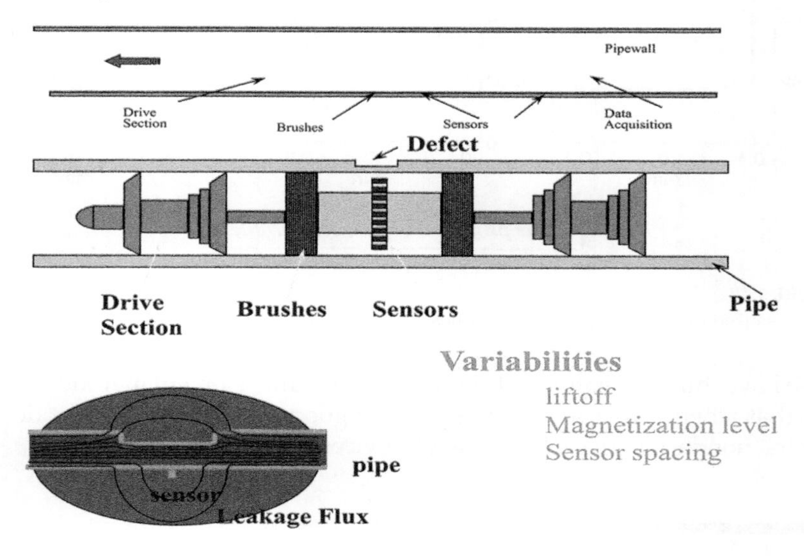

FIGURE 1.16
Gas pipeline with a sensor moved through it at speed.

* Figures 1.16 through 1.18 are courtesy of Prof Lalita Udpa of Michigan State University.

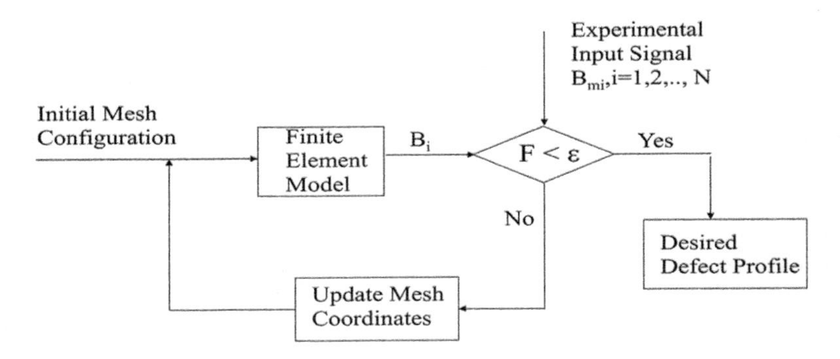

FIGURE 1.17
The computational process to determine defect profile instead of design.

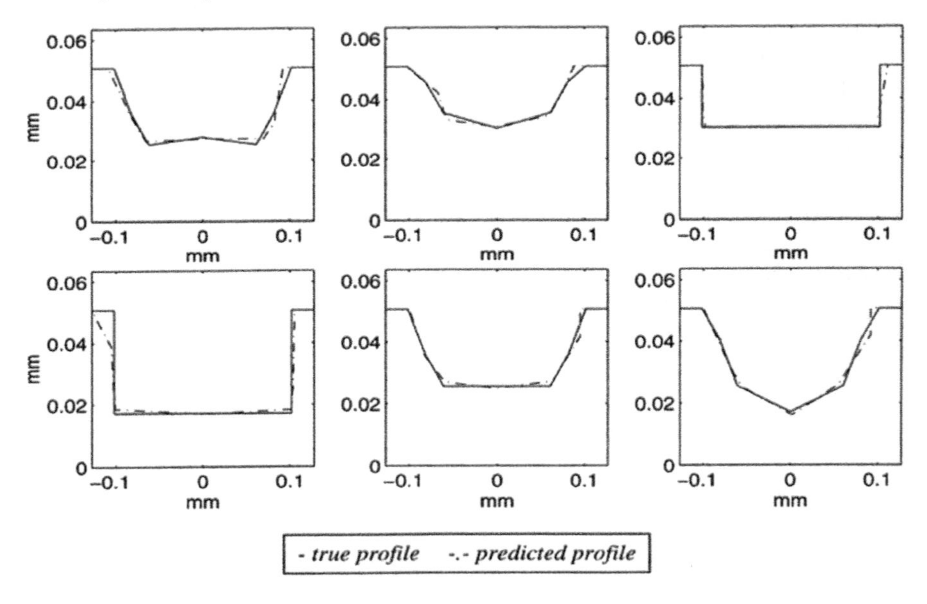

FIGURE 1.18
The measured and predicted defect profiles.

We have thus presented, in this chapter, the finite element method for analysis that when properly recast, is equally good for the synthesis of designs and the nondestructive evaluation of devices.

1.3 What This Book Is About

The preceding section has shown the role of synthesis – real design rather than the analysis that passes for synthesis – and how it can aid in meeting design goals. The process involves the formulation of the design goals as an optimization problem and then doing iterative steps of numerical analysis until the design goal is realized.

How to do this ourselves is what this book is about. So as to understand the principles of synthesis, we will confine ourselves for the most part to simple two-dimensional problems and go through the finite element method (the most powerful of the numerical methods of analysis), optimization methods, and how these two are put together for real synthesis. While outwardly it will be one program doing the synthesis, and therefore may be called one-step synthesis, as we have noted, that one-step consists of several analysis steps.

To broaden our understanding of synthesis and give the reader intellectual independence to innovate, we will look at wider issues like the use of symbolic algebra, the application of new equation types, and so on. We will rely on MATLAB, although many in advanced research or dealing with large problems, will be able to write their own programs in more powerful languages like C++.

At the end of this course of study, readers, it is hoped, will have their own suite of synthesis programs and the intellectual wherewithal to tackle new problems not directly broached in this text but whose principles have been laid out.

2

Analysis in Electromagnetic Product Design

2.1 Numerical Methods

We have noted in the previous chapter that the process of synthesis takes many iterative steps of analysis in the optimization cycle. Therefore, we will examine the analysis process in this chapter. Insofar as analysis is concerned, it needs to be broad in terms of applicability and therefore we will focus on the widely applicable numerical methods, which are not confined to circles and rectangles as explicit solutions almost always are.

Numerical methods involve integral as well as differential methods – that is those solving integral and differential forms, respectively, of the equations governing the electromagnetic fields we seek so as to determine how the electromagnetic device under analysis and synthesis behaves. We will, for reasons adduced later, deal exclusively with differential equations and choose between the finite element and finite-difference methods of solving the differential forms of the equations describing electromagnetic fields, confining ourselves to the finite element method.

Now to the equations defining the key electromagnetic quantities. Electromagnetic products are of many sorts – actuators, electric motors, waveguides, etc. They are all governed by Maxwell's equations, which define the electric field strength $\bar{E}$, electric flux density $\bar{D}$, magnetic field strength $\bar{H}$, and magnetic flux density $\bar{B}$.

$$\nabla \times \bar{E} = -\frac{\partial \bar{B}}{\partial t} \tag{2.1}$$

$$\nabla \times \bar{H} = J + \frac{\partial \bar{D}}{\partial t} \tag{2.2}$$

$$\nabla \cdot \bar{D} = \rho \tag{2.3}$$

$$\nabla \cdot \bar{B} = 0 \tag{2.4}$$

where ρ is the electric charge density and the constitutive material properties, permittivity ϵ and permeability μ relate the field intensities to the flux densities

$$\bar{D} = \epsilon \bar{E} \tag{2.5}$$

$$\bar{B} = \mu \bar{H} \tag{2.6}$$

This book assumes the familiarity of the reader with the basic concepts of electromagnetics.

Any analysis of how a device will perform, a predictive exercise rather than a design exercise on which we will see more in the next chapter, means solving these four Maxwell equations. Once that is done, for example the magnetic flux ψ through a coil, obtained by integration of the flux density over the coil cross section S,

$$\psi = \oint\!\!\oint_S \bar{B} \cdot \overline{dS} \tag{2.7}$$

may be computed. This will yield the inductance of the coil, among other things. Other quantities like force may also be computed once we know the electric and magnetic field intensities. To these ends, we must know what the electric and magnetic fields in a device are.

To be clear, these are analytical exercises. Given a particular postulated design, our analysis informs us how it will behave and whether it will perform as we expect – in such matters as inductance, capacitance, and force produced. Indeed, for such purposes, even in devices subject to dynamics, that is, transient performance, it often suffices to do a static analysis. For example, in an electric machine subject to transient-fault conditions like a sudden short, we may use coupled inductances to determine the fault currents. The inductances themselves may be computed using steady DC conditions. After setting up a coil system with N turns using steady DC conditions sourced by a current I and computing ψ for that current I, the inductance L would be given by

$$L = \frac{N\psi(I)}{I} \tag{2.8}$$

This inductance may then be used for analyzing transient behavior using circuit concepts.

However, a typical engineering electromagnetics text would give various closed form (i.e., mathematically explicit) solutions for simple geometries that are infinitely long or circular or spherical. An occasional rectangle will be thrown in. The reality, however, is that real engineering devices are of far more complex shape than these. To determine the fields in such devices, no simplifying assumptions on geometry are permissible.

The only way out is to bring in numerical solutions which for realistic devices are practicable only on digital computers. The most flexible and powerful of these numerical methods is the finite element method (although those whose careers are tied-up to the competing boundary-integral method will disagree). The finite element method involves:

i. Discretization of the geometry into subdomains (called elements) with nodes. Although more mathematically general forms are also occasionally seen, they do not need to detain us here.
ii. A trial function which expresses the field within each element in terms of the unknown nodal values (this is a simplified description).
iii. A second-order optimization function, often energy-based, that is expressed in terms of these unknown nodal values. Thus, the optimization function is a quadratic in the unknowns which when minimized with respect to the unknowns, yields linear equations relating the unknown quantities. This condition therefore yields a matrix equation, ideally symmetric and sparse to make the solution of the matrix equation speedier than otherwise.
iv. A matrix-solution process for that equation from the previous step.
v. Visualization of the fields everywhere using the nodal values and the trial function.

In its most common form, in electromagnetics, we solve some form of the Poisson equation. In electrostatics, since Equation (2.1) becomes under static consitions with $\partial / \partial t = 0$:

$$\nabla \times \bar{E} = 0. \tag{2.9}$$

Comparing and combining this with the vector identity

$$\nabla \times \nabla \varphi = 0, \tag{2.10}$$

we assert that the static electric field $\bar{E}$ is the gradient of an electric potential φ:

$$\bar{E} = -\nabla \varphi. \tag{2.11}$$

Substituting in Equation (2.3) using Equation (2.5) we get the Poisson equation for the electrostatic potential:

$$\nabla \cdot (-\epsilon \nabla \varphi) = -\epsilon \nabla^2 \varphi = -\epsilon \left[\frac{\partial^2 \varphi}{\partial x^2} + \frac{\partial^2 \varphi}{\partial y^2} \right] = \rho. \tag{2.12}$$

It is this equation that we often solve in many analysis problems. We note that the term $\partial^2\varphi/\partial z^2$ would get added within the square brackets if this were a three-dimensional problem,

Similarly, examining Equation (2.4) with the vector identity

$$\nabla \cdot \nabla \times \bar{A} = 0, \tag{2.13}$$

we get

$$\bar{B} = \nabla \times \bar{A} \tag{2.14}$$

yielding, in combination with Equation (2.1), for static magnetism,

$$\nabla \times \frac{1}{\mu} \nabla \times \bar{A} = \bar{J}. \tag{2.15}$$

Comparing and combining with the vector identity

$$\nabla \times \nabla \times \bar{A} = \nabla\left(\nabla \cdot \bar{A}\right) - \nabla^2 \bar{A}, \tag{2.16}$$

we get the vector form of the Poisson equation when we set $\nabla \cdot \bar{A} = 0$ as we are free to do by rules from vector analysis and the conditions for uniqueness:

$$-\frac{1}{\mu} \nabla^2 \bar{A} = \bar{J}. \tag{2.17}$$

In two-dimensional problems where both the current-density vector and the magnetic vector potential have only a single component, usually taken to be the z-component into the paper where the infinitely long device is depicted in its uniform cross section on the xy-plane, we work with the single component magnitudes of $\bar{A}$ and $\bar{J}$

$$-\frac{1}{\mu} \nabla^2 A = J, \tag{2.18}$$

which we note is again a Poisson equation. So, this too amounts to solving the Poisson equation for the magnitude of the vector potential.

For most of the rest of this text we will deal with this two-dimensional Poisson equation whether in magnetics or electrostatics. There will be occasion to examine more complex forms of this applicable in three-dimensions, for eddy current problems, etc.

2.2 Numerical Approximations versus Exact Methods

In today's context, numerical methods – as distinguished from explicit closed form classical methods, hereinafter referred to as classical methods – constitute the core of computer-aided design. Classical methods rather as

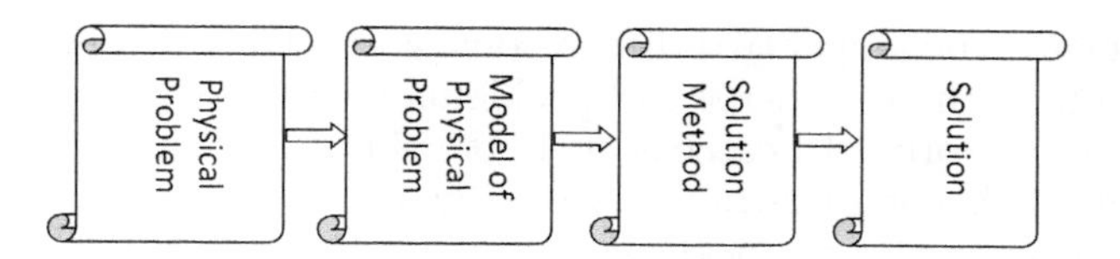

FIGURE 2.1
The solution process.

an almost sleight of hand or indeed of mind, have been referred to as exact methods while numerical methods are called approximate methods. These claims upon closer scrutiny do not stand up.

Classical methods are limited by the mathematics by which the solutions are expressed. As shown in Figure 2.1, for every problem, we first choose and select a model, and thereupon solve the problem posed by that model chosen rather than the actual problem. When the solution is not satisfactory, we reexamine the model with a view to improving it. For example, in the mechanics of projectiles our model may be an equation based on Newton's laws. This works most of the time. But when it comes to space travel at high speeds, the solutions may not match measurements and we will incorporate Einstein's laws to improve the model with better accuracy. Likewise, in electromagnetics, in classical electrodynamics modeling, say, of a transmission line, we assume symmetry in the direction of the line so that the problem is two-dimensional. Thereafter, we may further assume the transmission line to be circular in cross section. These assumptions permit the reduction of the problem to a simple one-dimensional differential equation of the form

$$-\frac{d\varphi}{dr} = \frac{1}{2\pi\epsilon}\frac{1}{r} \tag{2.19}$$

This model gives us a solution that is as exact only as much as the model is exact:

$$\varphi = -\frac{1}{2\pi\epsilon}\ln\frac{r}{r_0} \tag{2.20}$$

Should we need a more accurate solution, we need to improve the model. For instance, we may model the earth above which this infinitely long line runs by considering the image of the line. This makes the problem two-dimensional. With superposition, this too can be handled by our model and the corresponding solution method yielding an explicit solution in terms of the symbols for line dimensions. Should this also prove inadequate for purposes of the required accuracy, we go further and allow for a) a larger radius for the line that would invalidate the assumption of a point charge on the line, and must allow for charges redistributing on the line cross section and b) nearby transmission towers and their equipment like insulator strings, and the sag of the wire from tower to tower. At this point the model totally breaks down as the problem is no longer two-dimensional and is truly three-dimensional. Although many three-dimensional problems have classical solutions, they rely upon some form of neatness such as spherical shapes, homogeneous systems, etc. This complex-transmission line and its model of the line as it really is, has no known classical solution in explicit form.

At this point a numerical solution is a must. Numerical solutions by definition are approximate insofar as derivatives and integrals assume linear variation over small discrete distances. Examples of the approximations for the derivative and integral are:

$$\frac{d\varphi}{dx} = \lim_{x_2 \to x_1} \frac{\varphi_2 - \varphi_1}{x_2 - x_1} \approx \frac{\varphi_2 - \varphi_1}{x_2 - x_1} \tag{2.21}$$

$$\int_{x_1}^{x_2} \varphi dx \approx \left(\frac{\varphi_1 + \varphi_2}{2} \right) (x_2 - x_1) \tag{2.22}$$

Although a linear change of φ is assumed over the interval (x_1, x_2) this so called approximations of (2.21) and (2.22) are increasingly more accurate for smaller such intervals (as shown in early courses on calculus). That is, while classical solutions are only as accurate as the models on which they are predicated, the numerical approximations for derivatives and integrals can be made as accurate as we want by our resorting to finer subdivisions of the relevant intervals over the domain.

That is, the accuracy of numerical methods depends on the subdivisions resorted to. This involves data preparation (called preprocessing), then solving the problems, and, now faced with inordinate amounts of data corresponding to the numerous points of discretization, some form of computer-based depiction of the results to make meaning of the numbers, a stage called post-processing. As we will see, numerically-based computer methods involve three stages; namely preprocessing, solving, and post-processing.

2.3 Methods of Approximate Solution – Differential and Integral

In analyzing a device for how it would perform, we need to solve the equations governing its behavior. The kinds of analysis we employ depend on the equations we choose to solve, and these come in two forms. These two forms are differential and integral. The differential forms are expressed in terms of derivative operations, such as (2.1) to (2.4). The differentiation operators in terms of rate of change of the field quantities are expressions at a point in space. As derivatives may be expressed in terms of change of a field variable divided by the change in distance or time, the corresponding approximations of derivatives, express the relationship of the field at a point in space to that at a point close by. Approximations of the derivative equations are therefore said to be sparse insofar as only the relationship to fields at neighboring points is expressed. When the relationship between field values is expressed through approximations of differential relationships, therefore, the matrix equation will be very sparse – that is most places on the matrix will be unoccupied. This has strong implications for computational efficiency.

Every one of the differential equations (2.1) to (2.4) is derived from an integral equation. For example, we say that the divergence equation for flux density came from Coulomb's law on the field caused at a point in space by the charge density ρ. These express far away effects through a homogeneous medium. All field quantities become related when integral equations are approximated, leading to dense matrices.

Based on these considerations, we may say that there are two classes of numerical methods based on whether we are solving integral equations or differential equations. In electrostatics, for example, the Coulomb equation becomes the basis of the boundary-element method:

$$\varphi = \iiint_R \frac{1}{4\pi\epsilon} \frac{\rho}{r} dR \tag{2.23}$$

This equation is discretized and the potential φ is expressed in terms of the charge distribution ρ. However, on material boundaries, say between regions 1 and 2, there is a surface charge density σ according to the derivatives in the normal direction n at the interface separating regions 1 and 2:

$$\epsilon_1 \frac{\partial \varphi_1}{\partial n_1} - \epsilon_2 \frac{\partial \varphi_2}{\partial n_2} = \sigma \tag{2.24}$$

This simply states that the difference in normal-flux density across the interface is the surface-charge density – a form of Gauss' Law saying that the total flux out of a volume is the charge enclosed.

Thus, the integral Poisson equation relates φ at each point to the known charge distribution ρ and φ's normal derivative at all material interfaces. The charge distribution too may be unknown as the charges rearrange themselves because of Coulomb forces. Thus, every discrete point has an unknown related to the unknowns at all other discrete points. The resulting matrix equation is fully populated and not necessarily symmetric. Thus, although unlike in differential methods relating every point only to others in its neighborhood, thereby measuring rates of change, with integral equations every point in the domain has a relationship with all other points in the domain of solution. However, here with integral methods, we have unknowns only where there are conductors and material interfaces – that is, we have fewer unknowns. Yet the method is not too popular especially because of the nonsymmetric nature of the matrix associated with the solution equation, besides the fully-populated nature of the matrix equation.

The best-known differential methods are the finite-difference and finite element methods. The finite-difference method is best suited to homogeneous problems and therefore is very popular in solving the wave equation which usually involves modeling in free homogeneous space. It is not so popular in the many problems where inhomogeneous materials are present such as in a stuffed waveguide. While we will focus on the finite element method because of its wide applications, flexibility, and suitability to the most efficient matrix-solution methods, here we will briefly mention the finite-difference method.

We shall take as a general example for demonstrating the finite-difference method, the solution of the Poisson equation, which appears in many places in all specialties of engineering:

$$-\epsilon \nabla^2 \varphi = \rho \tag{2.25}$$

where in the case of electrostatic problems, ϵ is the permittivity, φ is the electric potential for which we will be solving, and ρ is the space charge density. We note that the operator (in two-dimensional problems) is

$$\nabla^2 = \frac{\partial^2}{\partial x^2} + \frac{\partial^2}{\partial y^2} \tag{2.26}$$

In the finite-difference method we work with a rectangular grid. Although there are numerous papers claiming to work with triangular and quadrilateral grids, they are really re-presentations of the finite element method to make the finite-difference authors' work of a lifetime not seem outdated. The real simplicity of the finite-difference method comes from the rectangular grid as also do its weaknesses. A rectangular grid means that when a device involves fine details, the finite-difference mesh needs to be fine there; but now uniformity of the mesh size demands that the same fineness be applied everywhere, including places where fineness is not required.

We need to approximate the Poisson equation at Node 0. To this end, we shall estimate the first derivative at points A and B, which are the midpoints between Nodes 3 and 0 and Nodes 0 and 1, respectively. Approximately, we may say that at the two midpoints A and B, as shown in Figure 2.2,

$$\left.\frac{\partial \varphi}{\partial x}\right|_A = \frac{\Delta \varphi}{\Delta x} = \frac{\varphi_0 - \varphi_3}{h} \tag{2.27}$$

$$\left.\frac{\partial \varphi}{\partial x}\right|_B = \frac{\Delta \varphi}{\Delta x} = \frac{\varphi_1 - \varphi_0}{h} \tag{2.28}$$

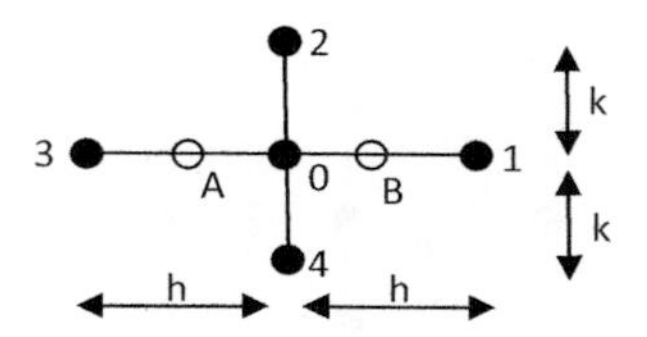

FIGURE 2.2
An hxk finite difference mesh with typical Node 0 surrounded by Nodes 1, 2, 3, 4.

Therefore, recognizing that the change in x between the points A and B is also *h*,

$$\frac{\partial^2 \varphi}{\partial x^2} = \frac{\Delta \frac{\partial \varphi}{\partial x}}{\Delta x} = \frac{\frac{\varphi_1 - \varphi_0}{h} - \frac{\varphi_0 - \varphi_3}{h}}{h} = \frac{\varphi_1 + \varphi_3 - 2\varphi_0}{h^2} \tag{2.29}$$

Similarly,

$$\frac{\partial^2 \varphi}{\partial y^2} = \frac{\Delta \frac{\partial \varphi}{\partial y}}{\Delta y} = \frac{\varphi_2 + \varphi_4 - 2\varphi_0}{k^2} \tag{2.30}$$

so that the approximation of the Poisson equation becomes (see Figure 2.3.1)

$$-\varepsilon\left(\frac{\varphi_1 + \varphi_3 - 2\varphi_0}{h^2} + \frac{\varphi_2 + \varphi_4 - 2\varphi_0}{k^2}\right) = \rho_0 \tag{2.31}$$

Frequently a square mesh with $h = k$ is used in which case we have

$$4\varphi_0 - \varphi_1 - \varphi_2 - \varphi_3 - \varphi_4 = \frac{h^2 \rho_0}{\varepsilon} \tag{2.32}$$

Generally, an equation can be fitted to every interior node. Where the node 0 is on a boundary with the potential specified (called the Dirichlet boundary condition) there is no need to have an equation written for it. Where the node is on the boundary where the normal gradient of potential is zero (the so-called Neumann boundary condition), the node that would lie outside the boundary becomes a reflection of the interior node so that the above equation would involve the three nodes on the boundary and the interior node twice.

For example, if the node 0 is on a left vertical boundary (see Figure 2.3), the Neumann condition can also be expressed in finite differences to deal with the nonexistent node 3, which if it exists would lie outside the domain of solution:

$$\left.\frac{\partial \varphi}{\partial n}\right|_{\text{At } 0} = \left.\frac{\partial \varphi}{\partial x}\right|_{\text{At } 0} = \frac{\varphi_1 - \varphi_3}{h} = 0 \rightarrow \varphi_1 = \varphi_3 \tag{2.33}$$

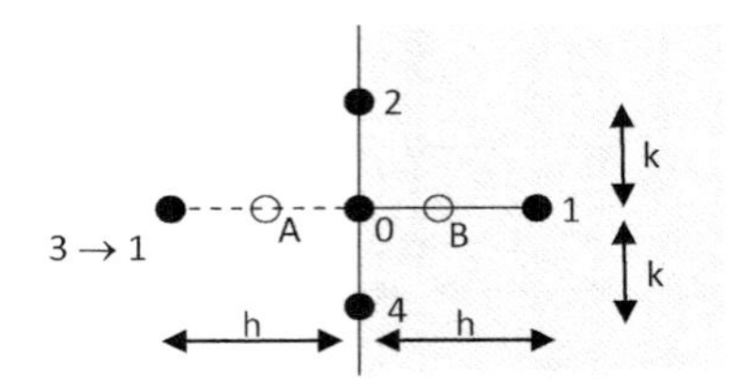

FIGURE 2.3
Node on left vertical boundary of solution.

Substituting, the relevant finite difference approximation for node 0, eliminating node 3, becomes

$$-\varepsilon\left(\frac{2\varphi_1 - 2\varphi_0}{h^2} + \frac{\varphi_2 + \varphi_4 - 2\varphi_0}{k^2}\right) = \rho_0 \tag{2.34}$$

We will stop there on the finite-difference method saying that the Neumann boundary condition makes the matrix of coefficients nonsymmetric and that inhomogeneities also do that besides imposing the requirement that material interfaces must run vertically or horizontally. The latter requirement necessitates modeling slanted boundaries by short vertical and horizontal lines making the mesh very fine, thereby increasing the matrix size.

We will therefore look instead at the far more powerful finite element method, which is state of the art in terms of the numerical solution of partial differential equations. Matrices remain symmetric even if we had Neumann boundaries. Graded meshes can be used to model parts in finer detail.

2.4 A Note on Matrix Representation of Polynomials

It is convenient first to express second-order polynomials in matrix form and establish some terminology required for a proper discussion of the finite element method. In particular, we want to establish the rules for adding and differentiating polynomials through matrix representation. This makes the development of the finite element method elegant and succinct in terms of the use of symbols and algebra

The variables $x_1, x_2, \ldots, x_n$ may be expressed as the components of the column vector $\underline{x}$. Thus, the scalar s, a first-order polynomial expression in the components of $\underline{x}$ becomes

$$s = c_1 x_1 + c_2 x_2 + \ldots + c_n x_n = \underline{c}^t \underline{x} = \underline{x}^t \underline{c} \tag{2.35}$$

Defining

$$\frac{\partial s}{\partial \underline{x}} = \frac{\partial \underline{x}^t \underline{c}}{\partial \underline{x}} = \begin{Bmatrix} \frac{\partial s}{\partial x_1} \\ \vdots \\ \frac{\partial s}{\partial x_n} \end{Bmatrix} = \begin{Bmatrix} c_1 \\ \vdots \\ c_n \end{Bmatrix} = \underline{c} \tag{2.36}$$

A second-order polynomial p, it may be verified, is

$$\begin{aligned} p = a_{11}x_1^2 + 2a_{12}x_1x_2 + \cdots + 2a_{1n}x_1x_n + a_{22}x_2^2 + 2a_{23}x_2x_3 \\ + \cdots + 2a_{2n}x_2x_n + \cdots + a_{nn}x_n^2 = \underline{x}^t \left[A\right] \underline{x} \end{aligned} \tag{2.37}$$

where the symmetric matrix

$$\left[A\right] = \begin{bmatrix} a_{11} & \cdots & a_{1n} \\ \vdots & \ddots & \vdots \\ a_{n1} & \cdots & a_{nn} \end{bmatrix} \tag{2.38}$$

If we considered a second polynomial

$$\begin{aligned} q = b_{11}x_1^2 + 2b_{12}x_1x_2 + \cdots + 2b_{1n}x_1x_n + b_{22}x_2^2 + 2b_{23}x_2x_3 \\ + \cdots + 2b_{2n}x_2x_n \cdots + b_{nn}x_n^2 = \underline{x}^t \left[B\right] \underline{x} \end{aligned} \tag{2.39}$$

then we have the rule for adding second-order polynomials

$$p + q = \underline{x}^t \left[A\right] \underline{x} + \underline{x}^t \left[B\right] \underline{x} = \underline{x}^t \left[A + B\right] \underline{x}$$

The rule for differentiation comes from differentiating p:

$$\frac{\partial p}{\partial \underline{x}} = \begin{Bmatrix} \frac{\partial p}{\partial x_1} \\ \vdots \\ \frac{\partial p}{\partial x_n} \end{Bmatrix} = \begin{Bmatrix} 2a_{11}x_1 + 2a_{12}x_2 + \cdots + 2a_{1n}x_n \\ \vdots \\ 2a_{1n}x_1 + 2a_{2n}x_2 + \cdots + 2a_{nn}x_n \end{Bmatrix} = 2\left[A\right] \underline{x} \tag{2.40}$$

2.5 The Finite Element Method

The finite element method is today the most commonly used numerical method in the area of the solution of partial differential equations. Unlike integral methods and the finite-difference method, it always yields symmetric sparse matrices, which enable the use of the most powerful and elegant matrix solvers. It enables us to use a fine mesh in parts where fine device detail has to be incorporated or high accuracy is required, while a sparse mesh may be used in other parts. Inhomogeneities are not a problem and are handled by ensuring that an element is always entirely in a homogeneous region.

To describe the finite element method it is good to note that it has certain ingredients and to spell them out:

Ingredient 1

A trial or test function of the geometric coordinates on the domain of solution (the xy-plane in our two-dimensional demonstration) consisting of many interpolation functions α_i with $i=1 \ldots$ n. The test function would be a sum of all the α_is and the αs are functions that we prudently choose so as to best model the solution. This means that we have some knowledge of the solution. In this form, the method was known as Ritz's method and was not quite successful since we do not usually know, *a priori*, the form of the solution. So, our postulates based on our choice of αs is that the solution is of the form:

$$\varphi = \sum_{i=1}^{n} c_i \alpha_i \tag{2.41}$$

Our task then would be to determine the numbers c_i so as to yield the best fit of the actual solution that can be made by the trial functions chosen.

Ingredient 2

The second required ingredient is a mesh of finite elements on the domain of solution, involving a tessellation of the area of solution. Commonly, the domain of solution is subdivided into triangles or quadrilaterals. Such division of the domain is termed a tessellation. In this lies the brilliance of the method. Once we divide the domain into small regions, in place of the complicated functions, our αs, we may go for simple functions such as linear or quadratic polynomials. In simple terms, to approximate a squiggly function y(x) on a large domain, we may need to add several αs of different forms; but if we divide the domain into small bits, a straight line in one-dimension and a flat plane in two-dimensions on each bit or element would suffice. The smaller the elements, the more accurate our model. Thus, on a triangular linear finite element, we have

$$\varphi = a + bx + cy \tag{2.42}$$

where the functions 1, x, and y take the place of the three αs we need for the small triangular domain as shown in Figure 2.4. As we divide the domain into smaller and smaller triangles, the modeling of the actual shape of the unknown would become more and more accurate, as the approximation of Equation (2.42) also will.

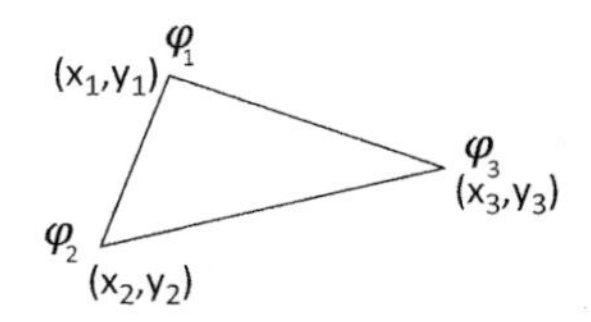

FIGURE 2.4
A triangle with three nodes.

This linear model is an approximation of the potential because φ may not vary linearly and may have a complex relationship with x and y. However, on the small element φ is linear because x and y are linear first order functions. This is the key to determining a, b, and c in a triangle.

Usually from one triangle to its neighbor we would need continuity conditions so that the trial functions match on the edge common to two adjacent elements. For this the a, b, and c of one triangle need to have a relationship with the a, b, and c of any adjacent triangle. This relationship being complicated, instead of a, b, and c we deal with the three vertex potentials φ_i, $i=1, 2, 3$. So at each vertex i of a triangle (x_i,y_i) at potential φ_i the value φ_i would be known at the end of the solution process because we are solving the Poisson equation in a specific domain divided into a specific mesh of known vertices connected into a triangular mesh for the vertex potentials rather than for a, b, and c. This means that a, b, and c need to be replaced by the three potentials φ_i, $i=1, 2, 3$. Applying the linear approximation:

$$\varphi_1 = a + bx_1 + cy_1 \tag{2.43a}$$

$$\varphi_2 = a + bx_2 + cy_2 \tag{2.43b}$$

$$\varphi_3 = a + bx_3 + cy_3 \tag{2.43c}$$

Or in matrix form

$$\begin{bmatrix} \varphi_1 \\ \varphi_2 \\ \varphi_3 \end{bmatrix} = \begin{bmatrix} 1 & x_1 & y_1 \\ 1 & x_2 & y_2 \\ 1 & x_3 & y_3 \end{bmatrix} \begin{bmatrix} a \\ b \\ c \end{bmatrix} \tag{2.44}$$

Solving by Cramer's rule

$$a = \frac{\begin{vmatrix} \varphi_1 & x_1 & y_1 \\ \varphi_2 & x_2 & y_2 \\ \varphi_3 & x_3 & y_3 \end{vmatrix}}{\begin{vmatrix} 1 & x_1 & y_1 \\ 1 & x_2 & y_2 \\ 1 & x_3 & y_3 \end{vmatrix}} \tag{2.45}$$

$$b = \frac{\begin{vmatrix} 1 & \varphi_1 & y_1 \\ 1 & \varphi_2 & y_2 \\ 1 & \varphi_3 & y_3 \end{vmatrix}}{\begin{vmatrix} 1 & x_1 & y_1 \\ 1 & x_2 & y_2 \\ 1 & x_3 & y_3 \end{vmatrix}} = \frac{1}{\Delta}\left\{\varphi_1\left(y_2 - y_3\right) + \varphi_2\left(y_3 - y_1\right) + \varphi_3\left(y_1 - y_2\right)\right\}$$

$$= \sum_{i=1}^{3} \varphi_i \frac{\left(y_{i1} - y_{i2}\right)}{\Delta} = \sum_{i=1}^{3} b_i \varphi_i \tag{2.46}$$

$$c = \frac{\begin{vmatrix} 1 & \varphi_1 & y_1 \\ 1 & \varphi_2 & y_2 \\ 1 & \varphi_3 & y_3 \end{vmatrix}}{\begin{vmatrix} 1 & x_1 & y_1 \\ 1 & x_2 & y_2 \\ 1 & x_3 & y_3 \end{vmatrix}} = \frac{1}{\Delta}\left\{\varphi_1\left(x_3 - x_2\right) + \varphi_2\left(x_1 - x_3\right) + \varphi_3\left(x_2 - x_1\right)\right\}$$

$$= \sum_{i=1}^{3} \varphi_i \frac{\left(x_{i2} - x_{i1}\right)}{\Delta} = \sum_{i=1}^{3} c_i \varphi_i \tag{2.47}$$

where Δ is twice the area of the triangle provided that the nodes 1, 2, and 3 are numbered counterclockwise (and the negative of that if the nodes go clockwise); the numbers i, $i1$, and $i2$ are the numbers 1, 2, and 3 in cyclic order (that is, the integers i, $i1$, and $i2$, respectively are 1, 2, 3 if i is 1, or 2, 3, 1 if i is 2, or 3, 1, 2 if i is 3; or put another way $i1 = i \bmod 3 + 1$ and $i2 = i1 \bmod 3 + 1$ where the mod 3 function (also often referred to as the modulo 3 function) gives the remainder when divided by 3); and

$$b_i = \frac{y_{i1} - y_{i2}}{\Delta} \tag{2.48}$$

and

$$c_i = \frac{x_{i2} - x_{i1}}{\Delta} \tag{2.49}$$

Ingredient 3

The third key ingredient to the finite element method is the identification of a functional $\mathcal{L}(\varphi)$ (a function of a function, in this case of φ), minimizing which is the same as satisfying the differential equation. We note that only some differential equations may be identified with a functional. The variational finite element method is general. There is an alternative approach called the Galerkin formulation, that is described in Section 6.2, relying on the theory of function spaces going into minimizing the residual when our trial function is put into the differential equation being solved; but we will not get into that here. A limitation of the approach we describe, the variational calculus approach, is that to take the functional route we take here, we must have a functional corresponding to the solution of the differential equation whose solution is sought. Suffice it to say that the Poisson equation does have a functional corresponding to it:

$$\mathcal{L}(\varphi) = \iiint_R \left[\tfrac{1}{2}\epsilon (\nabla\varphi)^2 - \rho\varphi \right] dR \tag{2.50}$$

We can justify this intuitively. Since the electric field strength $\bar{E} = -\nabla\varphi$ and the flux density $\bar{D} = \epsilon\bar{E} = -\varepsilon\nabla\varphi$, we see that the first term is the stored energy density $\bar{D}\cdot\bar{E}$ times elemental volume dR. Since by definition the potential φ is the work done in bringing a unit charge to where it is, the second term is the work done in bringing the charge ρdR, the charge density times elemental volume, to where it is. The essence of the functional is that it says that the difference between the two co-energies ought to be a minimum when the Poisson equation is satisfied. But let us see this more precisely in mathematical terms.

Let the potential φ take a small excursion to $\varphi + \kappa\eta(x, y, z)$, where $\kappa$is a constant and $\eta(x, y, z)$ is a small arbitrary function. That it is termed arbitrary is critical and we shall make recourse to this soon. Also, the term z may be dropped in two-dimensions but this proof is valid for two- and three-dimensions. The change in the functional then is

$$\begin{aligned} \mathcal{L}(\varphi + \kappa\eta) = \mathcal{L} + \delta\mathcal{L} &= \iiint_R \left[\tfrac{1}{2}\epsilon (\nabla\varphi + \kappa\nabla\eta)^2 - \rho(\varphi + \kappa\eta) \right] dR \\ &= \iiint_R \left[\tfrac{1}{2}\epsilon \left(\nabla\varphi\cdot\nabla\varphi + 2\kappa\nabla\varphi\cdot\nabla\eta + \kappa^2\nabla\eta\cdot\nabla\eta \right) - \rho(\varphi + \kappa\eta) \right] dR \end{aligned} \tag{2.51}$$

Subtracting (2.50) from (2.51)

$$\delta\mathcal{L} = \iiint_R \left[\epsilon(\kappa\nabla\varphi\cdot\nabla\eta) - \rho\kappa\eta \right] dR + O^2(\kappa\eta) \tag{2.52}$$

where for small changes, the terms of order 2 (designated O^2) in $\kappa\eta$ may be dropped for being negligibly small but are retained to make a point we will encounter soon. To proceed we need to use a vector identity expressing the divergence of a vector $\vec{V}$ scaled by a scalar s:

$$\nabla \cdot \left(s\bar{V}\right) = \nabla s \cdot \bar{V} + s\nabla \cdot \bar{V} \tag{2.53}$$

together with the divergence theorem:

$$\iiint_R \nabla \cdot \bar{V} dR = \iint_S \bar{V} \cdot \overline{dS} \tag{2.54}$$

where the surface S bounds the domain R. Integrating the vector identity just encountered and applying the divergence theorem to it

$$\iiint_R \nabla \cdot s\bar{V} dR = \iiint_R \left(\nabla s \cdot \bar{V} + s\nabla \cdot \bar{V}\right) dR \tag{2.55}$$

or

$$\iint_S s\bar{V} \cdot \overline{dS} = \iiint_R \left(\nabla s \cdot \bar{V} + s\nabla \cdot \bar{V}\right) dR \tag{2.56}$$

Now set $s = \eta$ and $\bar{V} = \nabla\varphi$

$$\iint_S \eta\nabla\varphi \cdot \overline{dS} = \iiint_R \left(\nabla\eta \cdot \nabla\varphi + \eta\nabla^2\varphi\right) dR \tag{2.57}$$

We note further that the elemental surface vector points in the normal direction so that $\overline{dS} = dS\bar{u}_n$ where $\bar{u}_n$ is a unit vector in the normal direction. Further, since

$$\nabla\varphi = \bar{u}_x \frac{\partial\varphi}{\partial x} + \bar{u}_y \frac{\partial\varphi}{\partial y} + \bar{u}_z \frac{\partial\varphi}{\partial z} \tag{2.58}$$

and therefore,

$$\bar{u}_x \cdot \nabla\varphi = \frac{\partial\varphi}{\partial x} \tag{2.59}$$

and since the direction x is arbitrary according to how we orient the axes, we may write

$$\bar{u}_n \cdot \nabla\varphi = \frac{\partial\varphi}{\partial n} \tag{2.60}$$

So, Equation (2.57) upon rearranging becomes

$$\iiint_R (\nabla\eta \cdot \nabla\varphi) dR = \iint_S \eta \frac{\partial\varphi}{\partial n} dS - \iiint_R \left(\eta\nabla^2\varphi\right) dR \tag{2.61}$$

Using this in the expression for the change in the Lagrangian functional in (2.52)

$$\delta\mathcal{L} = \left\{ \iint_S \epsilon\kappa\eta \frac{\partial\varphi}{\partial n} dS - \iiint_R \left(\eta\nabla^2\varphi\right) dR \right\} \iiint_R [-\rho\kappa\eta] dR + O^2(\kappa\eta) \quad (2.62)$$

Now the boundary S has either Dirichlet or Neumann conditions on each point of it. Where the Dirichlet condition applies $\eta = 0$ since the potential is specified and cannot vary. On those parts of S where the Neumann condition is specified, $\partial\varphi/\partial n = 0$. Therefore, the surface integral vanishes because the integrand is zero at every point on S, thereby leading to

$$\delta\mathcal{L} = \left\{ \iiint_R \kappa\eta\left(-\epsilon\nabla^2\varphi - \rho\right) dR \right\} + O^2(\kappa\eta) \quad (2.63)$$

Now we are in a position to say

a) If the functional is at an extremum, whether a minimum or a maximum, the left-hand side is zero since changes are flat at a point of extremum. Further, the second-order terms $O^2(\kappa\eta)$ may be neglected for small changes $\kappa\eta$ so that

$$\iiint_R \kappa\eta\left(-\epsilon\nabla^2\varphi - \rho\right) dR = 0 \quad (2.64)$$

But since the change in potential $\kappa\eta$ is arbitrary, this is possible only if the other factor of the integrand

$$-\epsilon\nabla^2\varphi - \rho = 0 \quad (2.65)$$

or

$$-\epsilon\nabla^2\varphi = \rho \quad (2.66)$$

meaning that the Poisson equation is satisfied at the extremum.

b) We need one more consideration to show that the extremum is a minimum. At that extremum since the residual of the Poisson equation $-\epsilon\nabla^2\varphi - \rho = 0$, we must have

$$\delta\mathcal{L} = O^2(\kappa\eta)$$

which is a strictly positive quantity so that any changes in φabout the extremum of $\mathcal{L}$will make $\mathcal{L}$ rise so that that the extremum has to be a minimum.

2.6 Uniqueness

For completeness we will show that the Poisson equation has a unique solution when the potential φ or its normal gradient is defined at every point on the boundary S and φ at least at one point. Practically, this is a requirement for the well posedness of a problem. If a problem is not well posed, the solution will not be unique and the matrix equation we form will a) have a noninvertible-coefficient matrix if we are using a direct method of solution; or b) not have a solution to converge to in the solution space if we are using an iterative solution method and may have multiple solutions to converge to. To this end of proving the conditions for uniqueness, consider two solutions to the Poisson equation φ_1 and φ_2 satisfying the boundary conditions:

$$-\epsilon\nabla^2\varphi_1 = \rho \text{ in } R \tag{2.67}$$

$$-\epsilon\nabla^2\varphi_2 = \rho \text{ in } R \tag{2.68}$$

with

$$\varphi_1 = f \text{ on } S_1 \text{ and } \frac{\partial\varphi_1}{\partial n} = g \text{ on } S_2 \tag{2.69}$$

$$\varphi_2 = f \text{ on } S_1 \text{ and } \frac{\partial\varphi_2}{\partial n} = g \text{ on } S_2 \tag{2.70}$$

where S_1 is the Dirichlet part of S and S_2 is the Neumann part, and f and g are the specified values. Let us now examine $\psi = \varphi_1 - \varphi_2$ to see under what circumstances it will be zero, the condition for uniqueness – that is, there being no difference between two solutions. Subtracting the equations in R and S

$$-\epsilon\nabla^2\psi = 0 \text{ in } R \tag{2.71}$$

or simply

$$\nabla^2\psi = 0 \text{ in } R \tag{2.72}$$

with

$$\psi = 0 \text{ on } S_1 \text{ and } \frac{\partial\psi}{\partial n} = 0 \text{ on } S_2 \tag{2.73}$$

In the vector identity

$$\nabla\cdot\left(s\bar{V}\right) = \nabla s\cdot\bar{V} + s\nabla\cdot\bar{V} \tag{2.74}$$

If we let $s = \psi$ and $\bar{V} = \nabla\psi$

$$\nabla \cdot (\psi \nabla \psi) = \nabla\psi \cdot \nabla\psi + \psi \nabla^2 \psi \tag{2.75}$$

Noting that $\nabla^2\psi$ is zero, and integrating over R

$$\iiint_R \nabla\psi \cdot \nabla\psi \, dR = \iiint_R \nabla \cdot (\psi \nabla \psi) \, dR, \tag{2.76}$$

and applying the divergence theorem noting further that the dot-product of the vector $\nabla\psi$ with itself is the square of its magnitude:

$$\iiint_R (\nabla\psi)^2 \, dR = \iint_S \psi \nabla\psi \cdot \overline{dS} = \iint_2 \psi \frac{\partial \psi}{\partial n} dS = 0, \tag{2.77}$$

since the surface integral is zero because ψ or $\partial\psi/\partial n$ is zero on every point on S. And now since $(\nabla\psi)^2$ is either positive or zero, for its integral to be zero we must have

$$\nabla\psi = 0 \text{ everywhere in } R, \tag{2.78}$$

which means that ψ is a constant. So, we make the additional stipulation that φ ought to be specified at least at one point on S, giving us a reference potential. At that point $\psi = 0$ so that constant is zero giving us the requisite condition for uniqueness:

> *The potential or its normal gradient should be specified at every point on the boundary* S *with the potential specified at least at one point on* S.

2.7 Natural Boundary Conditions

It may be shown that the variational formulation of the finite element method has the zero Neumann boundary condition implicit to it – that is, on boundaries where no condition is imposed, it will naturally turn out to be imposed in the solution.

Consider the problem shown on the left in Figure 2.5. Say that the problem is prepared for a finite element solution with free potentials $\varphi_1, \ldots, \varphi_n$. The functional $\mathcal{L} = \mathcal{L}(\varphi_1, \ldots, \varphi_n)$. The left-hand side end at $x = x_1$ has a Dirichlet condition while the right-hand side end at $x = x_2$ is free in the sense that no condition is imposed there. As of now we do not know if this problem as posed has a unique solution if tackled by the finite element method.

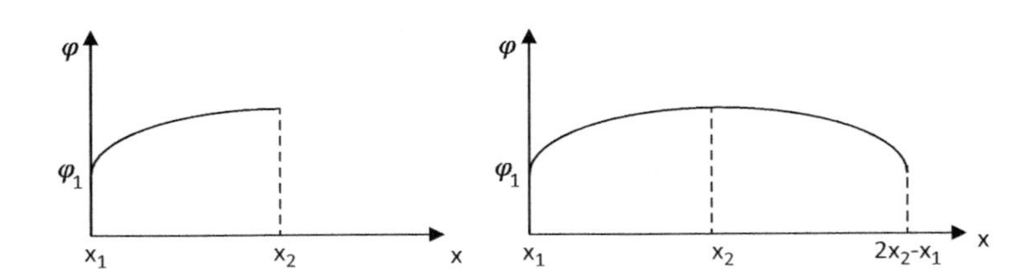

FIGURE 2.5
A finite element problem with the right-end free and a symmetric doubled problem.

Let us now double the problem by symmetrically reflecting it about the right end x_2. All nodes in the left half too are reflected and those in the right half have the same variable values as the corresponding nodes in the left half. That is, the unknowns are still $\varphi_1, \ldots, \varphi_n$ as before. The variables ε and ρ too are symmetrically reflected in the extended half of the problem. The new right end of the doubled problem will be at coordinate location $x_2 + (x_2 - x_1) = 2x_2 - x_1$ and has the same value of φ, φ_1, imposed as at the left end. Now this problem satisfies the condition that the ends ought to have either the Neumann or Dirichlet condition imposed so it has a unique solution. But the new functional by the way the doubled problem has been formulated is twice $\mathcal{L}(\varphi_1, \ldots, \varphi_n)$. Therefore, the original problem also has not only a unique solution but, in the left half, the same solution as the doubled problem. By symmetry, at the midpoint, $d\varphi/dx = 0$. Therefore, we may assert that when we ignore boundary conditions, the variational formulation automatically imposes a Neumann condition and yields unique solutions.

This argument may be repeated in two-dimensions with some limitations because multiple portions of the Neumann boundary or curved portions cannot be reflected as done here in one-dimension. Getting into that, however, is beyond the scope of this work.

Often, we have the freedom to impose the Neumann condition rather than ignoring it. This is called a strongly imposed condition as opposed to what we call a weakly imposed condition when it is allowed to occur naturally by ignoring the condition. In a one-dimensional situation, say imposing it strongly on the right boundary would mean making the potential values in the last two nodes equal while being free. This would ensure that the φ curve would be flat. But then it would be flat throughout the last element on the right which means that the equal potential values in the last two nodes cannot bend so as to accommodate the differential equation being solved. This then is the classic poser: Is it more important to satisfy the equation set (consisting of the differential equation and boundary conditions) at the end point by strong imposition or overall through weak imposition? Clearly the latter is the better and indeed more convenient approach to implement.

2.8 One-Dimensional Linear Finite Elements

Let us approach the implementation of the finite element method through a simple one-dimensional problem in demonstration of the principles. In the domain [0,10] let the electric potential be governed by the one-dimensional Poisson equation

$$-\varepsilon\frac{d^2\varphi}{dx^2}=\rho \tag{2.79}$$

with $(\varepsilon,\rho)=(2,0)$ in the interval (0,5) and (1,1) in the interval (5,10). In addition, φ is 30 V at the two-end points $x=0$ and $x=10$ (i.e., Dirichlet conditions at both boundaries). Solve for φ.

The element in one-dimension x is a straight line (see Figure 2.6). A linear function in x, $\varphi=a+bx$, has two constants a and b, which need to be replaced by values of φ at the two nodes.

Applying the interpolation to the two nodes and solving

$$a+bx_1=\varphi_1 \tag{2.80a}$$

$$a+bx_2=\varphi_2 \tag{2.80b}$$

Or in matrix form

$$\begin{bmatrix}1 & x_1\\ 1 & x_2\end{bmatrix}\begin{Bmatrix}a\\ b\end{Bmatrix}=\begin{Bmatrix}\varphi_1\\ \varphi_2\end{Bmatrix}, \tag{2.81}$$

yielding

$$\begin{Bmatrix}a\\ b\end{Bmatrix}=\frac{1}{x_2-x_1}\begin{bmatrix}x_2 & -x_1\\ -1 & 1\end{bmatrix}\begin{Bmatrix}\varphi_1\\ \varphi_2\end{Bmatrix} \tag{2.82}$$

Therefore, noting that $x_2-x_1=L$, the length of the element

$$\begin{aligned}\varphi=a+bx&=\frac{1}{L}\left[x_2\varphi_1-x_1\varphi_2+x\left(\varphi_2-\varphi_1\right)\right]\\ &=\varphi_1\frac{x-x_1}{L}+\varphi_2\frac{x_2-x}{L}=\alpha_1\varphi_1+\alpha_2\varphi_2\end{aligned} \tag{2.83}$$

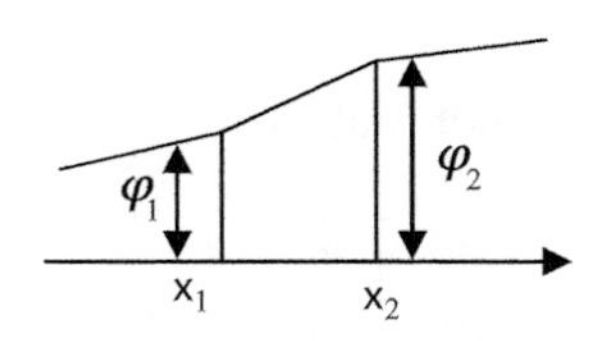

FIGURE 2.6
The first-order line element.

where the interpolation functions

$$\alpha_1 = \frac{x - x_1}{L}; \text{ and } \alpha_2 = \frac{x_2 - x}{L} \tag{2.84}$$

We can also see that

$$\frac{d\varphi}{dx} = \frac{\varphi_2 - \varphi_1}{L} \tag{2.85}$$

The functional in one-dimension is now

$$\begin{aligned}
\mathcal{L} &= \int_R \left\{ \frac{1}{2}\varepsilon \left[\frac{d\varphi}{dx}\right]^2 - \rho\varphi \right\} dx = \sum_{\text{All elements}} \int_{\text{Element}} \left\{ \frac{1}{2}\varepsilon \left[\frac{d\varphi}{dx}\right]^2 - \rho\varphi \right\} dx \\
&= \sum_{\text{All elements}} \int_{x_1}^{x_2} \left\{ \frac{1}{2}\varepsilon \left[\frac{\varphi_2 - \varphi_1}{L}\right]^2 - \rho\varphi \right\} dx \\
&= \sum_{\text{All elements}} \left\{ \frac{1}{2}\varepsilon \left[\frac{\varphi_2 - \varphi_1}{L}\right]^2 L - \rho_0 \frac{\varphi_1 + \varphi_2}{2} L \right\}.
\end{aligned} \tag{2.86}$$

Here we have used the fact that the integrand of the first integral term is constant so that, taking it out of the integral sign, the integral is simply x which becomes L when limits are applied; and for the second integral we have assumed that ρ is a constant so that the integral of φ, varying as a straight line (see Figure 2.8.1), is the area under the curve, which is the area of a trapezium (which in some places is called a trapezoid). Expressing this polynomial expression for energy in matrix terms

$$\mathcal{L} = \sum_{\text{All elements}} \left\{ \frac{1}{2}\underline{\varphi}^t \begin{bmatrix} \frac{\varepsilon}{L} & -\frac{\varepsilon}{L} \\ -\frac{\varepsilon}{L} & \frac{\varepsilon}{L} \end{bmatrix} \underline{\varphi} - \underline{\varphi}^t \begin{Bmatrix} \frac{\rho_0 L}{2} \\ \frac{\rho_0 L}{2} \end{Bmatrix} \right\} = \sum_{\text{All elements}} \left\{ \frac{1}{2}\underline{\varphi}^t [P] \underline{\varphi} - \underline{\varphi}^t \underline{Q} \right\} \tag{2.87}$$

where what we call the Dirichlet matrix

$$[P] = \begin{bmatrix} \frac{\varepsilon}{L} & -\frac{\varepsilon}{L} \\ -\frac{\varepsilon}{L} & \frac{\varepsilon}{L} \end{bmatrix} \tag{2.88}$$

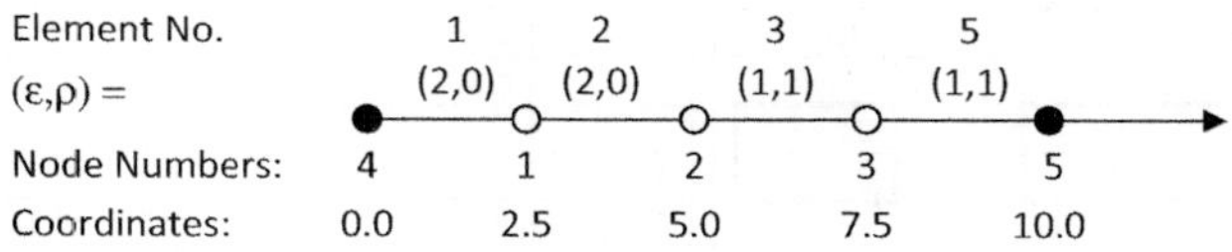

Element No.		1	2	3	5	
(ε,ρ) =		(2,0)	(2,0)	(1,1)	(1,1)	
Node Numbers:	4	1	2	3		5
Coordinates:	0.0	2.5	5.0	7.5		10.0

FIGURE 2.7
Data for sample problem and its mesh.

and

$$\underline{Q} = \begin{Bmatrix} \dfrac{\rho_0 L}{2} \\ \dfrac{\rho_0 L}{2} \end{Bmatrix} \tag{2.89}$$

Let us divide the domain into four elements of equal length as shown in Figure 2.7. The two known nodes are shown in black and the three unknowns in white. Note well that the two known nodes (that is whose potential values are known) have been numbered last. The reason for this will become apparent as we go through this example.

There being five potentials $\varphi_1, \ldots, \varphi_5$ the functional will be a polynomial in these, formed by a 5×5 matrix. Let us say that the final form of the functional is

$$\mathcal{L} = \frac{1}{2}\underline{\varphi}_g^t \left[P_g\right]\underline{\varphi}_g - \underline{\varphi}_g^t \left\{Q_g\right\} \tag{2.90}$$

These matrices $[P_g]$ and $\{Q_g\}$ are referred to as global matrices because they contain the functional with the contributions of the local matrices from the elements.

The so-called local matrices [P] and {Q} for element no. 1 of length = 2.5 are

$$\left[P\right] = \begin{bmatrix} \dfrac{2}{2.5} & -\dfrac{2}{2.5} \\ \dfrac{2}{2.5} & -\dfrac{2}{2.5} \end{bmatrix} = \begin{bmatrix} 0.8 & -0.8 \\ -0.8 & 0.8 \end{bmatrix} \text{ and } \left\{Q\right\} = \begin{Bmatrix} 0 \\ 0 \end{Bmatrix} \tag{2.91}$$

because ρ is zero. The same goes for the second element. But for the third and fourth elements,

$$\left[P\right] = \begin{bmatrix} \dfrac{1}{2.5} & -\dfrac{1}{2.5} \\ \dfrac{1}{2.5} & -\dfrac{1}{2.5} \end{bmatrix} = \begin{bmatrix} 0.4 & -0.4 \\ -0.4 & 0.4 \end{bmatrix} \text{ and } \left\{Q\right\} = \begin{Bmatrix} \dfrac{1 * 2.5}{2} \\ \dfrac{1 * 2.5}{2} \end{Bmatrix} = \begin{Bmatrix} 1.25 \\ 1.25 \end{Bmatrix} \tag{2.92}$$

Figure 2.8 shows the global matrices being built up from the four local matrices and four local vectors.

Elem. 1	PLocal		QLocal	PGlobal	1	2	3	4	5	QGlobal
Vertices	4	1		1	0.8			-0.8		0
4	0.8	-0.8	0	2						
1	-0.8	0.8	0	3						
				4	-0.8			0.8		0
				5						

Elem. 2	PLocal		QLocal	PGlobal	1	2	3	4	5	QGlobal
Vertices	1	2		1	0.8+0.8 =1.6	-0.8		-0.8		0+0=0
1	0.8	-0.8	0							
2	-0.8	0.8	0	2	-0.8	0.8				0
				3						
				4	-0.8			0.8		0
				5						

Elem. 3	PLocal		QLocal	PGlobal	1	2	3	4	5	QGlobal
Vertices	2	3		1	1.6	-0.8		-0.8		0
2	0.4	-0.4	1.25	2	-0.8	0.8+0.4	-0.4			0+1.25
3	-0.4	0.4	1.25	3		-0.4	0.4			+1.25
				4	-0.8			0.8		0
				5						

Elem. 4	PLocal		QLocal	PGlobal	1	2	3	4	5	QGlobal
Vertices	3	5		1	1.6	-0.8		-0.8		0
3	0.4	-0.4	1.25	2	-0.8	1.2	-0.4			1.25
5	-0.4	0.4	1.25	3		-0.4	0.4+0.4		-0.4	1.25+1.25
				4	-0.8			0.8		0
				5			-0.4		0.4	+1.25

FIGURE 2.8
Contributions of the local matrices P and Q as they add to the global matrices PG, QG.

Therefore,

$$\mathcal{L} = \frac{1}{2}\underline{\varphi}_g^t \begin{bmatrix} 1.6 & -0.8 & 0 & -0.8 & 0 \\ -0.8 & 1.2 & -0.4 & 0 & 0 \\ 0 & -0.4 & 0.8 & 0 & -0.4 \\ -0.8 & 0 & 0 & 0.8 & 0 \\ 0 & 0 & -0.4 & 0 & 0.4 \end{bmatrix} \underline{\varphi}_g - \underline{\varphi}_g^t \begin{bmatrix} 0 \\ 1.25 \\ 2.5 \\ 0 \\ 1.25 \end{bmatrix} \tag{2.93}$$

The three potentials φ_1, φ_2, and φ_3, are as small as possible, but the two potentials φ_4 and φ_5 are fixed and cannot vary. Therefore, of the derivative vector $\partial\mathcal{L}/\partial\underline{\varphi}_g$, only the first three rows may be set to zero:

$$\begin{bmatrix} 1.6 & -0.8 & 0.0 & -0.8 & 0.0 \\ -0.8 & 1.2 & -0.4 & 0.0 & 0.0 \\ 0.0 & -0.4 & 0.8 & 0.0 & -0.4 \end{bmatrix} \begin{Bmatrix} \varphi_1 \\ \varphi_2 \\ \varphi_3 \\ \varphi_4 \\ \varphi_5 \end{Bmatrix} - \begin{Bmatrix} 0 \\ 1.25 \\ 2.5 \end{Bmatrix} = \begin{Bmatrix} 0 \\ 0 \\ 0 \end{Bmatrix} \tag{2.94}$$

Now we see why the unknown nodes were numbered first. This ensured that we could simply drop the last few rows of PG corresponding to known nodes and shift the last columns of PG corresponding to known nodes to the right. Otherwise, we would have had rows in the middle of the matrix needing to be eliminated. Of course, in modern programs where we progressively refine the mesh as a solution is found to be too crude to be acceptable, nodes are numbered arbitrarily but then mapped on to variables that are structured unknowns first and knowns last.

Returning to our problem, recognizing that the last two values, φ_4 and φ_5, are 30 V, these may be shifted to the right-hand side of the three equations:

$$\begin{bmatrix} 1.6 & -0.8 & 0.0 \\ -0.8 & 1.2 & -0.4 \\ 0.0 & -0.4 & 0.8 \end{bmatrix} \begin{Bmatrix} \varphi_1 \\ \varphi_2 \\ \varphi_3 \end{Bmatrix} = \begin{Bmatrix} 0+0.8*30 \\ 1.25 \\ 2.5+0.4*30 \end{Bmatrix} = \begin{Bmatrix} 24 \\ 1.25 \\ 14.5 \end{Bmatrix} \tag{2.95}$$

Solving, we have

$$\varphi_1 = 32.0433,\ \varphi_2 = 34.1667,\ \varphi_3 = 35.2083 \tag{2.96}$$

We have seen how the polynomials are added to the global matrices using the corresponding node numbers. But hand calculations are prone to error and tedious. Therefore let us look at a problem to see how this process is implemented on the computer.

Defining the demonstrative example, the domain is [0,10]. The left and right ends are at potential 10. ε and ρ are 1 and 2. This is a test problem for learning purposes and has an analytical solution so it is a problem for which we would normally not need a computer:

$$-\frac{d^2\varphi}{dx^2} = 2 \tag{2.97}$$

Integrating twice

$$\frac{d\varphi}{dx} = -2x + c \tag{2.98}$$

$$\varphi = -x^2 + cx + d \tag{2.99}$$

Applying the boundary conditions, $d=10$ and $c=10$. Therefore, the exact solution is

$$\varphi = -x^2 + 10x + 10 \tag{2.100}$$

To see if we get the same solution by the finite element method, the domain length is divided into 100 elements so we have 101 nodes with all elements of equal length = 10/100 = 0.1 m. Since the known nodes are numbered last, the numbering from left to right is 101, 1, 2, …, 100. So, element 1 has nodes 101–1, element 2 is 1–2, element 3 is 2–3, etc., to element 100 which connects nodes 99–100. All elements have the same local matrices:

$$P = \frac{\varepsilon}{L}\begin{bmatrix} 1 & -1 \\ -1 & 1 \end{bmatrix} = \begin{bmatrix} 10 & -10 \\ -10 & 10 \end{bmatrix} \quad Q = \begin{Bmatrix} \frac{\rho L}{2} \\ \frac{\rho L}{2} \end{Bmatrix} = \begin{Bmatrix} 0.1 \\ 0.1 \end{Bmatrix} \tag{2.101}$$

The MATLAB code is shown boxed as Code Box 2.1. Note that in placing the local matrices P and Q in the global matrices PG and QG, if the node number is of one of the knowns, then the row of the global matrix is not formed. If the

CODE BOX 2.1 One-D FEM with Dirichlet Conditions

```
clc; clear all; %Clear memory and screen
n=100;
%n= No. of elements. No. of nodes=n-2
m=n-1; %no. of unknowns
PG=zeros(m,m); %Initializing
QG=zeros(m,1);
L=10/n; %Domain Length/No. of Elem
y=zeros(n,1);
%Solution in y. Setting last 2 known values
y(n+1)=10;
y(n)=10;
P=[1, -1;-1,1]/L;
%Same Local P,Q for all elements
Q=[1;1]*L;

for e=1:n %e is element number
  if e==1
    v(1)=n+1;
    v(2)=1;
  else
    v(1)=e-1;
    v(2)=e;
  end;%if
  for r=1:2
    if v(r)<=m %Otherwise do nothing
      QG(v(r))=QG(v(r))+Q(r);
      for c=1:2
        if v(c)<=m %P(r,c) goes into PG
          PG(v(r),v(c))= PG(v(r),v(c))+P(r,c);
        else %P(r,c)needs to go into QG
```

```
          QG(v(r))=QG(v(r))-P(r,c)*y(v(c));
        end;
      end;
    end;
  end;
end;
% The equation PG x xx = QG has been formed
xx=inv(PG)*QG;
%Here MATLAB inversion function is being used
for i=1:n+1
  xaxis(i,1)=(i-1)*L;
  yexact(i,1)=10+10*xaxis(i,1)-(xaxis(i,1))^2;
  if i==1
    yaxis(i,1)=10;
  else
    if i==n+1
      yaxis(i,1)=10;
    else
      yaxis(i,1)=xx(i-1,1);
    end;
  end;
end;
%plot (xaxis, yexact);
plot (xaxis, yaxis,'r', xaxis, yexact, 'b');
```

row is unknown, the corresponding Q goes into QG; and if the column of P is also of an unknown, then the corresponding P value goes into PG. But if the row of P corresponds to an unknown while the column of P corresponds to a known, then that P value times the known potential value is moved to the right-hand side of the equation (i.e., the multiple is taken out of QG). Upon formation of PG and QG, the solution is obtained through inversion here (although in more sophisticated programs more efficient matrix solution schemes will be used.

The analytical and numerical solutions are shown plotted in Figure 2.9 and they are seen to match exactly as best as the eye can see. It is easy to see how the method is very powerful when different values of ε and ρ come into different segments of the problem where an analytical solution would be very difficult.

To test and see for ourselves how naturally the Neumann boundary condition works when ignored, let us revisit the one-dimensional finite element problem with the right-end boundary condition turned into a Neumann condition.

$$-\frac{d^2\varphi}{dx^2} = 2 \tag{2.102}$$

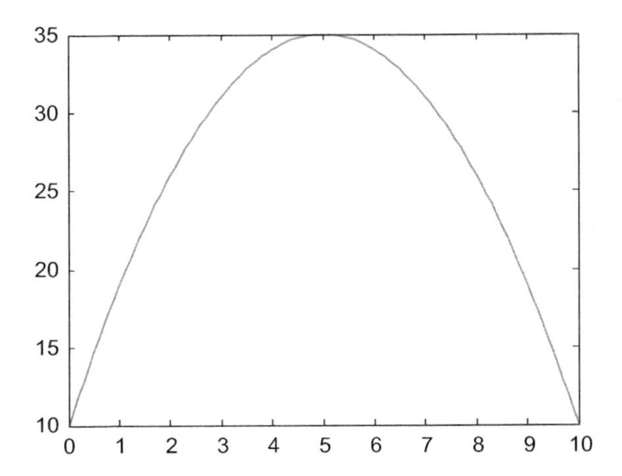

FIGURE 2.9
Explicit and numerical solutions to one-dimensional problem.

Integrating twice

$$\frac{d\varphi}{dx} = -2x + c \tag{2.103}$$

$$\varphi = -x^2 + cx + d \tag{2.104}$$

Applying the boundary conditions on the right $c = 20$ and now on the left, $d = 10$ so that the exact solution is

$$\varphi = -x^2 + 20x + 10 \tag{2.105}$$

Now, however, we have 100 unknowns, the right-end node also corresponding to an unknown potential. The MATLAB code is given in Code Box 2.2. Figure 2.10 shows the explicit and numerical solutions overlapping as best as the eye can see with the right end of the curve flat, that is the zero Neumann boundary condition turning out naturally.

2.9 Two-Dimensional Linear, Triangular Finite Elements

Just as we did this on one-dimension, we will broach this section on two-dimensions too through an example. Figure 2.11 shows a two-dimensional system governed by the Poisson equation $-\sigma\nabla^2 T = q$ with zero Neumann conditions along the upper and lower boundaries and Dirichlet conditions on the left and right boundaries; and $\sigma = 4$ and $q = 1$ in the shaded region, and

CODE BOX 2.2 One-D FEM with Dirichlet and Neumann Conditions

```
clc; clear all;
n=100;
m=n;%no. of unknowns
PG=zeros(m,m);
QG=zeros(m,1);
L=10/n;
y=zeros(n,1);
y(n+1)=10;
P=[1, -1;-1,1]/L;
Q=[1;1]*L;
for e=1:n
  if e==1
    v(1)=n+1;
    v(2)=1;
  else
    v(1)=e-1;
    v(2)=e;
  end;%if
  for r=1:2 %
    if v(r)<=m
      QG(v(r))=QG(v(r))+Q(r);
      for c=1:2
        if v(c)<=m
          PG(v(r),v(c))= PG(v(r),v(c))+P(r,c);
        else
          QG(v(r))=QG(v(r))-P(r,c)*y(v(c));
        end;
      end;
    end;
  end;
end;
xx=inv(PG)*QG;
for i=1:n+1
  xaxis(i,1)=(i-1)*L;
  yexact(i,1)=10+20*xaxis(i,1)-(xaxis(i,1))^2;
  if i==1
    yaxis(i,1)=10;
    else
      yaxis(i,1)=xx(i-1,1);
    end;
end;
%plot (xaxis, yexact);
plot (xaxis, yaxis,'r', xaxis, yexact, 'b');
```

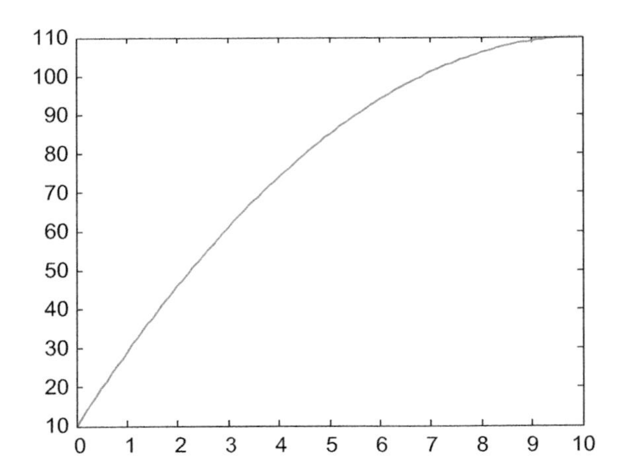

FIGURE 2.10
Explicit and numerical solutions to one-dimensional problem with natural boundary condition.

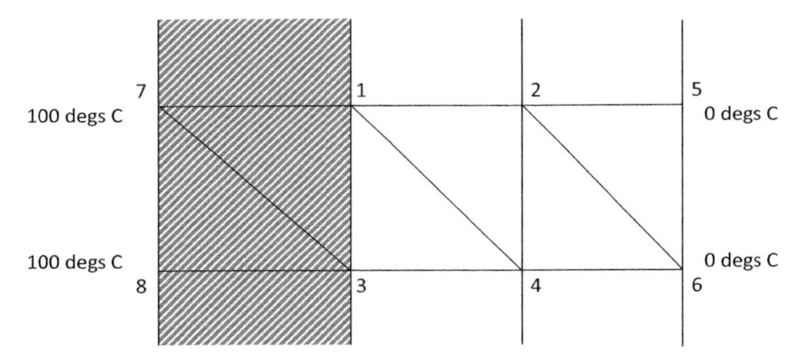

FIGURE 2.11
Heat flow problem.

$\sigma = 1$ and $q = 0$ elsewhere. Although this is a problem governing the flow of heat, the equations are similar to those developed for an electrostatic system with $\sigma = \varepsilon$and $q = \rho$. Solve for T everywhere using hand-calculations

We have by our linear approximations or trial functions

$$T = a + bx + cy \tag{2.106}$$

We have already seen that the first-order approximation $\varphi = a + bx + cy$ on a triangle leads to

$$b = \sum_{i=1}^{3} b_i \varphi_i \tag{2.107}$$

$$c = \sum_{i=1}^{3} c_i \varphi_i \tag{2.108}$$

Now

$$\nabla T = \vec{u}_x \frac{\partial T}{\partial x} + \vec{u}_y \frac{\partial T}{\partial y} = \vec{u}_x b + \vec{u}_y c \tag{2.109}$$

The functional

$$\begin{aligned}\mathcal{L}(T) &= \iint_R \left(\left[\frac{1}{2}\sigma(\nabla T)^2 - qT \right] \right) dR = \sum_{\text{Elements}} \iint_{\text{Triangle}} \left[\frac{1}{2}\sigma\left(b^2 + c^2\right) - qT \right] dR \\ &= \sum_{\text{Elements}} \left\{ \frac{1}{2}\sigma\left(b^2 + c^2\right) A - q_0 \frac{T_1 + T_2 + T_3}{3} A \right\}\end{aligned} \tag{2.110}$$

where, as before in one-dimension, we have used the fact that $\sigma(b^2 + c^2)$ is a constant to pull it out of the integral sign so that the integral of dR is simply the triangle area A. Likewise, by assuming that q is a constant at q_0 in a triangle, that too comes out of the integral sign so that the integral of the linearly varying T is the average of T at the three vertices times a measure of the integration area which is A; this is a variant of Simpson's rule in one-dimension for two-dimensions. Thus we have, substituting for b and c, while noting that

$$b = \sum_{i=1}^{3} b_i T_i = \underline{T}^t \underline{b} \tag{2.111}$$

$$c = \sum_{i=1}^{3} c_i T_i = \underline{T}^t \underline{c} \tag{2.112}$$

with notation for $\underline{T}$, $\underline{b}$ and $\underline{c}$ defined by correspondence for terms. Now noting that a scalar b is its own transpose, we have

$$b^2 = b.b = b.b^t = \left(\underline{T}^t \underline{b}\right)\left(\underline{b}^t \underline{T}\right) = \underline{T}^t \underline{b}\,\underline{b}^t \underline{T} = \underline{T}^t \begin{bmatrix} b_1^2 & b_1 b_2 & b_1 b_3 \\ b_2 b_1 & b_2^2 & b_2 b_3 \\ b_3 b_1 & b_3 b_2 & b_3^2 \end{bmatrix} \underline{T} \tag{2.113}$$

where the brackets have been dispensed with because of there being multiplicative compatibility. Similarly,

$$c^2 = c.c = c.c^t = \left(\underline{T}^t \underline{c}\right)\left(\underline{c}^t \underline{T}\right) = \underline{T}^t \underline{c}\,\underline{c}^t \underline{T} = \underline{T}^t \begin{bmatrix} c_1^2 & c_1 c_2 & c_1 c_3 \\ c_2 c_1 & c_2^2 & c_2 c_3 \\ c_3 c_1 & c_3 c_2 & c_3^2 \end{bmatrix} \underline{T} \tag{2.114}$$

Therefore,

$$\mathcal{L}(T) = \sum_{\text{Elements}} \left\{ \tfrac{1}{2}\sigma\left(b^2 + c^2\right)A - q_0 \tfrac{T_1+T_2+T_3}{3} \right\}$$

$$= \sum_{\text{Elements}} \left\{ \tfrac{1}{2}\underline{T}^t \sigma A \begin{bmatrix} b_1^2 + c_1^2 & b_1b_2 + c_1c_2 & b_1b_3 + c_1c_3 \\ b_2b_1 + c_2c_1 & b_2^2 + c_2^2 & b_2b_3 + c_2c_3 \\ b_3b_1 + c_3c_1 & b_3b_2 + c_3c_2 & b_3^2 + c_3^2 \end{bmatrix} \right\} \underline{T} \quad (2.115)$$

$$-\left(\underline{T}\right)^t \begin{Bmatrix} \frac{q_0A}{3} \\ \frac{q_0A}{3} \\ \frac{q_0A}{3} \end{Bmatrix}$$

where, in corresponding notation, the local matrices

$$\left[P\right] = \sigma A \begin{bmatrix} b_1^2 + c_1^2 & b_1b_2 + c_1c_2 & b_1b_3 + c_1c_3 \\ b_2b_1 + c_2c_1 & b_2^2 + c_2^2 & b_2b_3 + c_2c_3 \\ b_3b_1 + c_3c_1 & b_3b_2 + c_3c_2 & b_3^2 + c_3^2 \end{bmatrix} \quad (2.116)$$

and

$$\underline{Q} = \begin{Bmatrix} \frac{q_0A}{3} \\ \frac{q_0A}{3} \\ \frac{q_0A}{3} \end{Bmatrix} \quad (2.117)$$

From here onwards the procedure is exactly as in the one-dimensional example we worked on before. Consider the first triangle from the left of the mesh of Figure 2.11, namely, triangle 8-3-7 presented in Figure 2.12. Where we start

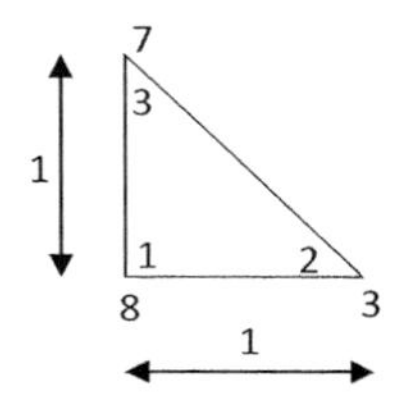

FIGURE 2.12
Triangle with internal generic numbers 1, 2, 3 and corresponding actual numbers 8, 3, 7.

numbering the nodes (whether 8-3-7 or 3-7-8 or 7-8-3 does not matter but that the order ought should be counterclockwise is important).

For this triangle

$$\Delta = \begin{vmatrix} 1 & x_1 & y_1 \\ 1 & x_2 & y_2 \\ 1 & x_3 & y_3 \end{vmatrix} = \begin{vmatrix} 1 & 0 & 0 \\ 1 & 1 & 0 \\ 1 & 0 & 1 \end{vmatrix} = 1 \tag{2.118}$$

Therefore, area $A = ½$. We could have said Area $= ½ \times \text{base} \times \text{height} = ½$ but this is not general enough an approach for computer implementation where triangles whose sides are not parallel to the Cartesian axes or perpendicular to each other must be included in the algorithms for generality. We have

$$\underline{b} = \begin{Bmatrix} \frac{y_2 - y_3}{\Delta} \\ \frac{y_3 - y_1}{\Delta} \\ \frac{y_1 - y_2}{\Delta} \end{Bmatrix} = \begin{Bmatrix} \frac{0-1}{1} \\ \frac{1-0}{1} \\ \frac{0-0}{1} \end{Bmatrix} = \begin{Bmatrix} -1 \\ 1 \\ 0 \end{Bmatrix} \tag{2.119}$$

$$\underline{c} = \begin{Bmatrix} \frac{x_3 - x_2}{\Delta} \\ \frac{x_1 - x_3}{\Delta} \\ \frac{x_2 - x_1}{\Delta} \end{Bmatrix} = \begin{Bmatrix} \frac{0-1}{1} \\ \frac{0-0}{1} \\ \frac{1-0}{1} \end{Bmatrix} = \begin{Bmatrix} -1 \\ 0 \\ 1 \end{Bmatrix} \tag{2.120}$$

Observe that only relative coordinates are important since we are dealing with differences in coordinate values which make the actual values of the coordinates irrelevant. Now

$$\begin{aligned} [P] &= \sigma A \begin{bmatrix} b_1^2 + c_1^2 & b_1 b_2 + c_1 c_2 & b_1 b_3 + c_1 c_3 \\ b_2 b_1 + c_2 c_1 & b_2^2 + c_2^2 & b_2 b_3 + c_2 c_3 \\ b_3 b_1 + c_3 c_1 & b_3 b_2 + c_3 c_2 & b_3^2 + c_3^2 \end{bmatrix} \\ &= 4 \times 0.5 \begin{bmatrix} 1+1 & -1+0 & 0-1 \\ -1+0 & 1+0 & 0+0 \\ 0-1 & 0+0 & 0+1 \end{bmatrix} = \begin{bmatrix} 4 & -2 & -2 \\ -2 & 2 & 0 \\ -2 & 0 & 2 \end{bmatrix} \end{aligned} \tag{2.121}$$

$$\underline{Q} = \begin{Bmatrix} \frac{1 \times 0.5}{3} \\ \frac{1 \times 0.5}{3} \\ \frac{1 \times 0.5}{3} \end{Bmatrix} = \begin{Bmatrix} \frac{1}{6} \\ \frac{1}{6} \\ \frac{1}{6} \end{Bmatrix} \tag{2.122}$$

The local matrices $[P]$ and $\underline{Q}$ may be computed for general implementation by the Algorithms 2.1 and 2.9.2.

Algorithm 2.1: Computing First-Order Differentiation Matrices

```
Procedure Triangle (b, c, A, x, y)
{Function: To compute the first order differentiation
matrices
Outputs: b,c = The 3×1 first-order differentiation
matrices defined in eqns. (2.119) to (2.122);
A = Area of triangle
Inputs: x,y = 3×1 vectors containing the coordinates
of the three vertices}
Begin
  Delta ← x[2]*y[3] - x[3]*y[2] + x[3]*y[1] -
  x[1]*y[3] + x[1]*y[2] - x[2]*y[1] !Eqn.(7.5.17)
 A ← Abs(Delta)/2          !Eqn.(7.5.17)
 For i ← 1 To 3 Do
 i1 ← Mod(i, 3) + 1 {Or i Mod 3 + 1}
 i2 ← Mod(i1, 3) + 1
 b[i] ← (y[i1] - y[i2])/Delta                  !Eqn.(2.119)
 c[i] ← (x[i2] - x[i1])/Delta                  !Eqn.(2.120)
End
```

Algorithm 2.2: Forming Local Matrices for First-Order Triangle

```
Procedure FirstOrderLocalMats (P, q, x, y, Eps, Rho)
{Function: To compute the first-order differentiation
matrices
Outputs: P = The 3×1 first-order local element matrix
- Eqn. (2.121); q = The 3×1 local
right-hand side vector - Eqn. (2.122)
Inputs: x,y = 3×3 vectors containing the coordinates
of the three vertices; Eps = The constant
material value ε in the triangle - Eqn. (1.2.22); Rho
= 3×1 vector giving source ρ at the three
vertices (or a constant if ρ is a constant)
Required: Procedure Triangle}
Begin
```

```
Triangle (b, c, A, x, y)
For i ← 1 To 3 Do
  q[i] ← A*(2*Rho[i] + Rho[i1] + Rho[i2])/12
   {Or if Rho is a constant q[i] ← A*Rho/3
   in place of the previous two lines}
  For j ← 1 To 3 Do
   P[i, j] ← Eps*A*(b[i]*b[j] + c[i]*c[j])
                                          {Eqn. (2.121)}
End
```

We observe (and leave it as an exercise to the reader to check this out) that if we number the next triangle as 1-7-3 to have shape correspondence to the previous triangle, the local matrices will be exactly the same. This is a common way to reduce hand computations although most computations are done today on a computer so that how we number triangles does not matter. Likewise, for the next two triangles except for differences in σ and q, which can be easily accounted for through scaling. So, the local matrices are as shown in Figure 2.13.

The four rows and columns of the matrix [PG] are filled up as shown in Figure 2.14. For clarity, each position of the global matrix [PG] is divided into six, each sub-location corresponding to the contribution from one of the six elements. Similarly, the column vector $\underset{\sim}{Q}G$ shows its contributions from the six elements as it is built up.

The numbers 200 in the right-hand side column vector $\underset{\sim}{Q}G$ come from the two additional columns of [PG] corresponding to the known T values at 100.

Ele. 1	PMatrix				QMatrix	Ele. 2	PMatrix				QMatrix
Verts	8	3	7			Verts	1	7	3		
8	4	-2	-2		1/6	1	4	-2	-2		1/6
3	-2	2	0		1/6	7	-2	2	0		1/6
7	-2	0	2		1/6	3	-2	0	2		1/6
Ele. 3	PMatrix				QMatrix	Ele. 4	PMatrix				QMatrix
Verts	3	4	1			Verts	2	1	4		
3	1	-0.5	-.5		0	2	1	-0.5	-.5		0
4	-0.5	0.5	0		0	1	-0.5	0.5	0		0
1	-0.5	0	0.5		0	4	-0.5	0	0.5		0
Ele. 5	PMatrix				QMatrix	Ele. 6	PMatrix				QMatrix
Verts	4	6	2			Verts	5	2	6		
4	1	-0.5	-.5		0	5	1	-0.5	-.5		0
6	-0.5	0.5	0		0	2	-0.5	0.5	0		0
2	-0.5	0	0.5		0	6	-0.5	0	0.5		0

FIGURE 2.13
The local matrices of the six elements of Figure 2.11.

The 4x4 Matrix [PG]									$\underline{QG}$ of size 4x1		
	4				-2				1/6	200+1/6	
0.5	0.5		-0.5	-0.5							
	-0.5		1				-0.5				
		0.5	0.5			-0.5					
	-2			2	2				200+1/6		
-0.5				1		-0.5					
						-0.5					
							1				
			-0.5	-0.5							
		-0.5				1					

FIGURE 2.14
The local matrices of Figure 2.13 added into global matrices (each matrix location has six boxes for each contributing element).

The numbers zero when they need to be added are not reflected. Thus we have, from Figure 2.14,

$$\begin{bmatrix} 5 & -0.5 & -2.5 & 0 \\ -0.5 & 2 & 0 & -1 \\ -2.5 & 0 & 5 & -0.5 \\ 0 & -1 & -0.5 & 2 \end{bmatrix} \begin{Bmatrix} T_1 \\ T_2 \\ T_3 \\ T_4 \end{Bmatrix} = \begin{Bmatrix} \frac{601}{3} \\ 0 \\ \frac{1201}{3} \\ 0 \end{Bmatrix} \tag{2.123}$$

Solving

$$\underline{T}^t = \begin{bmatrix} 2000 & 2000 \end{bmatrix} \tag{2.124}$$

2.10 Cholesky's Factorization

Consider the Cholesky scheme of matrix solution which is most suitable and efficient for the symmetric matrices we encounter in finite element analysis. It is better than the competitive Conjugate Gradients Algorithm for matrix sizes less than 1000x1000 – a rough statement because an iterative scheme where the solution is progressively improved, may be made to converge faster or slower by playing with the convergence criterion, whereas in a direct scheme, we finish only when all steps are over.

Cholesky's Method for solving the equation

$$[A]\{x\} = \{B\} \tag{2.125}$$

requires the splitting or decomposition of [*A*] into lower and upper triangular factors [*L*] and [*U*]

$$[A] = [L][U] \tag{2.126}$$

This is Cholesky's factorization, alternatively called El-Yoo factorization. The equation to be solved is therefore,

$$[L][U]\{x\} = \{B\} \tag{2.127}$$

Letting

$$[U]\{x\} = \{z\} \tag{2.128}$$

we have

$$[L]\{z\} = \{B\} \tag{2.129}$$

Once [*L*] and [*U*] are computed it is a simple matter to compute {*z*} in the order $z_1, z_2, \ldots, z_n$ by forward elimination from [*L*]{*z*}={*Q*} and thus {*x*} in the reverse order $x_n, x_{n-1}, \ldots, x_1$ by back substitution from [*U*]{*x*}={*z*}. The method becomes efficient, and therefore attractive for symmetric matrices [*A*] because in that case

$$[A] = [A]^t \tag{2.130}$$

so that

$$[L][U] = [[L][U]]^t = [U]^t[L]^t \tag{2.131}$$

Since the transpose of an upper-triangular matrix is a lower triangular matrix and *vice versa*,

$$[L] = [U] \tag{2.132}$$

meaning that we need to compute only half the numbers as when we need to compute both [*L*] and [*U*]. In this case the procedure for computing [*L*] is found by examining $[A]=[L][L]^t$

$$\begin{bmatrix} L11 & & & & & & \\ L21 & L22 & & & & & \\ L31 & L32 & L33 & & & & \\ etc & etc & etc & & & & \\ Lj1 & Lj2 & Lj3 & etc & Ljj & & \\ etc & etc & etc & etc & etc & & \\ etc & etc & etc & etc & etc & & \\ Li1 & Li2 & Li3 & etc & Lij & etc & Lii \end{bmatrix} \mathrm{X} \begin{bmatrix} L11 & L21 & L31 & etc & Lj1 & etc & Li1 & etc \\ & L22 & L32 & etc & Lj2 & etc & Li2 & etc \\ & & L33 & etc & Lj3 & etc & Li3 & etc \\ & & & & etc & etc & etc & etc \\ & & & & Ljj & etc & etc & etc \\ & & & & & & Lii & etc \\ & & & & & & & etc \\ & & & & & & & \end{bmatrix}$$

and equating the element at 11 with the product for that location

$$A_{11} = L_{11}^2 \text{ or } L_{11} = \sqrt{A_{11}} \tag{2.133}$$

Proceeding thus to the second-row columns 1 and 2, third-row columns 1, 2, and 3, etc.

$$A_{21} = L_{21}L_{11} \text{ or } L_{21} = \frac{1}{L_{11}}A_{21} \tag{2.134}$$

$$A_{22} = L_{21}^2 + L_{22}^2 \text{ or } L_{22} = \sqrt{A_{22} - L_{21}^2} \tag{2.135}$$

$$A_{31} = L_{31}L_{11} \text{ or } L_{31} = \frac{1}{L_{11}}A_{31} \tag{2.136}$$

etc., giving us in general

$$L_{ij} = \begin{cases} \text{When } j = 1: A_{ij} = L_{i1}L_{11} \text{ or } L_{i1} = \dfrac{1}{L_{11}}A_{i1} \\ \text{When } j \neq 1: L_{i1}L_{j1} + L_{i2}L_{j2} + \ldots + L_{ij}L_{jj} \text{ or } L_{ij} = \dfrac{1}{L_{jj}}\left(A_{ij} - \sum_{k=1}^{j-1} L_{ik}L_{jk}\right) \end{cases} \tag{2.137}$$

Similarly,

$$L_{ii}\left(i \neq 1\right) = \sqrt{L_{11}^2 + L_{21}^2 + \ldots + L_{ii}^2} \text{ or } L_{ii} = \sqrt{\left(\sum_{k=1}^{i-1} L_{ik}^2\right)} \tag{2.138}$$

We first note that A_{ij} is required only once and that is to compute L_{ij}. So as each L_{ij} is computed it is stored as A_{ij} whose memory is not needed thereafter.

This equation (2.137) with (2.138) is the basic algorithm. Special terms need to be accounted for in programming when the summation of Equation (2.137) is not involved. For example, terms such as L_{i1} which need no summation.

The El-Yoo function has already been presented in Code Box 2.4.

2.11 A Two-Dimensional Finite Element Program through an Example

We will consider without too much explanation a two-dimensional problem and a MATLAB program to solve it – not too much to absorb so early in this

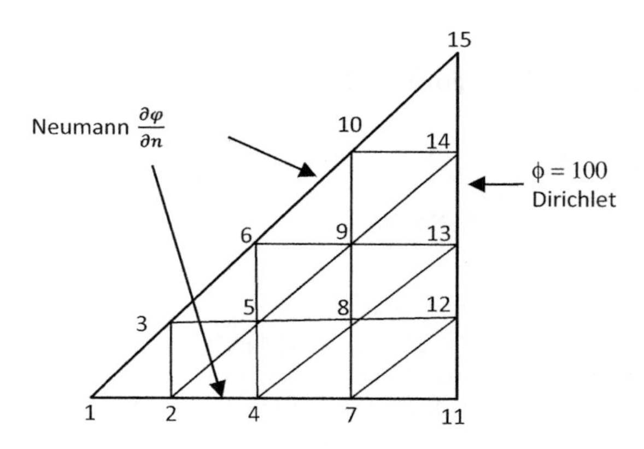

FIGURE 2.15
An eighth of a cable core prepared for finite element analysis.

text because the principles are an extension of the work already done and demonstrated in one-dimension.

Figure 2.15 shows an eighth of a square cable with a square core surrounded by a sheath at 100 V. It has been prepared for finite element analysis with 16 elements and 15 nodes. The sheath is a Dirichlet boundary and the two lines of symmetry are Neumann boundaries. So, the known nodes, nodes 11 to 15, are numbered last. There are 16 elements and 15 nodes of which 10 need their potential values computed. Left of the line 4-5-6, $(\varepsilon,\rho) = (2,1)$. To the right of the line the corresponding value of the number pair is (1,0).

Code Boxes 2.3 to 2.7 give the MATLAB solution to this problem. In the main program (see Code Box 2.3) the coordinates of the 15 nodes are defined in XG and YG. Phi contains dummy values for the potentials of the ten nodes for which it is to be computed and the last five values of Phi are the known values. DatE contains the element data – for each element the three vertices and its ε and ρ values. The program then, after initializing PG and QG, the global matrices, takes up each element, forms its local matrices P and q using the function Triangle (see Code Box 2.6) and then places them in the global matrices using the function GlobalPlace (see Code Box 2.5). Once all elements' contributions are done, the function ElYoo (see Code Box 2.4) solves the equation using Cholesky's factorization (also called LU-decomposition), forward elimination, and back-substitution. Now the solution is ready. To process it, the MATLAB function surf is used. It plots a vector z defined at coordinate locations in the vectors x and y on a rectangular grid. The function Nearest (see Code Box 2.7), of no general value but written for this problem, identifies the nodes of the finite element grid nearest to the nodes in x,y and gets the solution plotted.

CODE BOX 2.3 For Problem of Figure 2.15 by 2-D FEM

```
clc; clear all;
n= 10; %matrix size
NNodes = 15; %No. of Nodes
NElems = 16;
PG = zeros (10,10);
QG= zeros(10,1);
XG = [0;1;1;2;2;2;3;3;3;3;4;4;4;4;4]; %These are the
x-coordinates of nodes 1 to 10
YG = [0;0; 1;0;1;2;0;1;2;3;0;1;2;3;4];%These are the
y-coordinates of nodes 1 to 10
DatE= [1,2,3, 2, 10; %v1,v2,v3,eps,rho
  2, 4, 5, 2, 10;
  2, 5, 3, 2, 10;
  3, 5, 6, 2, 10;
  4, 7, 8, 1, 0;
  4, 8, 5, 1, 0;
  5, 8, 9, 1, 0;
  5, 9, 6, 1, 0;
  6, 9, 10, 1, 0;
  7, 11, 12, 1, 0;
  7, 12, 8, 1, 0;
  8, 12, 13, 1, 0;
  8, 13, 9, 1, 0;
  9, 13, 14, 1, 0;
  9, 14, 10, 1, 0;
  10, 14, 15, 1, 0];
Phi = [0.0; 0.0; 0.0; 0.0; 0.0; 0.0; 0.0; 0.0; 0.0;
0.0;100.0;100.0;100.0;100.0;100.0];
%Last 5 positions known
for Element = 1: NElems
  for i=1:3
    V(i)=DatE(Element,i);
  end;
  eps=DatE(Element,4);
  rho=DatE(Element,5);
  for i = 1:3
    x(i) = XG(V(i));% x-coordinates of 3 vertices
    y(i) = YG(V(i));% y-coordinates of 3 vertices
  end;
  [P,q] = Triangle(x,y,eps,rho);
  [PG,QG] = GlobalPlace(PG,QG,P,q,n,V,Phi);
end; %for Element
Solution=ElYooFunction(PG,QG,n);
for i=1:n
  Phi(i)=Solution(i);
end;
```

```
for i=1:5
  for j=1:i
    xx(i,1)=i-1;
    yy(i,1)=j-1;
    k=Nearest(xx(i,1),yy(i,1), XG,YG,NNodes);
    zz(i,j)=Phi(k);
    if j<i
      zz(j,i)=Phi(k);
    end;
  end;
end;
surf(xx,yy,zz);
end;
x
```

CODE BOX 2.4 The Function ElYoo for LU-Decomposition – that is, Cholesky's Method

```
function x= ElYooFunction(A,B,n)
L(1,1)= sqrt(A(1,1));
for i=2:n
  %First
compute all Ls up to the diagonal
  for j=1:i-1
    sum=A(i,j);
    for k=1:j-1
      sum=sum-L(j,k)*L(i,k);
    end;
    L(i,j)=sum/L(j,j);
  end;
  %Then compute L at the diogonal
  sum=A(i,i);
  for k=1:i-1
    sum=sum-L(i,k)*L(i,k);
  end;
  L(i,i)=sqrt(sum);
end;
L
U=L';
P=L*U
Z(1)=B(1)/L(1,1);
for i=2:n
  sum=B(i);
```

```
  for k=1:i-1
    sum=sum-L(i,k)*Z(k);
  end;
  Z(i)=sum/L(i,i);
end;
Q=L*Z'
x(n,1)=Z(n)/L(n,n);
for i=1:n-1
  ir=n-i;
  sum=Z(ir);
  for ic=ir+1:n
    sum=sum-L(ic,ir)*x(ic);
  end;
  x(ir,1)=sum/L(ir,ir);
```

CODE BOX 2.5 Function GlobalPlace Placing Local Matrices in Global Matrices

```
GlobalPlace.m
function [PG,QG] =
GlobalPlace(PGI,QGI,P,q,MatSize,V,Phi)
  PG=PGI;
  QG=QGI;
  for i = 1:3
    if V(i)<=MatSize %Then Goes into PG
      NR = V(i);
      QG(NR) = QG(NR) + q(i);
      for j = 1:3
        NC = V(j);
        if NC <= MatSize
          PG(NR,NC) = PG(NR,NC) + P(i,j);
        else
          QG(NR) = QG(NR) - P(i,j)* Phi(NC);
        end; %if statement end;
      end;% Loop for j ends
    end;
  end;
```

CODE BOX 2.6 Function Triangle Computes Local Matrices for a Triangle

```
function [P,q] = Triangle(x,y,eps,rho);
Del=(x(2)*y(3)-x(3)*y(2)) + (x(3)*y(1)-x(1)*y(3))+ ...
(x(1)*y(2)-x(2)*y(1));
  Area=Del/2;
  for i=1:3
    i1=mod(i,3)+1;
    i2=mod(i1,3)+1;
    b(i)=(y(i1)-y(i2))/Del;
    c(i)=(x(i2)-x(i1))/Del;
  end;
  const1=rho*Area/3;
  const2=Area*eps;
  for i=1:3
    q(i,1)=const1;
    for j=1:3
      P(i,j)=const2*(b(i)*b(j)+c(i)*c(j));
    end;
  end;
```

CODE BOX 2.7 Function Nearest Identified Nodes Nearest to Coordinate (x,y)

```
function k=Nearest(x,y, XG,YG,n)
Error=10000;
k=0;
for i=1:n
  SquareDistance=(XG(i)-x)^2+(YG(i)-y)^2;
  if Error > SquareDistance
    Error=SquareDistance;
    k=i;
  end;
end;
```

2.12 Other Equations

Besides the Poisson equation, two other equations commonly appear in electrical engineering. These are the wave equation and the diffusion equation for eddy current problems. Functionals need to be found for these.

The wave equation for waveguides is for the component of the non-transverse electric or magnetic field φ,

$$\nabla^2\varphi = -k^2\varphi \tag{2.139}$$

where ∇^2 operates on the transverse plane and k is defined in terms of the propagation constant γ, frequency ω and velocity of light c:

$$k^2 = \gamma^2 + \frac{\omega^2}{c^2} \tag{2.140}$$

Here we will use the homogeneous triangular coordinates $(\zeta_1, \zeta_2, \zeta_3)$, where in a triangle with vertices 1-2-3, the coordinate point (x,y) is replaced by $(\zeta_1, \zeta_2, \zeta_3)$. Each triangular coordinate, ζ_i is the height h_i of the point (x,y) above the base opposite the vertex i scaled by the height H_i of vertex i above the opposite base i:

$$\zeta_i = \frac{h_i}{H_i}. \tag{2.141}$$

Since two independent coordinates cannot be replaced by three independent coordinates, it may be shown (Hoole, 1989b) that

$$\zeta_1 + \zeta_2 + \zeta_3 = 1. \tag{2.142}$$

Our trial function interpolates the value of φ from the three vertices of the triangle

$$\varphi = \zeta_1\varphi_1 + \zeta_2\varphi_2 + \zeta_3\varphi_3 = \check{\alpha}\check{\varphi}^t \tag{2.143}$$

with corresponding notation and the accent standing for a row vector.

From our considerations, we know that

$$\mathcal{L} = \tfrac{1}{2}\iint_R (\nabla\varphi)^2\, dR \tag{2.144}$$

when minimized will satisfy $-\nabla^2\varphi = 0$. We also can see that $(k^2/2)\varphi^2$ upon differentiation with respect to φ will yield $k^2\varphi$. Therefore, the functional we seek is

$$\mathcal{L} = \tfrac{1}{2}\iint_R \left\{(\nabla\varphi)^2 - \frac{k^2}{2}\varphi^2\right\} dR \tag{2.145}$$

Putting in the approximations and noting that over a triangle of area A (Hoole, 1989b)

$$\iint_{\text{Triangle}} \zeta_1^i \zeta_2^j \zeta_3^k dR = \frac{i!j!k!2!}{(i+j+k+2)!} A \tag{2.146}$$

it can be shown that the relevant equation is

$$\sum_{\text{All elements}} \left\{ \tfrac{1}{2} \underline{\varphi}^t [P] \underline{\varphi} - \tfrac{k^2}{2} \underline{\varphi}^t [Q] \underline{\varphi} \right\} = 0 \tag{2.147}$$

where the local matrices are, drawing from Equation (2.116)

$$[P] = A \begin{bmatrix} b_1^2 + c_1^2 & b_1 b_2 + c_1 c_2 & b_1 b_3 + c_1 c_3 \\ b_2 b_1 + c_2 c_1 & b_2^2 + c_2^2 & b_2 b_3 + c_2 c_3 \\ b_3 b_1 + c_3 c_1 & b_3 b_2 + c_3 c_2 & b_3^2 + c_3^2 \end{bmatrix} \tag{2.148}$$

And, using Equation (2.146)

$$\begin{aligned} [Q] &= \iint_{\text{Triangle}} \check{\alpha}^t \check{\alpha}\, dR = \iint_{\text{Triangle}} \begin{bmatrix} \zeta_1^2 & \zeta_1\zeta_2 & \zeta_1\zeta_3 \\ \zeta_2\zeta_1 & \zeta_2^2 & \zeta_2\zeta_3 \\ \zeta_3\zeta_1 & \zeta_3\zeta_2 & \zeta_3^2 \end{bmatrix} dR \\ &= \frac{A}{12} \begin{bmatrix} 2 & 1 & 1 \\ 1 & 2 & 1 \\ 1 & 1 & 2 \end{bmatrix}. \end{aligned} \tag{2.149}$$

This leads to the linearized eigen value problem of the form

$$[P]\underline{\varphi} = \lambda [Q] \underline{\varphi} \tag{2.150}$$

In the case of eddy current problems in two-dimensions, the governing equation is

$$-\frac{1}{2} \nabla^2 \tilde{A} = \tilde{J}_0 - j\omega\alpha\tilde{A} \tag{2.151}$$

where $\tilde{A}$ is the complex, single-component vector potential. By similar considerations as to the wave equation, we may arrive at the functional

$$\mathcal{L} = \iint_R \left\{ \frac{1}{2\mu} \left(\nabla \tilde{A} \right)^2 + \frac{1}{2} j\omega\sigma \tilde{A}^2 - \tilde{A}\tilde{J}_0 \right\} dR \tag{2.152}$$

where the first and third terms together are from the Poissonian part of the diffusion equation and the middle term arises in like considerations to the wave equation.

3

Optimization in Product Design – Synthesis

3.1 An Introduction

Typically, performance requirements on a device to be synthesized through optimization are cast as minimizing an object function defined in terms of the parametric description of the device being designed. These parameters may be the various lengths of the parts of the device as well as material values. As presented in Chapter 1, the general problem is defined as implementing

$$\text{Minimize } f=f(p_1,p_2,\ldots,p_n),$$
$$\text{subject to the inequality constraints } g_i(p_1,p_2,\ldots,p_n)\leq 0 \text{ for } i=1,\ldots,m$$
$$\text{and the additional constraints } h_i(p_1,p_2,\ldots,p_n)=0 \; i=m+1,\ldots,l$$

Generally, the field of optimization was initially an outgrowth of mathematical departments and mathematicians who came up with neat and useful methods. Often, they worked with closed form solutions – that is, known functions expressible in terms of the variables.

Engineering optimization is a field that was pioneered by structural engineers (Hsu, 1994), a field led by such intellectual giants as O.C. Zienkiewicz and G.H. Gallagher (Atrek et al., 1984) and Vanderplaats (1984). In engineering optimization, however, the functions to be minimized are numerical insofar as they have no explicit form in terms of the parameters of design, $p_1, p_2, \ldots, p_n$. Thus, to minimize heat in an electrical machine of complex geometry, the heat dissipated needs to be computed by a numerical method and the object function will depend on the physical dimensions of the machine, the values of the currents that activate and enervate the machine, and materials of which the machine is made. What this means is that gradients of the object function f are difficult to compute, and such computation must be approached by slight variations of the parameters of device description one at a time to compute the gradients by finite difference. Thus, for n variables, $n+1$ field computations must be performed, once for the present set of values of the variables, and once each as the variables are varied one at a time with a slight change to each. With finite elements, there is a way to compute the gradients from one solution, but it is accompanied by restrictions on the mesh as shown by Hoole et al. (1991).

What this means is that in engineering optimization, we must avoid optimization methods that need derivatives. As such, there being three classes of optimization methods – zeroth, first order, and second order, requiring, respectively, no derivative information, first derivative information, and second derivate information – we tend to prefer zeroth-order methods. Despite the good results for first- and second-order methods, most engineers therefore restrict themselves to zeroth-order methods. Yet, first-order methods are also used by engineers because they converge faster. Second-order methods, such as Newton's, are rarely resorted to by engineers.

Therefore, this chapter will deal mainly with zeroth order and some first-order methods.

3.2 One-Dimensional Optimization

3.2.1 One-Dimensional Search

The simplest way to understand optimization is to start with a function of a single variable. Let us consider the function $f(x) = x^2$. In our one-dimensional optimization, we want to minimize $f(x)$ with respect to x. There are no constraints.

It is important to understand this because often even in many-dimensional systems, we reduce the problem to searching for the minimum along a line. The big problem is reduced to multiple sequential line searches.

3.2.2 Bisection Search

Bisection search is a search algorithm that makes use of first derivative information to compute the minimum. It takes in a closed interval $[a_0, b_0]$, and an object function $f(x)$ such that $f(x)$ is unimodal and continuously differentiable along $[a_0, b_0]$ and defines a tolerance of ϵ_0. The implementing algorithmic function outputs x_{opt} s.t. x_{opt} is no more than ϵ_0 away from the minimizer of the object function $f(x)$. The key principle behind this algorithm, as depicted in Figure 3.1, is that the gradient always points away from the minimum. Thus, at any point $a_0 < x_i < b_0$

$$\text{if } f'(x_i) > 0, \text{ then } a_0 < x_{opt} < x_i$$

Conversely,

$$\text{if } f'(x_i) < 0, \text{ then } x_i < x_{opt} < b_0$$

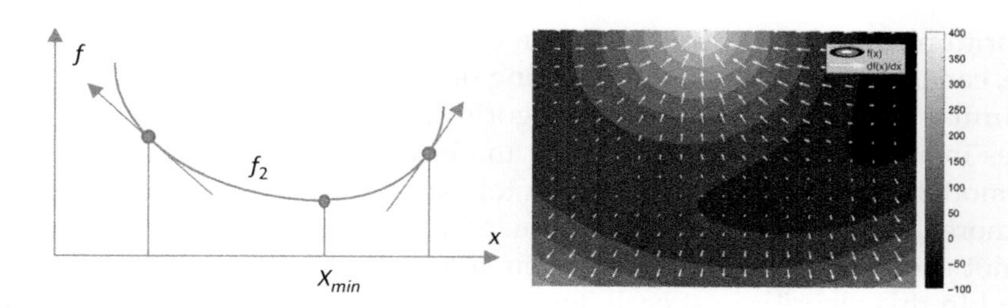

FIGURE 3.1
Gradient points away from minimum: Minimum to the left if gradient is positive and to the right if negative.

The bisection algorithm, presented as Algorithm 3.1, makes use of this property to halve the search space at each iteration. x_i is picked such that $x_i = (b_0 - a_0)/2$. At some point, $b_0 - a_o < \epsilon_0$ and the algorithm will converge.

Algorithms of this class which make use of first derivatives are termed first-order search algorithms. Note that if there are multiple minima in the interval, there is no guarantee that convergence will be to the lowest of the minima.

Algorithm 3.1: Bisection Search

Input: A closed interval $[a_0, b_0]$
$f(x)$ a function unimodal and continuously differentiable along $[a_0, b_0]$
ϵ_0 Output tolerance
Output: $x_{opt} \subset [a_0, b_0]$ s.t. x_{opt} is no more than ϵ_0 away from the minimizer of f(x)

1 if $b_0 - a_0 < \epsilon_0$ **then**
2 | **return** a_0
3 $x_{opt} = (b_0 - a_0)/2$;
4 **if** $f'(x_i) > 0$ **then**
5 | *BisectionSearch*$([a_0, x_{opt}], f(x), \epsilon_0)$
6 **else if** $f'(x_i) < 0$ **then**
7 | *BisectionSearch*$([x_{opt}, b_0], f(x), \epsilon_0)$
8 **else**
9 | **return** x_i

3.2.3 Golden Section Search

The first derivative information, however, is not always available, and even when available it is too costly to have meaningfully rapid convergence. For such cases, a zeroth-order search algorithm may be used. Zeroth-order algorithms only make use of function evaluations, which while still expensive, can be far cheaper than calculating derivative information.

Similar to the bisection search algorithm, the golden search algorithm takes in a closed interval $[a_0,b_0]$, and an object function $f(x)$ such that $f(x)$ is unimodal along $[a_0,b_0]$ and employs a tolerance of ϵ_0. It outputs x_{opt} s.t. x_{opt} is no more than ϵ_0 away from the minimizer of the object function $f(x)$. Since we do not use first derivative information, $f(x)$ need not be differentiable along $[a_0,b_0]$ In this algorithm, at each iteration, two probe points are calculated a_i and b_i s.t. $a_o < a_i, b_i, < b_0$. Since the function is unimodal, we know that the following property holds.

$$\text{if } f(a_i) < f(b_i), \text{ then } a_0 < x_{opt} < a_i$$

Conversely,

$$\text{if } f(a_i) > f(b_i), \text{ then } b_i < x_{opt} < b_0$$

While a_i,b_i may be picked at random, the golden section search requires that the following constraint be met to ensure faster convergence:

$$a_i - a_0 = b_0 - b_i \tag{3.1}$$

and that

$$\frac{a_i - a_0}{b_0 - a_0} = \frac{b_0 - b_i}{b_0 - a_0} = \varphi = \frac{\sqrt{5}-1}{2} \approx 0.618 \tag{3.2}$$

This ratio is based on empirical, quasi-formal reasoning. To maintain this ratio, a_i and b_i are selected such that

$$a_i = a_0 + \varphi(b_0 - a_0) \tag{3.3}$$

Using this we can search in successively smaller areas until $b_0 - a_0$ for a given iteration is smaller than ϵ_0. The algorithm is presented in Algorithm 3.2.

Algorithm 3.2 Golden Section Search

Input: A closed interval $[a_0, b_0]$
$f(x)$ a function unimodal and continuously differentiable along $[a_0, b_0]$
ϵ_0 Output tolerance

Output: $x_{opt} \subset [a_0, b_0]$ s.t. x_{opt} is no more than ϵ_0 away from the minimizer of $f(x)$

```
1 if b₀ - a₀ < ϵ₀ then
2 └return a₀
3 ρ=(√5 - 1)/2;
4 d=ρ * (b₀ - a₀);
5 x₁=b₀ - d;
6 x₂=a₀ + d;
7 ϵ=inf;
8 while ϵ > ϵ₀ do
9  │ if f(x₁) > f(x₂) then
10 │  │ b₀=x₂;
11 │  │ x₂=x₁;
12 │  │ d=ρ * (b₀ - a₀);
13 │  │ x₁=b₀ - d
14 │ else if f(x₁) < f(x₂) then
15 │  │ a₀=x₁
16 │  │ x₁=x₂
17 │  │ d=ρ * (b₀ - a₀);
18 │  │ x₂=a₀+d;
19 │  else
20 │  │ a₀=(x₁+x₂)/2;
21 │  └ b₀=a₀;
22 └ ϵ=2 * |b₀ - a₀|/(a₀+b₀)
23 x_opt=(x₁+x₂)/2
24 return x_opt
```

Figure 3.2 shows the golden section algorithm converging as we search for the minimum of *sin x*. It is converging to $x = 4.71239$ radians (270 degrees) with the *sin x* value of –1,

3.2.4 The Line Search or Univariate Search

Univariate search is the simplest of search methods. Univariate search takes as input an initial point and a set of search directions. At each iteration, we search along one of the search directions for a minimum. This minimum becomes our new point. From here we search along the next search direction (see Figure 3.3).

Also called a line search, the univariate search is a method of searching for the minimum of a function *f* along a direction – a one-dimensional search along a line. It is an integral part of reducing a multi-dimensional search to one direction – so that finding the minimum in a multi-dimensional space is reduced to searching sequentially in different directions. This is an integral component of the steepest descent method as well as the conjugate gradients method.

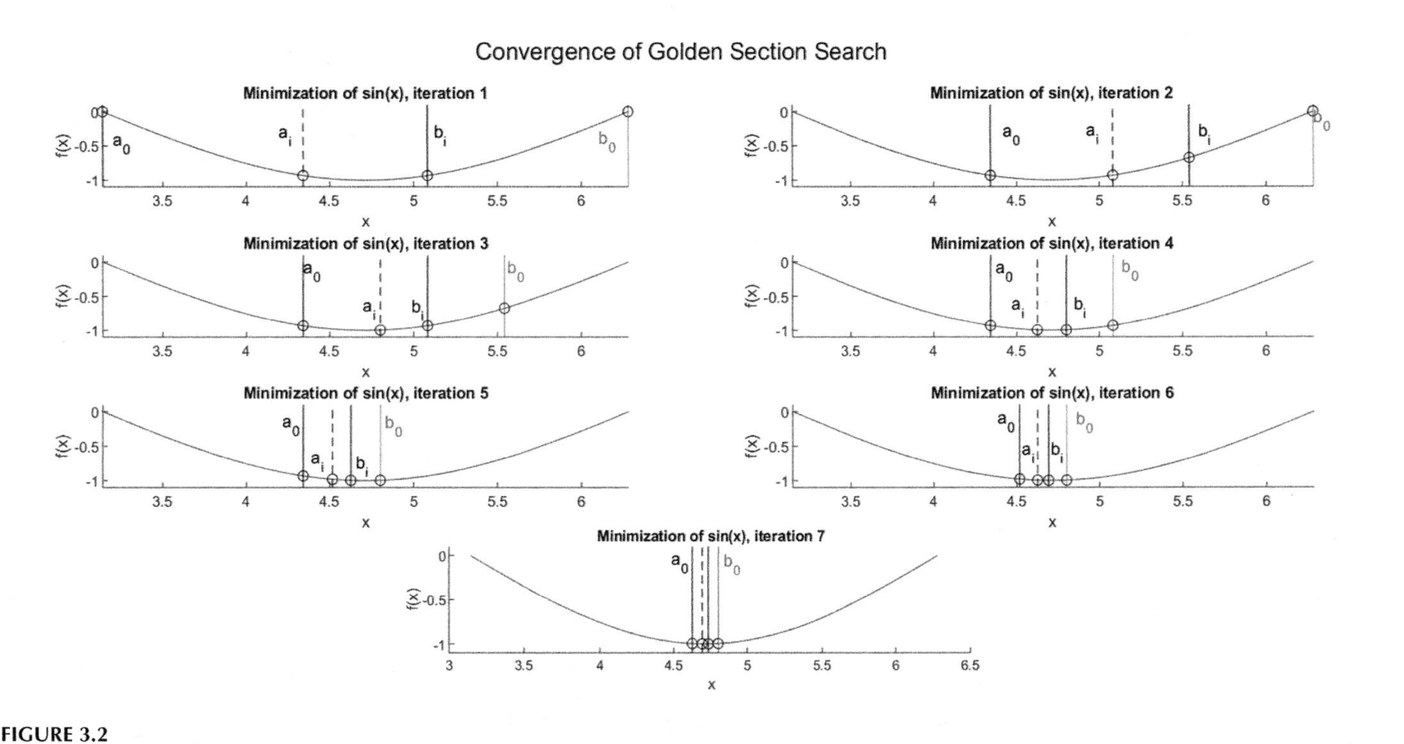

FIGURE 3.2
Convergence of the golden section search algorithm as the search intervals narrow in finding the minimum of the function *sin x*.

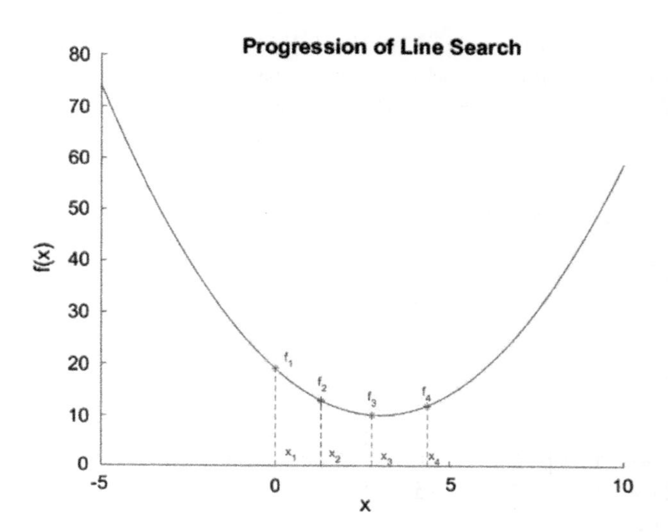

FIGURE 3.3
Curve fitting for the minimum after passing it from the last three points of steepest descent.

Many methods take this approach to searching. As an example, consider a line search starting from x = 0 for the function

$$f = 10 + (x-3)^2 = 19 + x * x - 6 * x \tag{3.4}$$

The derivative of the function is $2x-6$. Setting this to zero, we know that the minimum is at $x = 3$. The algorithm accomplishes finding the minimum numerically by starting at a guess (in this case $x = 0$ is a good start), incrementing x by α, then by 1.1α and again by 1.1 times the new α and so on until f begins to rise. At this point we take the last three values of (x,f) and curve fit

$$f = a + bx + cx^2 \tag{3.5}$$

to get

$$\begin{bmatrix} f_1 \\ f_2 \\ f_3 \end{bmatrix} = \begin{bmatrix} 1 & x_1 & x_1^2 \\ 1 & x_2 & x_2^2 \\ 1 & x_3 & x_3^2 \end{bmatrix} \begin{bmatrix} a \\ b \\ c \end{bmatrix} \tag{3.6}$$

Solving for a, b, and c, we know that the actual minimum, upon differentiation of Equation (3.2) and setting it to 0, is at

$$x_{\min} = -\frac{b}{2c} \tag{3.7}$$

That algorithm is summarized by the following MATLAB code where z is the column vector of *a*, *b*, and *c*, which is found by inverting the matrix of Equation (3.6) and the minimum then follows from Equation (3.7):

```
clear;

f=@ (x) 10+(x-3).^2;

x(1)=inf;

a=1.2;

x(1)=0;
i=1;

while i == 1 || f (x(i)) < f(x(i-1))
   x(i+1)=x(i)+(1.1)^(i) * a;
   i=i+1;
end

B=f(x(1:length(x) - 1));
A=[1, x(1), x(1)^2; 1, x(2), x(2)^2; 1,
x(3), x(3)^2];
z=inv(A) * B';
x_min=-z(2)/(2*z(3))

Output
x_min=3
```

3.3 N-Dimensional Zeroth-Order Optimization

3.3.1 Powell's Method

Powell's method is one such method employing the univariate search to reduce searching for the minimum in a multi-dimensional space to a sequence of one-dimensional univariate searches. It may be shown to be a zeroth-order approximation of a first-order method but that need not detain us here (Vanderplaats, 1984).

In Powell's method we start with an input of an initial point, and a set of search directions. A common starting set for search directions is made up by the unit vectors in the n-dimensional space so that our search directions are $S_{0,}$ which is along the x_1 axis and $S_{1,}$ which is along the x_2 axis. Referring to Figure 3.4, during the first iteration we start at X_0 and search along the given search directions for a minimum. We will move from X_0

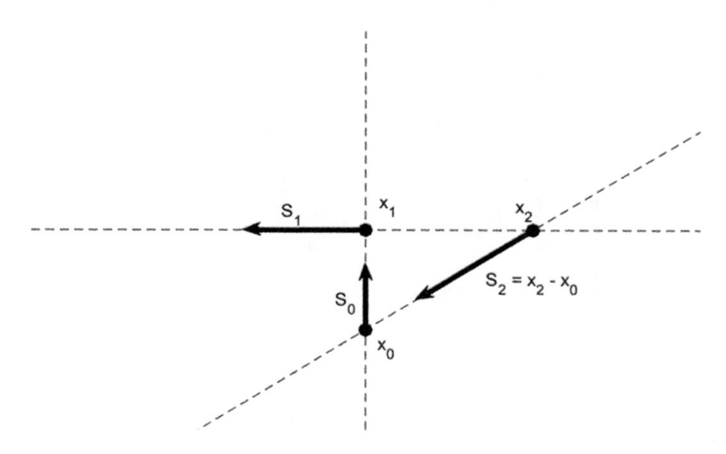

FIGURE 3.4
Search directions of Powell's method.

to X_1 searching along S_0 and then to X_2 searching along S_1. At each new point we search along the search direction and find the minimum in that direction. By the univariate search method, we would continue along in this fashion alternating between (in the case of this example) the two search directions.

Powell's method adds a twist to this methodology. We now generate a third direction given by $S_2 = X_2 - X_0$. After searching along S_2 we complete the iteration. For the next iteration, we now update our search directions to represent this change: we set $S_0 = S_1$, $S_1 = S_2$.

To demonstrate, consider the minimization of a spring system's energy which will give the values of (X_1, X_2) shown in Figure 3.5.

The stretched spring problem of Figure 3.5 (with values of forces and spring constants shown) is an ideal problem whose energy-based object function is given by the work stored in the spring minus the work expended by the two forces,

$$\begin{aligned}\mathcal{L} &= \tfrac{1}{2}\mathbf{K}_1\left\{\sqrt{\mathbf{X}_1^2+(\mathbf{l}_1-\mathbf{X}_2)^2}-\mathbf{l}_1\right\}^2\\ &\quad+\tfrac{1}{2}\mathbf{K}_2\left\{\sqrt{\mathbf{X}_1^2+(\mathbf{l}_2-\mathbf{X}_2)^2}-\mathbf{l}_2\right\}^2-\mathbf{P}_1\mathbf{X}_1-\mathbf{P}_2\mathbf{l}_2\end{aligned} \tag{3.8}$$

where **K** is the spring constant producing the force given by **K** times the stretch of its length representing the energy stored *Kxdx* as the spring is stretched by length *dx*, and force **P** times distance is the work done in stretching the spring, leading to the energy balance of Equation (3.8). Putting in the material values shown in Figure 3.5.

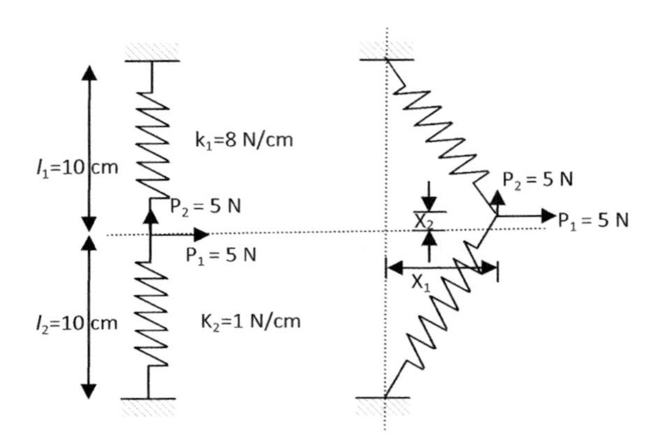

FIGURE 3.5
Stretched spring.

$$\mathcal{L} = 4\left\{\sqrt{X_1^2 + (10 - X_2)^2} - 10\right\}^2 + \tfrac{1}{2}\left\{\sqrt{X_1^2 + (10 - X_2)^2} - 10\right\}^2 - 5X_1 - 20 \quad (3.9)$$

This function is from Vanderplaats (1984) who has shown its properties to bring out the strengths of the various methods. The equal-F lines plotted by us are shown in Figure 3.6. This object function allows us to see how the different methods work with different starting points. Venkataraman (2009) gives several programs with his book so that with the MATLAB toolbox (Davis, 2011) on optimization, students, and researchers have an invaluable aid to problem solving.

Powell's algorithm always deals with n search directions and a univariate search in each direction. It begins with the n unit vectors in the space

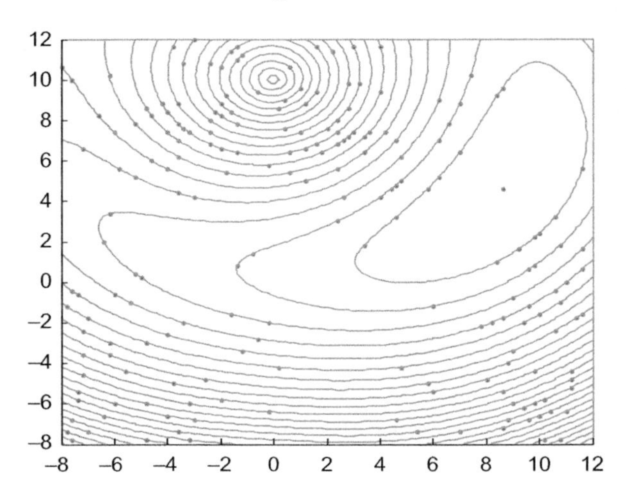

FIGURE 3.6
Object function of Equation (3.8) for the spring on the $X_1 - X_2$ plane.

as the search directions. Thus, for our two-dimensional space [X_1, X_2], the two search directions are $\underset{1}{\boldsymbol{S}} = [1, 0]$ and $\underset{2}{\boldsymbol{S}} = [1, 0]$. Let us say we begin at the n-dimensional point $\underset{0}{\boldsymbol{X}}$. Then searching in the directions $\underset{1}{\boldsymbol{S}}, \underset{2}{\boldsymbol{S}}, \ldots, \underset{n}{\boldsymbol{S}}$ say we arrive at $\underset{1}{\boldsymbol{X}}$. Then finally for this iteration we do a search in the direct $\underset{1}{\boldsymbol{X}} - \underset{0}{\boldsymbol{X}}$ to complete the present iteration.

For the next iteration, the new search directions are: $\underset{2}{\boldsymbol{S}}, \underset{3}{\boldsymbol{S}}, \ldots, \underset{n-1}{\boldsymbol{S}}, \underset{1}{\boldsymbol{X}} - \underset{0}{\boldsymbol{S}}$ and then in the direction from the starting point to the finishing point. So, there are $n+1$ searches per iteration.

This is graphically shown in Figure 3.7. For the first iteration the direction is $\underset{1}{\boldsymbol{S}} = [1, 0]$. That is parallel to the x-axis. Before searching with the line search, we must establish in which direction the function is falling. This being to the left, we go from X^0 to X^1. Now searching in the y-direction where the function is falling in the negative y-direction, we go from X^1 to X^2. To complete the iteration, we do a third search in the direction $X^2 - X^1$.

The three search directions for the next iteration are now [0, 1], taking us to X^4, then in the direction $X^2 - X^1$ taking us to X^5 and then finally in the

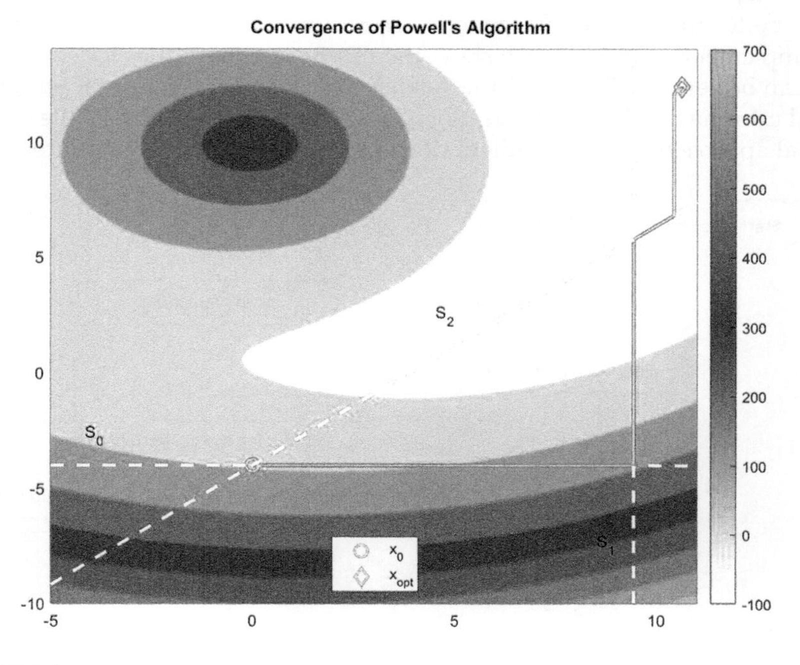

FIGURE 3.7
Progress of Powell's algorithm for the object function of Equation (3.9), with the first few search directions shown for reference.

direction $X^5 - X^3$ taking us to X^6 completing the third search of the second iteration of the two-dimensional space.

3.3.2 Genetic Algorithm

3.3.2.1 Broad Description of the Genetic Algorithm

The genetic algorithm (Sivanandan and Deepa, 2008) is a zeroth-order method of optimization suited to multi-dimensional problems, especially ones with multiple local minima. It is an optimization technique that mimics the process of evolution through natural selection. Given an initial population of solutions, the algorithm iteratively uses the fittest members of the population to breed the successive generation. As such, unlike classical algorithms, the genetic algorithm generates multiple points (that is, potential solutions) at each iteration rather than a single point while moving toward an optimum.

Additionally, while classical algorithms use a deterministic computation to advance to the next iteration, the genetic algorithm spawns the next iteration through probabilistic means.

The key stages of the genetic algorithm are representation, initialization, selection, crossover, mutation, and termination. These are shown below in Figure 3.8. The implementation of the genetic algorithm is in determining these steps.

To better understand the genetic algorithm, it is useful to walk through its implementation. Here, we show the minimization of the *Ackley function.* As can be seen in Figure 3.7, the Ackley function is riddled with numerous local minima, but has a unique global minimum of $f(0,0) = 0$. While a traditional approach such as gradient descent might prematurely declare a local

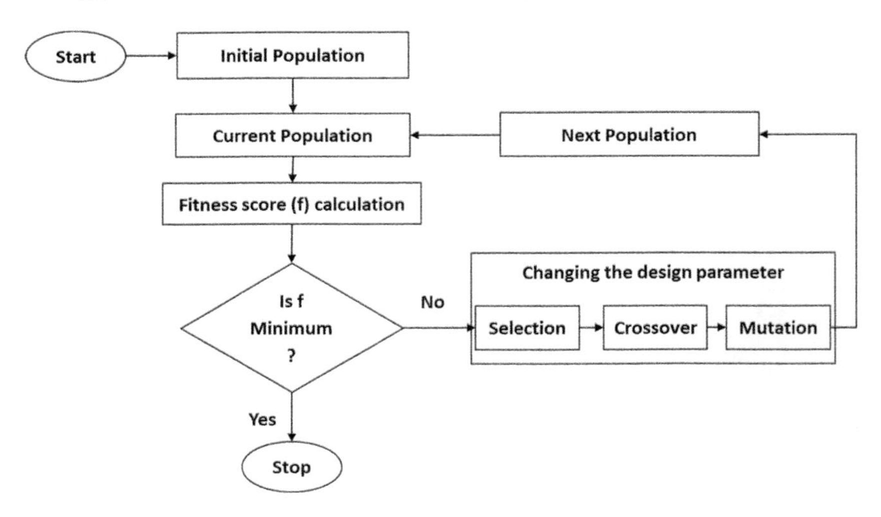

FIGURE 3.8
Optimization using the genetic algorithm.

minimum as the global minimum, we will show that the genetic algorithm uses an approach that in many cases is robust under such conditions. While it can often perform well, it is important to note, however, that the algorithm does not guarantee convergence at the global minimum due to its stochastic nature.

The Ackley function is defined as

$$f(\mathbf{x}) = -a\exp\left(-b\sqrt{\frac{1}{n}\sum_{i=1}^{n} x_i^2}\right) - \exp\left(-b\sqrt{\frac{1}{n}\sum_{i=1}^{n} \cos(cx_i)}\right) + a + \exp(1) \quad (3.10)$$

For our demonstrative example, we take $a = 20$, $b = 0.2$, and $n = 2$. Note that n specifies the dimensionality of the function. For $n = 2$, we have two variables x_1 and x_2; and we are testing on the two-dimensional Ackley function. For those with MATLAB's optimization toolbox, there is a nice implementation of genetic algorithm contained in it.

3.3.2.2 Representation in the Genetic Algorithm

The representation of solutions, genetic representation (Rothlauf, 2006), is highly problem specific. It is essentially the process of transforming our data into a form that maximizes its expressiveness, that is, how well it describes its features and inheritability or how well its characteristics can be passed on to successive generations. In genetic representation a solution is often referred to as an individual or a chromosome and individual features are referred to as genes (see Figure 3.9).

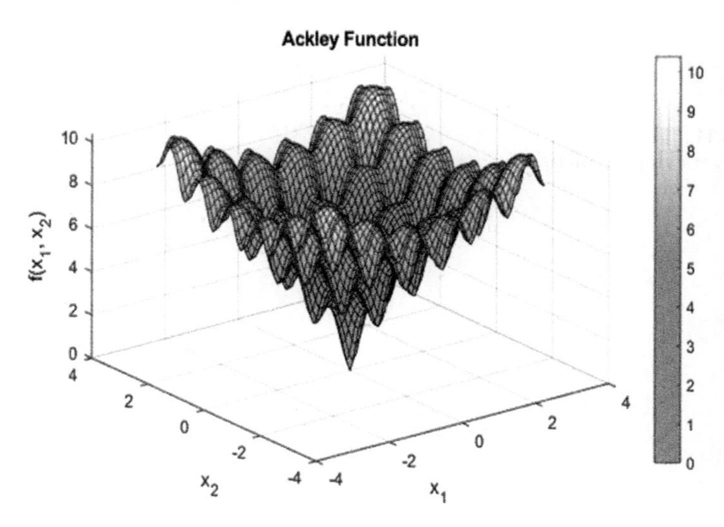

FIGURE 3.9
The Ackley function.

In the case of minimizing an n-dimensional object function $f(\boldsymbol{x})$ where $\boldsymbol{x} \subset \mathbb{R}^n$, the genetic representation can simply be an array of length n where each element i represents x_i. Thus, for our representation of the Ackley function, we have the representation:

x_1	x_2

It is important to note however that, there is no one representation that is best for all classes of problems. Consider the example of maximizing the flow of water through a network of pipes. For such a use case, our array is not expressive enough to describe such a rich problem space, since, for instance, we cannot represent capacity or interconnectivity of the pipes. In such an instance, it may be useful to represent our data as a graph with nodes and weighted edges. Here, each node can represent a pipe joint, and each weighted edge can represent the capacity of any given pipe.

Such a set of possible representations is provided from Figures 3.10 through 3.13. While many other representations exist, these provide a good starting point for newcomers to optimization (Rothlauf, 2006).

It is also important to note that the complexity of the representation has a direct impact on the computational cost of the genetic algorithm, and a more complex representation does not always lead to a better result.

1	0	1	1	0	1

FIGURE 3.10
Binary array representation.

1.23	2.22	3.11	-34.37	12.7	18

FIGURE 3.11
Standard array representation.

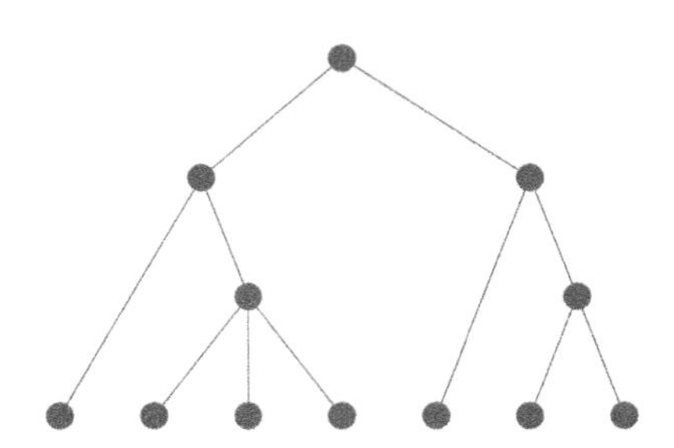

FIGURE 3.12
Binary tree representation.

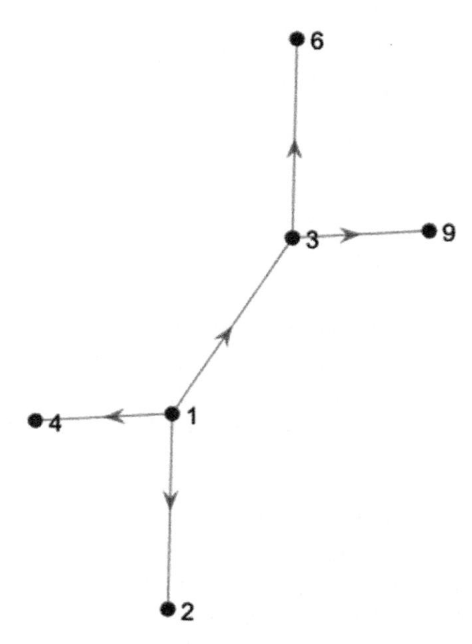

FIGURE 3.13
Directed acyclic graph (DAG) representation.

3.3.2.3 Initialization

In the initialization stage, an initial population of possible solutions is selected. A broad range of initial solutions helps the genetic algorithm to avoid termination at a local minimum. Looking at our biological analogy, this need for a broad range represents the need for genetic diversity. Genetic diversity increases the likelihood that individuals within the population will possess characteristics that are similar to those of the optimal solution.

The two prevailing methods are random initialization and heuristic initialization. In random initialization, a set of solutions is randomly sampled from the solution space. In cases where a known heuristic exists, heuristic initialization can be used. Here, the initial population can be seeded with solutions that match the heuristic. However, this process reduces the genetic diversity and may lead to premature convergence if done overzealously.

Since there are no constraints on the values of (x_1, x_2), random initialization involves randomly choosing numbers for x_1 and x_2. However, simply looking at the plot for the Ackley function shows that a global minimum might exist somewhere around (0,0). In fact, the global minimum is at (0,0). However, we will remain blissfully ignorant of that fact for this demonstration. Knowing this, we can use heuristic initialization to bias our initial population toward solutions near that range. A simple heuristic in this case might be to sample randomly 75% of the population, but for the remaining 25% to only choose values that satisfy the constraint $-5 \leq x_1, x_2 < 5$. This ensures that we explore a diverse search space, while also paying more attention to areas where we think the solution is likely to exist.

As mentioned above, it is important to be modest with our use of the heuristic since a direct consequence of added exploration in one area is less exploration in another.

3.3.2.4 Selection

Selection is the process of choosing the fittest individuals within an iteration, that is, the individuals that bring us closest to the optimum. This is accomplished using a fitness function, which determines how "good" or how "fit" a given candidate solution is. Since this function is evaluated for a number of candidates at each iteration, it is crucial that the fitness function is sufficiently inexpensive to calculate.

However, simply selecting the fittest individuals can easily lead to the domination of one well-fitting solution, thereby diminishing the genetic diversity of the population. To explore more of the problem space while still converging to an optimum, fitness proportionate selection is used.

To visualize fitness-proportionate selection, picture a roulette wheel. For each candidate i with fitness score f_i we assign a fraction $p_i = f_i / \sum_j f_j$ of the wheel (see Figure 3.14, and Table 3.1). A stopper sits at a fixed point around the wheel, and the wheel is spun. When the wheel stops, whichever piece the stopper points to, is selected to be the parent. The wheel is spun again to choose the second parent. An alternative approach is through tournament selection (see Figure 3.15). In tournament selection (Rothlauf, 2006), k candidates are chosen randomly from the current population. The best of these k candidates is chosen to be the parent.

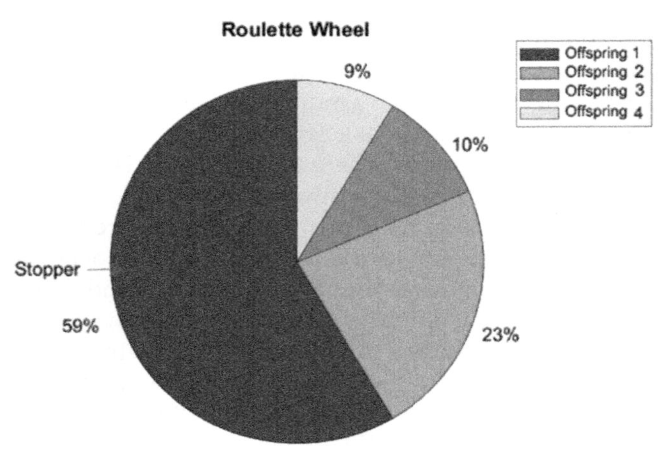

FIGURE 3.14
Roulette wheel for fitness proportional selection.

TABLE 3.1
Fitness Determination for Roulette Wheel

Offspring Name	Fitness	Proportional Fitness
Offspring 1	8.2	8.2/14=58%
Offspring 2	3.2	3.2/14=23%
Offspring 3	1.4	1.4/14=10%
Offspring 4	1.2	1.2/14=9%

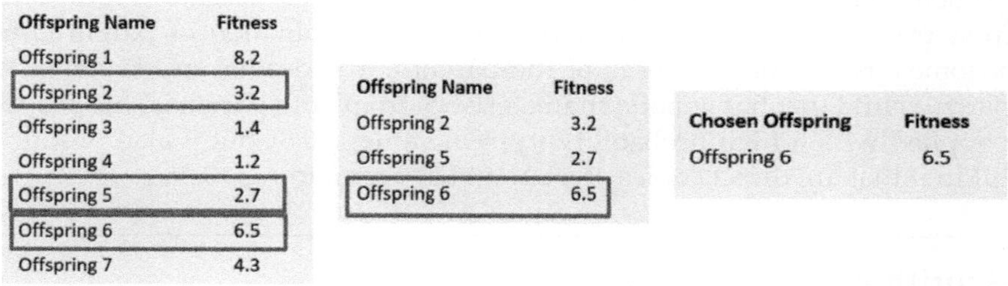

Offspring Name	Fitness
Offspring 1	8.2
Offspring 2	3.2
Offspring 3	1.4
Offspring 4	1.2
Offspring 5	2.7
Offspring 6	6.5
Offspring 7	4.3

Offspring Name	Fitness
Offspring 2	3.2
Offspring 5	2.7
Offspring 6	6.5

Chosen Offspring	Fitness
Offspring 6	6.5

FIGURE 3.15
Process of tournament selection with k=3.

However, due to the pseudo-random nature of such selection processes, the best solution may be lost by chance. To prevent this from happening, an approach called "elitism" may be invoked. In such a technique, the best individuals from a generation are automatically carried over to the next generation. While this reduces genetic diversity, it ensures that we do not have to rediscover the best solution over generations.

Going back to our minimization of the Ackley function, either of these methods may be used. Here the fitness function is simply the inverse of the function value at the specified solution:

$$\text{fitness}(x_1, x_2) = \frac{1}{f(x_1, x_2)} \tag{3.11}$$

Thus, solutions that lead to large values of f are penalized with low fitness scores whereas solutions that lead to small values of f are rewarded with a high fitness score.

3.3.2.5 Cross Over and Mutation

Cross over is the process of generating the next generation of offspring from a set of parents (Rothlauf, 2006). The size and diversity of the population play a large part in defining the success of the algorithm. Naturally, consistently large populations can substantially slow down the genetic algorithm and small populations can lead to diminished genetic diversity.

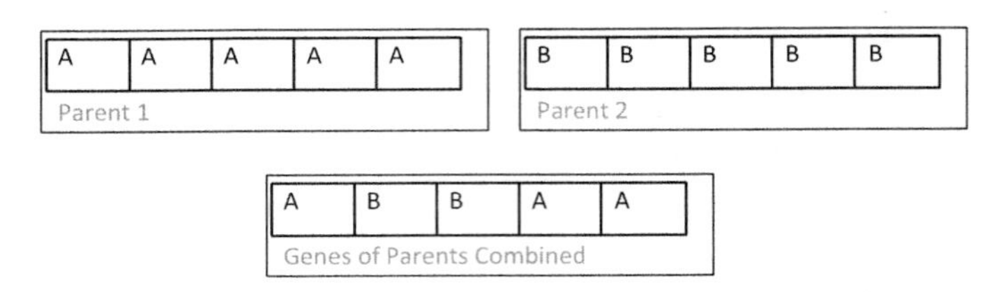

FIGURE 3.16
K-Point cross over.

When creating the succeeding generation, cross over is used (see Figure 3.16). Cross over is the genetic equivalent to the recombination of parent chromosomes. Here, subsequences of the parent chromosomes are swapped to spawn a child that has genetic characteristics from both parents. This process is applied with a high probability (typical values ~0.8), but it also produces children that are direct copies of a parent (The remaining ~0.2).

Algorithm 3.4: Cross over

```
   Input: Parents A set of parents with which to spawn
          the next generation
          P_CrossOver The probability of applying the Cross
          Over operator
          f(x) a fitness function that measures the
          quality of a solution
          Population Size the number of children to
generate
   Output: Children A set of solutions that contain a
           combination of genetic information from the
           parent set
1 Children=∅
2 R=1.2
3 while |Children| < PopulatianSize do
4 |    parent_1=Randomly select a parent from Parents
5 |    parent_2= =Randomly select a parent form Parents
6 |    parent_strong=argmax_parent1,parent2 f(x)
7 |    parent_weak=argmin_parent1,parent2 f(x)
8 |    if rand < P_crossOver where rand ⊂ U[0,1] then
9 |    |   Children =
  |    |    Children U parental+R * (parent_strong
  |    L   − parent_weak);
10|    else
11L    [  Children=Children ⋃ parent_strong
```

Intuitively, this process attempts to extract the characteristics of each parent that leads to a high fitness score and tries to make a child that is fitter than either parent.

Similar to genetic representation, there are a number of ways to implement cross over, the choice of which is problem specific. In the case of our minimization of the Ackley function, we use the implementation described in Algorithm 3.4. The algorithm essentially computes a line that passes through the two parent solutions and generates a child on that line that is close to the stronger parent. This is essentially an approximation of the gradient, similar to Powell's method described previously.

The key factor in exploring the problem space is the mutation operator. With a small probability, genes within a child chromosome are mutated within a range specified by the implementer. Mutation allows for the population to get out of a local minimum and explore a larger search space. Figure 3.16 shows a possible implementation of the mutation operator. It can be seen from the graph that the parent set seems to be converging to the region $x, y \geq 3$. The mutation operator in this scenario breaks the algorithm out of that region and forces the algorithm to explore a more diverse search space or our minimization of the Ackley function, we use a similar methodology where we add a randomly perturb x_1 and x_2 using a gaussian distribution (see Figure 3.17).

3.3.2.6 Termination

Termination occurs after a specified number of iterations or once we have reached a desired optimum. When a poor choice of representation or operators is made, the genetic algorithm is likely to get stuck at a non-optimal state

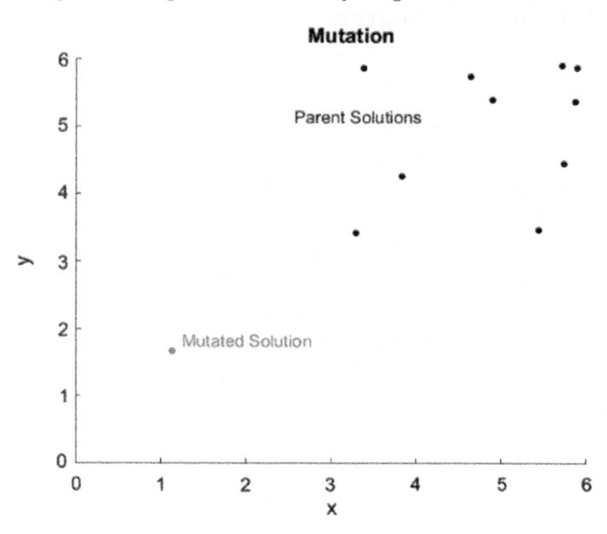

FIGURE 3.17
Mutation operator where parents converge to $x, y \geq 3$.

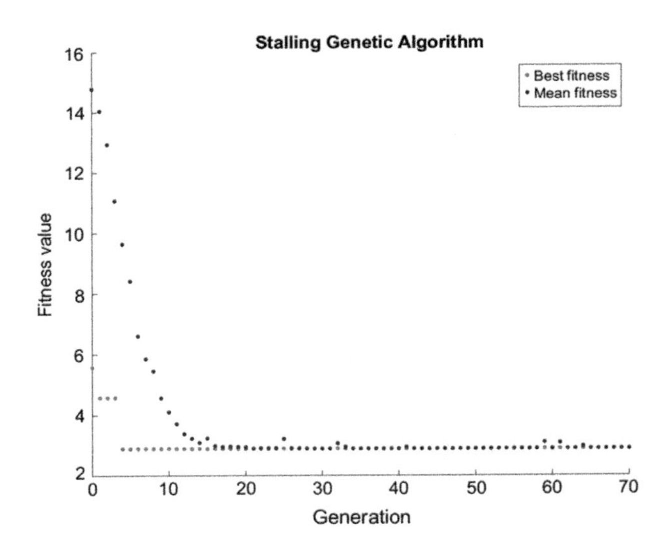

FIGURE 3.18
The genetic algorithm stalling due to lack of genetic diversity.

and we must terminate once we have run a specified number of iterations. This can be seen in Figure 3.18. Notice how the mean and best fitness values rarely change beyond a certain point. This is the outcome of poor genetic diversity within the population.

Algorithm 3.5 Genetic Search

Input: *PopulationSize* The size of the initial population *SolutionSpace* The constraints on where the solution can be $P_{CrossOver}$ The probability of applying the crossover operator $P_{Mutation}$ The probability of applying the mutation operator *Elitism* The number of best fit solutions to preserve *MaxGenerations* The maximum number of generations to spawn $f(x)$ A fitness function that measures the quality of a solution

Output: $x_{opt} \subset$ *SolutionSpace* s.t. x_{opt} is the best estimate for the global minimum over *MaxGenerations*

1 $Population_0$ = *InitializePopulation*(*SolutionSpace*, *PopulationSize*);

2 Generation = 0;

3 **while** *Generation* < *MaxGenerations* **do**

```
4  |   Parents=GetFittest(f(x), Population_Generation, 2);
5  |   Population_Generation+1 =
   |    CrossOver(Parents, P_CrossOver, f(x), PopulationSize);
6  |   Population_Generation+1 =
   |    Mutate(Population_Generation+1, P_Mutation);
7  |   if Elitism > 0 then
8  |   |  Population_Generation+1 =
   |   L Population_Generation+1 ∪ GetFittest(Population_Generation, 1);
9  L generation=generation+1;
10 return x_i
```

3.3.2.7 The Genetic Algorithm Applied to the Ackley Function

The progress of the genetic algorithm over each generation can be seen in Figure 3.19. Our implementation of the algorithm converged to $2.60819 * 10^{-8}$ at $x = (7.021954 * 10^{-9}, 5.977054 * 10^{-9})$ – close enough to (0.0,0.0).

3.3.3 Simulated Annealing

Like the genetic algorithm, simulated annealing (van Laarhoven and Aarts, 1987) is another stochastic algorithm that works well in the presence of many local minima. The methodology is drawn from the annealing process in metallurgy. Annealing is the process of heating up a metal and allowing it to cool to soften it.

At a given iteration, in classical algorithms we always move to a neighboring solution that is better than the current one. In simulated annealing however, there exists a probability that we will accept a worse solution. The probability of "up-hill" acceptance is usually, but not always, proportional to the error induced by accepting the solution. The worse the solution, the less likely the algorithm is to take the step. An example of a valid probability function is

$$P_i = \begin{cases} 1, \Delta E < 0 \\ \exp\left(\dfrac{-\Delta E}{kT}\right), \Delta E \geq 0 \end{cases} \tag{3.12}$$

Here E is the energy level, k is a constant analogous to the Boltzman constant in thermodynamics, and T is temperature. The energy level is the "fitness" of the function we want to minimize. The temperature T, as in annealing, starts out high, decreasing over time as iterations progress. Note the dependence on T for the $\Delta E \geq 0$ or the up-hill case. When T is high, P_i approaches 1, and as T approaches 0, P_i approaches 0. Thus, at higher temperatures, worse solutions are more likely to be accepted and as the temperature approaches 0 only solutions that decrease the energy are selected. Algorithm 3.6 presents the algorithm.

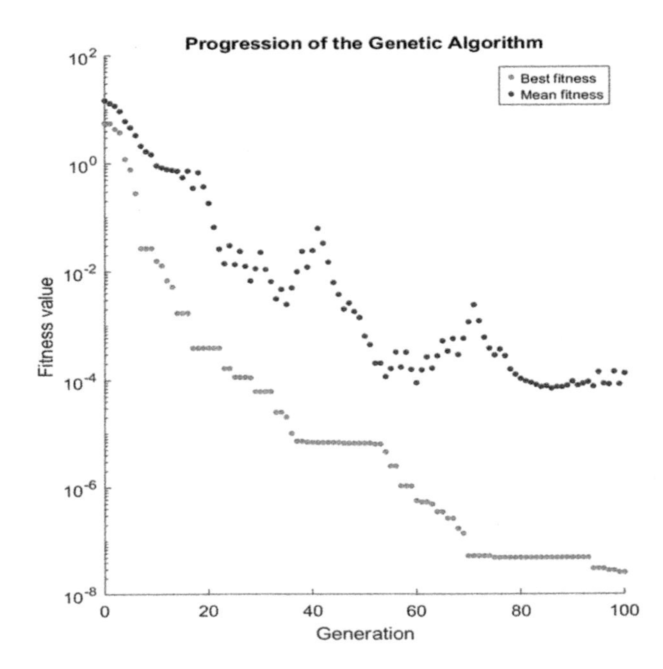

FIGURE 3.19
The Ackley function of Figure 3.9: convergence of the genetic algorithm to the minimum using best fitness and mean fitness.

Algorithm 3.6 Simulated Annealing

Input: *CoolingSchedule* A schedule describing the change in temperature over iterations $T_{Initial}$ the starting temperature T_{stop} the stopping temperature x_0 an initial guess at the solution $f(x)$ a function to minimize

Output: $x_{opt} \subset SolutionSpace$ s.t. x_{opt} is the best estimate for the global minimum over the number of iterations allowed by the cooling schedule

1 $S_0 = x_0$
2 $S_{best} = x_0$
3 $T_0 = T_{init}$
4 iteration = 0
5 **while** $T > T_{stop}$ **do**
6 | *iteration* = *iteration* + 1
7 | *proposed* = *GenerateNewSolution*($S_{iteration-1}$)) ;
8 | **if** $f(S_{iteration-1})$ - *proposed*) $> 1 * 10^{-6}$ **then**

```
9  |    S_iteration = proposed;
10 |    if f(S_iteration) > f(S_best) then
11 |    |  | S_best = S_iteration
12 |  else if R < e^(f(S_iteration - 1) - proposed)/(k_boltzman*T_iteration - 1),
   |  where
   |   R ⊂ U([0,1]) then
13 |    | S_iteration = proposed;
14 |  else
15 |    | S_iteration = S_iteration-1
16 |  T_iteration = CoolSchedule(T, iteration);
17 |  return S_best
```

3.4 N-Dimensional First-Order Optimization

3.4.1 Gradient Descent or Steepest Descent

Gradient descent is ubiquitous in machine learning applications. It essentially involves walking down the steepest slope until a minimum is found. The steepest slope is found by repeatedly taking a step proportional to the negative of the gradient at the current point. Recall from Figure 3.1 that the minimum is against the slope. The algorithm is based on the principle that the function decreases the fastest in the direction of the negative gradient.

As a one-dimensional example, consider the function

$$f(x) = x^4 - x^3 - x^2 + x \tag{3.13}$$

Before approaching this numerically, let us see if the solutions may be identified by hand calcultations. The gradient

$$\frac{df}{dx} = 4x^3 - 3x^2 - 2x + 1 \tag{3.14}$$

The minima must be the roots of $df/dx = 0$. It is readiy evident that at $x = 1$, df/dx is zero so $x = 1$ is a root where there lies a minimum of f(x) and (x–1) is a factor of df/dx. Dividing df/dx by (x–1) by polynomial long division we get $4x^2+x-1$ as the quotient with zero remainder. Solving

$$4x^2 + x - 1 = 0 \tag{3.15}$$

By the quadratic formula, we obtain x = –0.64 and x = 0.390 as the other two factors.

Figure 3.20 shows the convergence of gradient descent to the root at x = –0.64. To which root the algorithm converges depends on where we start and the proximity of the starting point to the root. Note in Figure 3.20 that

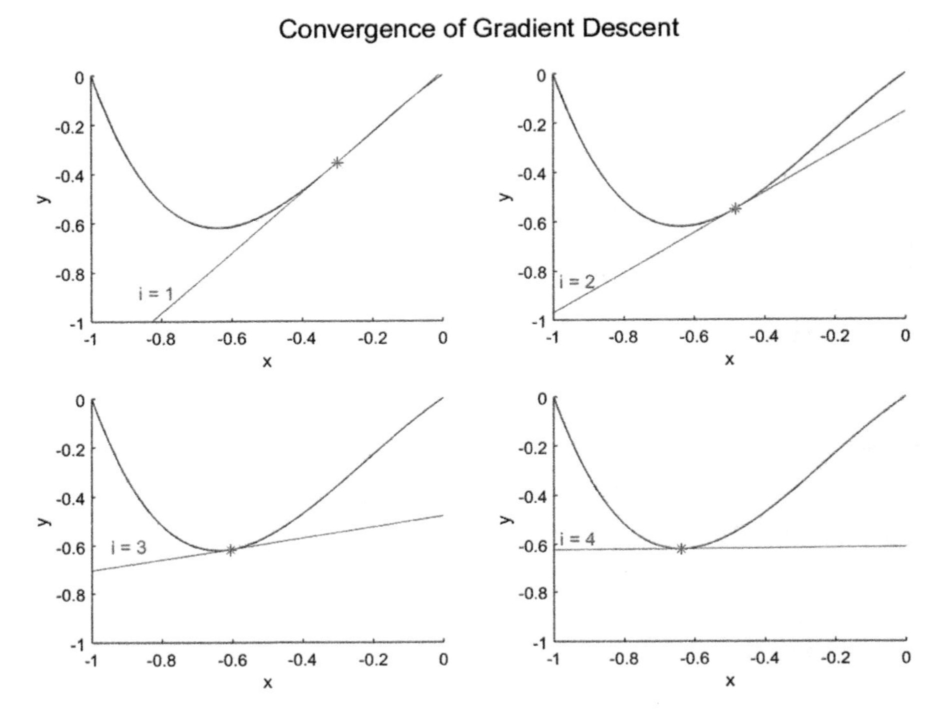

FIGURE 3.20
Convergence of gradient descent to zero slope at $x=-0.64$ of $y=x^4-x^3-x^2+x$.

the gradient gets smaller as we move through the iterations and gets to be practically zeo at convergence.

To examine the convergece by gradient descent of a two-dimensional consider the function

$$f(x,y)=(x-3)^2+xy+3y^2 \tag{3.16}$$

Figure 3.21 shows the descent toward the minimum. The broader algorithm for multi-dimensional systems is presented in Algrithm 3.7 where ∇f represents the gradient vector which is as long as there are variables.

Algorithm 3.7: Gradient Descent

Input: α A stop size
$f(x)$ A function to minimize
x_0 An initial guess
ϵ_0 tolerance
MaxIterations the number of steps to take

```
    Output: x_opt The The best estimate to the minimizer
                  of f(x)
1 i=0;
2 δ=∞
3 while i < MaxIterations Λδ > ϵ do
4  |  x_{i+1}=x_i - α∇f(x_i)
5  |  δ=|f(x_i) - f(x_{i+1})|
6  L  i=i+1
7 return x_{i-1}
```

As an additional exercise the reader is urged to consider finding the minimum of the two-dimensional object function

$$f(x_1.x_2) = x_1^2 + 2x_2^2 - 4x_1x_2 + 2x_1 + 2x_2 + 10 \tag{3.17}$$

To see where our algorithm should converge, at the minimum

$$\frac{\partial f}{\partial x_1} = 2x_1 - 4x_2 + 2 = 0 \tag{3.18}$$

$$\frac{\partial f}{\partial x_2} = 4x_2 - 4x_1 + 2 = 0 \tag{3.19}$$

Solving, we get

$$(x_1, x_2) = (2.0, 1.5) \tag{3.20}$$

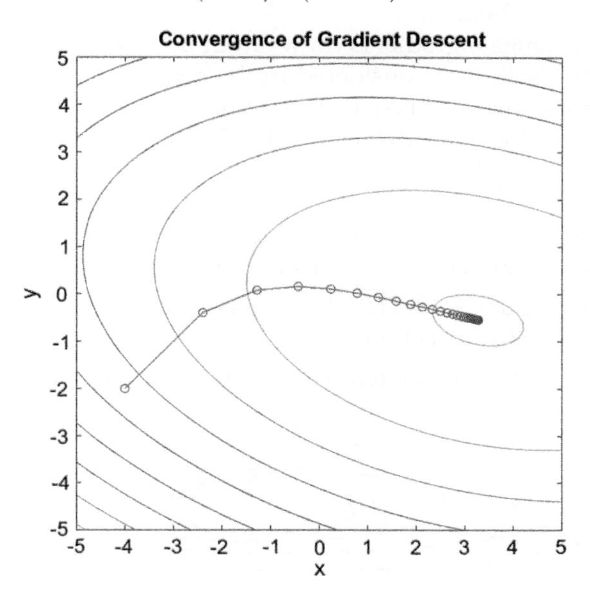

FIGURE 3.21
Convergence of gradient descent for $f(x, y) = (x-3)^2 + x*y + 3*y^2$.

Thus in Algorithm 3.7,

$$\nabla f = \begin{bmatrix} 2x_1 - 4x_2 + 2 \\ 4x_2 - 4x_1 + 2 \end{bmatrix} \tag{3.21}$$

3.4.2 Conjugate Gradients

One issue with gradient descent is that at progressive iterations, some of the work done in prior iterations is undone by the following step. This is particularly common in narrow valleys, where we overshoot the minimum and are forced to move in a back and forth manner as shown in Figure 3.22.

We can instead use the knowledge of the previous step to avoid backtracking. The conjugate gradient method does this by enforcing orthogonality on Q-orthogonal vectors.

Two vectors u, and v are said to be Q-orthogonal, with respect to Q, or conjugate if they satisfy the property

$$u^T Q v = 0 \tag{3.22}$$

where Q is a square $n \times n$ matrix for u, and v is of length n. There is a proposition that If **Q** is positive definite, and a set of nonzero vectors are Q-orthogonal, then these **vectors** are linearly independent. A more visual understanding of this definition is that in the space transformed by Q, vectors u^T and v are orthogonal.

The idea in conjugate gradients draws from the conjugate gradients algorithm for matrix solutions considered in the next chapter. It is the same as steepest descent with a slight change. Imagine we are improving the solution by searching in different search directions. In the steepest descent method, we will

1. $i = 1$; Start from x^0 and compute the search direction $p_1 = -\nabla f(x^0)$ and use the line search to find the next point x^1 along the direction p_1
2. $i = i + 1$; Compute $p_i = -\nabla f(x^{i-1})$. Use the line search to find the next point x^i along the direction p_i
3. If converged, flag convergence and stop. Otherwise go to Step 2.

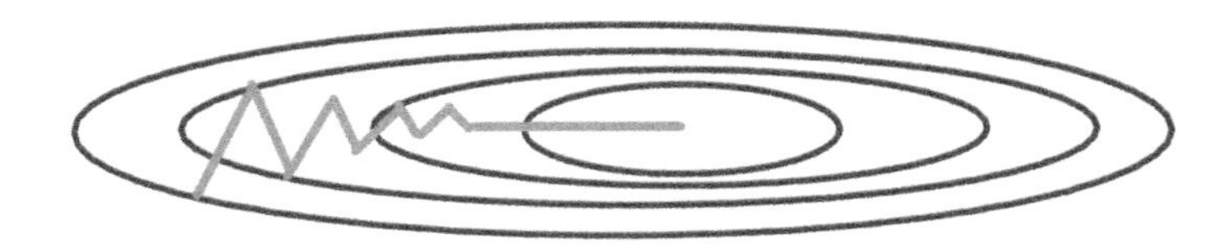

FIGURE 3.22
Progression of gradient descent in a sloped valley.

In conjugate gradients optimization, just like in conjugate gradients matrix solution, (see Section 4.5) we say at Step 2 above where we compute the new search direction $p_i = -\nabla f(x^{i-1})$, that having searched in the directions $p_1, p_2, p_3, \ldots, p_{n-1}$, there is no point in searching in directions already searched; therefore we must take off a little bit of p_{i-1} from p_i:

$$p_i = -\nabla f\left(x^{i-1}\right) - \beta p_{i-2} \tag{3.23}$$

It may be shown from the theory of Q-conjugate vectors that

$$\beta = -\frac{\left\|\nabla f\left(x^{i}\right)\right\|^2}{\left\|\nabla f\left(x^{i-1}\right)\right\|^2} \tag{3.24}$$

This leads to Algorithm 3.8.

Algorithm 3.8: Non-linear Conjugate Gradient Method

Input: $f(x)$ A function to minimize
x_0 An initial guess
ϵ_0 tolerance
Output: x_{opt} The best estimate to the minimizer of $f(x)$

1. $r = b - Ax_0$
2. $S_0 = -\nabla f/(x_0)$
3. **while** $\| f(x_{k+1} - f(x_k) \| < \epsilon_0$ **do**
4. $\quad \beta = \frac{\|\nabla f(x_k)\|^2}{\|\nabla f(x_k - 1)\|^2}$
5. $\quad P_{k+1} = -\nabla f(x_k) + \beta S_{k-1}$
6. $\quad \alpha_k = \text{argmin}_\alpha \ f(x_k + \alpha)$
7. $\quad x_{k+1} = x_k + \alpha_k S_k$
8. $\quad k = k+1$

3.5 A Good Test Problem from Magnetics – The Pole Face

3.5.1 Problem Description

To figure out how well a method is working and indeed to compare different methods, it is good to have a test problem with predictable outcome. If the method is easily implementable, and the result of the test problem's general predictable outcome is obtained matching our expectations of the solution through an optimization method, then we know that the optimization method appears correct and properly programmed. If the application of two different methods to a given problem gives the same optimized solution, then we may say that the two methods are validated; and if one of the methods has already been validated, we may add that the other method is validated with greater certainty now. We will therefore describe a problem that has been found suitable as a good test problem particularly for the zeroth-order methods that are often the most useful.

We note that in the optimization context, a good benchmark test problem will have multiple minima.

We describe the problem of determining the shape of a pole face that is required to produce constant flux density (see Figure 3.23). It is one of the oldest problems studied as electromagnetic devices were optimized (Pironneau, 1984) and has been developed since then (Weeber and Hoole, 1992a, 1992b). We have briefly encountered this problem in Section 1.2.2 and Figure 1.3. We define here in detail many relevant features of the pole-face problem. In the literature this problem is set up and then other different methods are brought to bear to analyze it to show the behavior under different methods. The multiple minima are shown here for interested readers to investigate. This problem is well accepted for having many local minima (Hoole et al., 2007).

Let us consider the benchmark problem described in Figure 3.24 which represents one repetitive section of Figure 3.23. The object is to achieve a uniform flux density distribution in the vertical direction in the air-gap of a pole face.

Figure 3.24 also gives the related dimensions, material properties, and field excitation values used. The symmetry of the magnetic fields with respect to the pole axis allows the modeling of just half the pole pitch, where the pole axis, which is the line of symmetry, is located at the right boundary to the finite element solution domain. The relative permeability of 20 for the magnetic circuit is deliberately set this low, so that the leakage flux through air at the left edge of the pole face is larger than for higher and more realistic permeability. The influence of this leakage flux requires

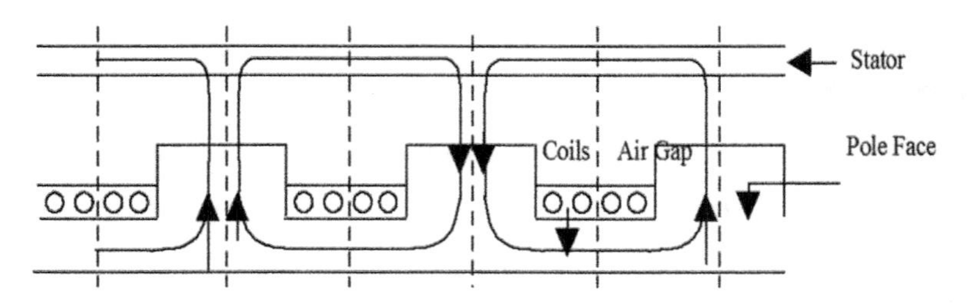

FIGURE 3.23
Benchmark problem – constant flux density on pole faces.

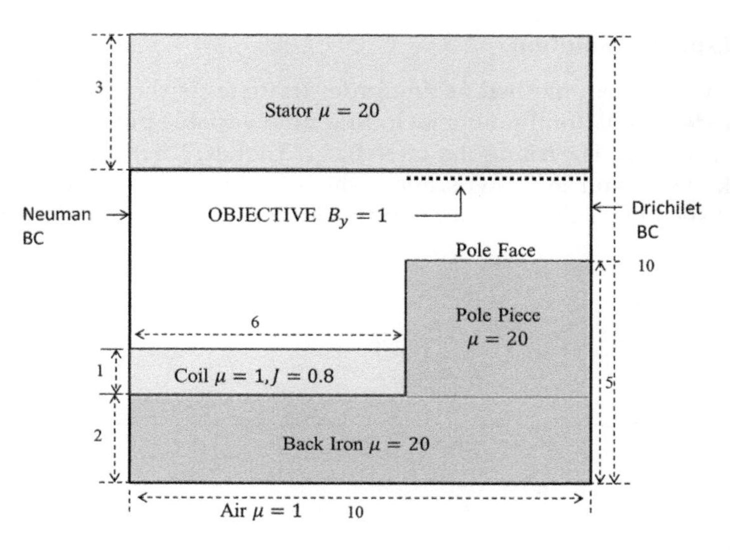

FIGURE 3.24
Numerical model for shape optimization.

significant correction in the shape of the pole face close to the left edge in order to achieve the desired constant flux density in the air-gap. As shown by Weeber and Hoole (1992a, 1992b), the constraints can be based on specification of minimum percentage of leakage flux or as shown by Subramaniam et al. (1994), the jaggedness of the resulting geometry. This example is frequently used in the demonstration of electromagnetic-optimization methods and it can be considered as a standard demonstration example (Subramaniam et al., 1994).

The pole face is split into eight segments and nine heights (see Figure 3.25 and the structure placed from x-coordinates 6 to 10 of Figure 3.26) at x-coordinates 5, 5.5, 6, 6.5, 7, 7.5, 8, 8.5, 9, 9.5, and 10 are measured from the bottom and are used for the design parameters.

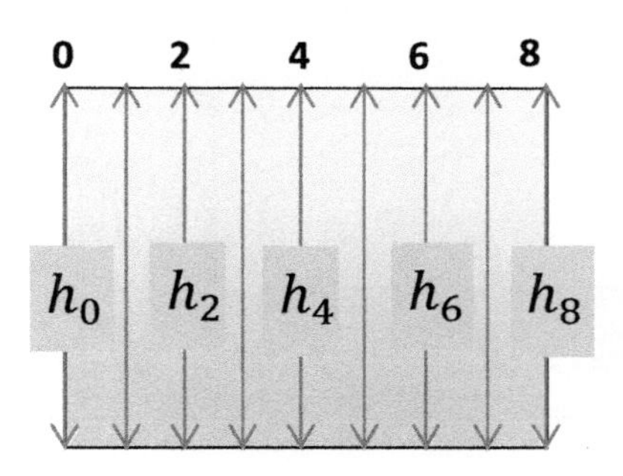

FIGURE 3.25
Segmentation and eight design parameters $h_0, h_2, \ldots, h_8$.

3.5.2 Expected Solution

The above problem meshed as first order triangles is shown in Figure 3.26. This model has 19 total points including nine variable points, 22 segments, and four regions. The triangular mesh has 133 nodes, 228 triangular elements, 105 unknowns, and 28 knowns, and is shown in Figure 3.26. The corresponding finite element solution prior to optimization is shown in Figure 3.27.

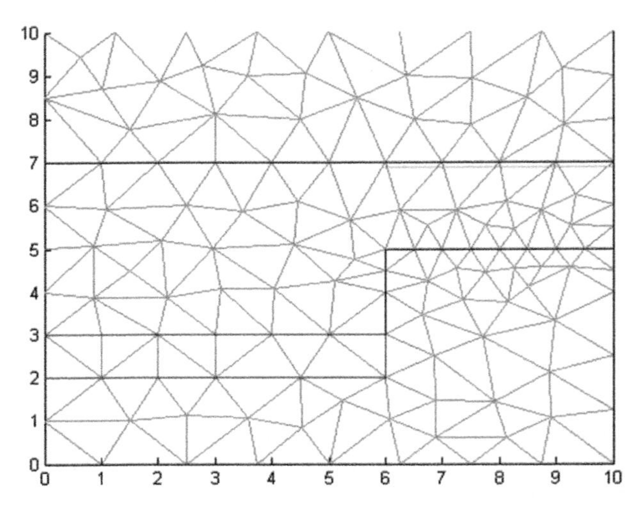

FIGURE 3.26
The finite element mesh for the benchmark problem.

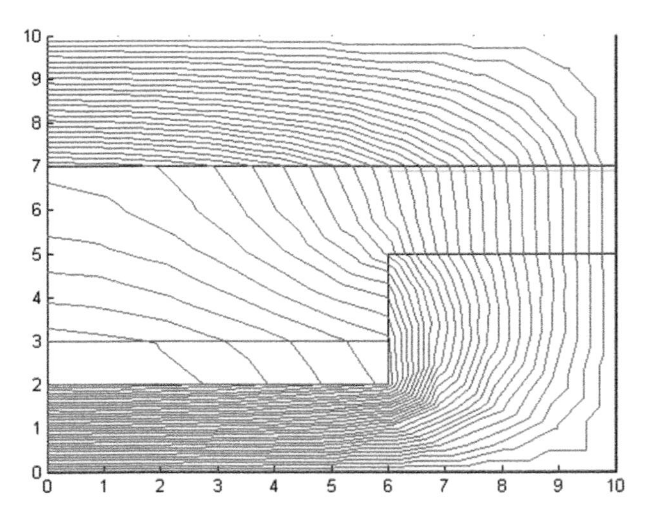

FIGURE 3.27
View of the equipotential lines computed through the model.

We note that when equipotential lines are spaced more apart, the flux density is lower. This is because the flux near the left corner of the pole face, taking the path of least resistance, chooses to flow leftwards rather than upwards. This is what is called fringing. So, the flux density is lower close to the corner. To increase flux density above the corner there to match that in parts to the right of the corner, the reluctance of the air-gap needs to be reduced above the corner. Since reluctance is proportional to the length of the path that flux takes, that corner has to rise closer to the stator face above the corner in the process of optimization. This is what we expect to see in the solution.

3.5.3 Choice of Optimization Method

The object here is to develop the characteristics of the proposed benchmark problem and not to investigate all the methods and make comparisons. When designing a general-purpose optimization software package, the optimization method must be selected in such a way that it can be used for the optimization of many general object functions, under many general constraints. Any FEA software requires solving large FEA matrix equations. So the optimization method should be selected to minimize the number of funtional evaluations.

For optimization purposes, there is an array of methods to pick from as described in Sections 3.2 and 3.4. We decided on implementing a zeroth-order method because higher order methods are best coupled with special methods to ensure C^1 continuity (Vanderplaats, 1984; Krishnakumar and Hoole, 2005) when the calculation of derivatives is involved.

As shown by Krishnakumar and Hoole (2005) special mesh generators are required with restrictions on free meshing. As a shape is modified during optimization and the mesh is deformed, as shown in Figure 3.28, a mesh will often violate the Delaunay criterion (Cendes et al., 1983) for a good mesh and change the nodal connections. In that figure there are two traingles P-S-Q and S-R-Q. They are Delaunay optimal in that as the Delaunay critierion

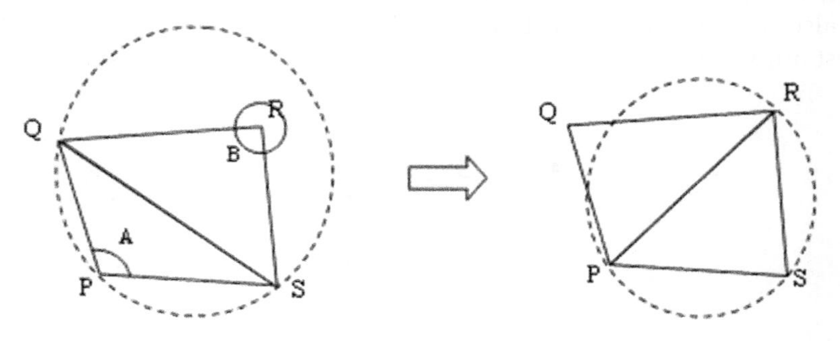

FIGURE 3.28
The Delaunay criterion for optimal triangular meshes.

demands for neighboring triangles, say as in Figure 3.28a, the circumcircle of triangle P-S-Q must exclude node R of the neighboring triangle S-R-Q. That means the sum of the angles A and B must exceed 180°. Note that if R is on the circumcircle, a theorem from plane geometry says angle A and B must sum to 180°.

As optimization proceeds, referring to Figure 3.29a, and node 4 moves toward node 1 (see Figure 3.29b), optimality may be lost as the circumcircle of triangle 1-2-3 may now include node 4. To return to optimality the two triangles must be deleted from the mesh data, and replaced by triangles 1-2-4 and 1-4-3 (see Figure 3.29c).

This sudden mesh change is therefore what introduces a jump in the object function (Weeber and Hoole, 1994b). Such sudden change in object function f as nodal connections change introduces a jump in f, threby violating the C^1 continuity of the object fuction. This will make gradient methods underperform as mesh-induced discontinuities in the object function are seen as local minima and the optimization scheme will get stuck there thinking it has found a minimum. This is why methods like the genetic algorithm are better suited to situations with multiple minima. This situation of multiple minima becomes very computationally expensive with the increase of the degrees of freedom of the object function. Besides, direct methods require knowledge about the derivatives of the constraints.

Statistical optimization methods such as the genetic algorithm and simulated annealing have a better chance of converging to the global minimum. However, they require many more object function evaluations than other numerical techniques. Since function evaluations are expensive in FEA (Beveridge and Schechter, 1970), they are not very suitable for this context

From among the zeroth-order methods we picked the Powell's method because it has been extensively tested by structural engineers (Beveridge and Schechter, 1970) and, being a zeroth-order approximation of a first-order method, it has the advantages of the conjugate-gradients method without the expense of gradients computation. As an alternative, the genetic algorithm was also implemented so as to ensure, if necessary, that we always got the lowest minimum.

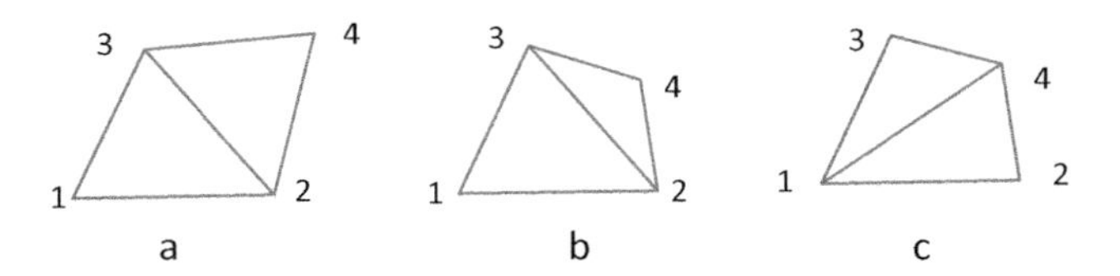

FIGURE 3.29
Object functions made non-C^1 continuous by the Delaunay condition.

3.5.4 Preprocessing the Pole Face

The in-house package we used provides a user-friendly visual interface for the problem definition. Figure 3.30 shows the problem described. The movable points move up and down to make the shape. So, the y-coordinates of these points are our parameters of design with the object of getting a flux density of 1 T in the air-gap at the measuring points.

In design optimization the material boundaries must be "movable" as shape is being optimized.

Note the movable points in Figure 3.30. These can be assigned to be variables as they are here, and can be used to define the design boundary for optimization. These points are called "movable points."

Once the problem is defined, we have to define the expression to be optimized – our object function *F*. Our requirement is to have a uniform flux density of 1 Tesla along the eight measuring points on the stator (see Figure 3.30). That means, all our measuring points must have their y direction flux density (x direction derivative of the vector potential) of 1 Tesla. This can be achieved by calculating the mean square error and minimizing that error. Therefore, our requirement is to minimize the square of the difference between the flux density and its desired value of 1 T, summed over the eight measuring points of Figure 3.30. This object function *F* may be written out as

$$F = \sum_{i=1}^{8} \frac{1}{2}(B_i - 1)^2 \tag{3.25}$$

As already established in his early work by Pironneau (1984), an impracticable jagged contour resulted from a flat starting configuration of the pole face with all heights equal. Similar to Pironneau's experience, the jagged shape of Figure 3.32 resulted for us from the alternative starting shape of Figure 3.31.

These results were both optimal in a mathematical sense in that the air-gap had a constant flux density of 1 T, but cannot be manufactured.

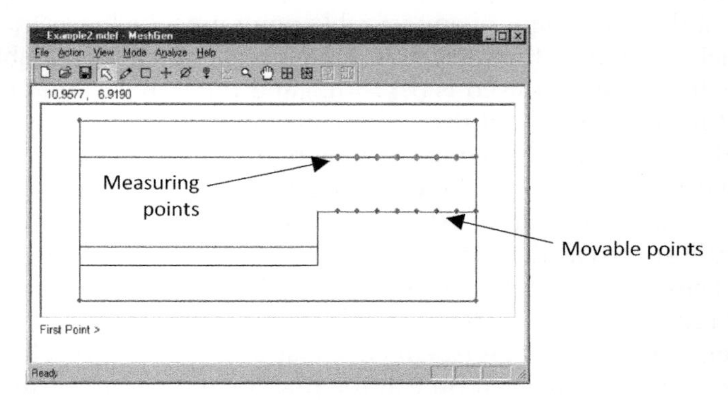

FIGURE 3.30
A screen capture of the application.

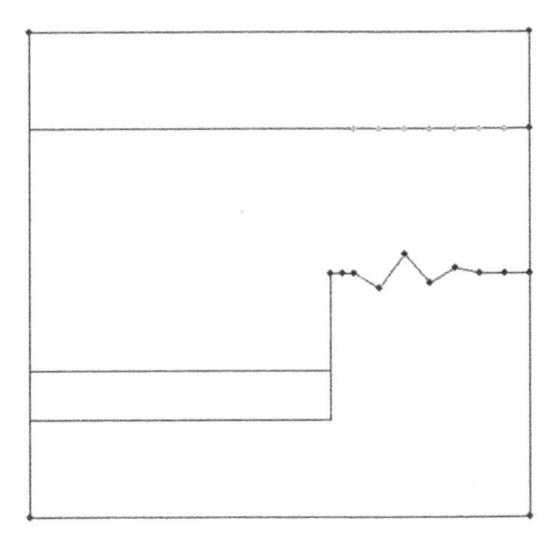

FIGURE 3.31
Alternative starting configuration.

Therefore contraints need to be imposed so as to ensure some regularity of the geometry.

These show the presence of multiple minima and the need for constraints to obtain a smooth manufacturable shape.

3.5.5 Powell's Method – Special Treatment and Constraints

Powell's method is well known (Vanderplaats, 1984) and has been described in Section 3.3.1. We wish to deal with a special case where it breaks down as it does for the lack of constraints here. This needs elaboration.

In Figure 3.32, to ensure non-jagged contours as in Figure 3.33, one of our constraint sets is that the height of the first parameter, that is the height of the left-most movable point (see Figure 3.25), is higher than that of the second, which in turn is higher than that of the third, and so on. We have applied as a constraint that the height of any variable point must be lower than all the other variable points to the right of it. This gives a practically usable result. Therefore, the constraint used is (h1>=h2) and (h2>=h3) … and (h8>=h9) (Note that h1 to h9 represent the parameters governing the vertical-coordinate position of the variable points as shown in Figure 3.25). With these constraints, the result of Figure 3.33 was obtained and it is seen to be acceptable in terms of manufacturability. Although there is fringing, the lines are seen to be equally spaced as they arrive at the top stator, making the flux density uniform. The constraints of the heights going down from left to right may be noted in Figure 3.33.

However, under these user-defined constraints, since Powell's method changes a single variable at a time at the first iteration, points cannot move down to meet the object without violating the constraints. This is because the optimum solution needs these parameters to be reduced. Therefore, only the last parameter will be changed during the first iteration of Powell's optimization. Only the last two parameters will be changed during the second iteration and so on. Therefore, it will take a long time for the real Hessian matrix to build up – the Hessian matrix being the matrix formed by arranging the initial search vectors one after the other (Vanderplaats, 1984). Until then this optimization process would be extremely inefficient. This is a good example to show that some problems cannot be solved as they are, if we wish to be efficient in our computations.

This problem is circumvented by having the parameters start at low values as in Figure 3.34 so that, as they rise to meet the object, no constraint

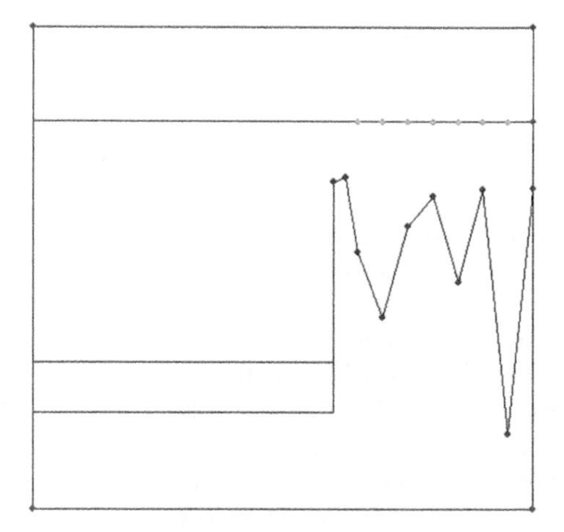

FIGURE 3.32
Optimum from different starting shape of Figure 3.31.

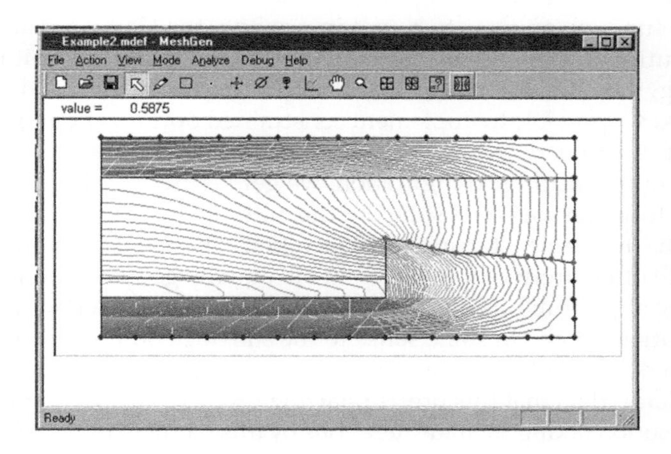

FIGURE 3.33
Optimized result with constraints.

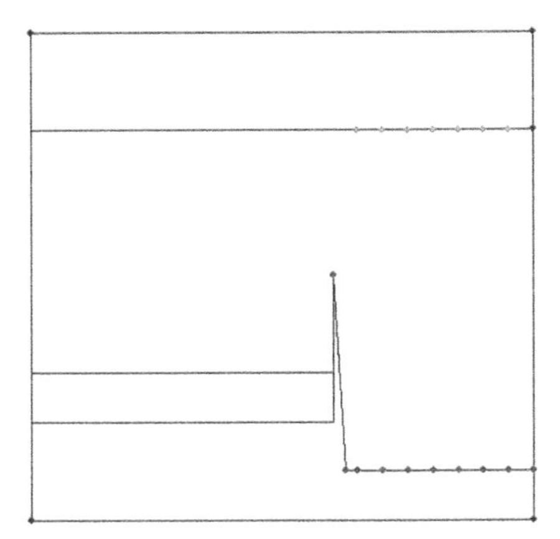

FIGURE 3.34
An initial shape for constrained optimization.

is violated. This need to inspect the constraints makes any package be not general purpose. So, we have solved this problem with steepest descent and conjugate gradients as alternative methods. We find that they work but are not always converging. This is because of the problem of mesh-induced local minima and the problem described with constraints. We may circumvent the problem of mesh-induced local minima by using a parameterized mesh generator (Krishnakumar and Hoole, 2005).

Improvement of the value of the object function may stop in Powell's method before achieving the optimum value due to two reasons. One is the search directions being parallel due to numerical imprecision of floating-point calculations in the computer. The other reason is that the Hessian matrix is calculated without considering the presence of constraints. Therefore, once the object function appears to be not improving, the Hessian matrix is reset to the identity matrix and a unidirectional search is started. If it is still not improving, then it is concluded that the optimum value is found. This technique makes it possible to implement constrained optimization for any given constraint function.

With another starting configuration, we get Figure 3.35. In both instances the object function was zero, showing the multiplicity of solutions. Likewise, even with the same starting configuration, the genetic algorithm and Powell's do often produce two different but equally good solutions – because the genetic algorithm hits on its solution statistically while Powell's tends to find a solution "close" in some sense to the starting configuration

Where a benchmark has a unique solution, it is presented to peer scientists through data – data that this precise starting problem will yield this precisely defined solution taking so many iterations by this method and so on. Given the multiplicity of solutions based on starting solutions and the method employed, that cannot be done here in view of the sheer volume of data being involved.

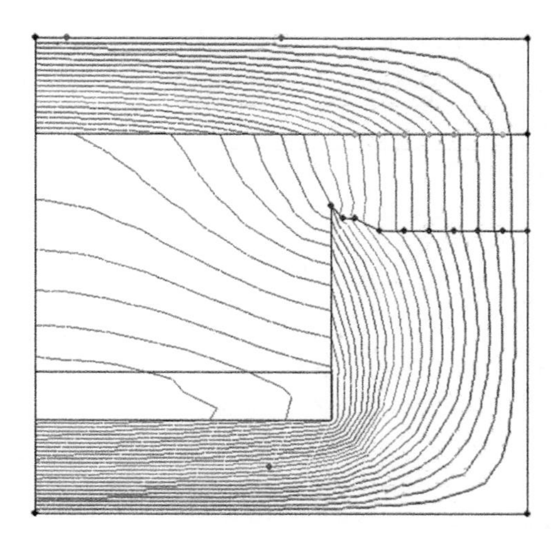

FIGURE 3.35
An optimized shape by the genetic algorithm with constraints.

3.5.6 Solution by the Genetic Algorithm

The genetic algorithm has already been descried in Section 3.3.2. The computational process follows the steps shown in Figure 3.8 and is the same even for other methods. What would be unique to each of the different methods is the box where the design is changed.

The computational process starts with initial design parameters $\boldsymbol{h}_1, \boldsymbol{h}_2, \ldots, \boldsymbol{h}_n$ (see Figure 3.25), generates the mesh, calculates the vector potential **A** by the finite element method, and from that the flux density **By**. From **By** at the measuring points, the object function is computed using Equation (3.25). We then check for whether the minimum value is reached. If it is minimum enough then the process comes to the end, otherwise the process loops back with the new design parameters.

Figure 3.8 shows the optimization process using the genetic algorithm. It works based on the fitness function and continues in loop until we get the minimum fitness. Each iteration changes the design parameters through the three continuous sub-processes, selection, cross over, and mutation one by one in this order.

First, for the starting configuration of Figure 3.30, **By** was computed at the ten measuring points – Table 3.2 shows the results.

The genetic algorithm (GA, described in Section 3.3.2) was then initialized with the values shown in Table 3.3 and used to optimize for the desired **By** value of 1 Tesla in the measuring points. The GA was run ten times and the heights with best minimum fitness value (**0.020956**) were chosen (see Table 3.4). In this case, we obtained **By** values almost at 1 T at every measuring point (see Table 3.5). The corresponding mesh is shown in Figure 3.36, and the equipotential plot in Figure 3.37.

TABLE 3.2

By Value Measured by Finite Element Solution

y	x	By
6.9	5.5	0.83229
6.9	6.0	1.013364
6.9	6.5	1.02816
6.9	7.0	1.163368
6.9	7.5	1.123182
6.9	8.0	1.184965
6.9	8.5	1.141732
6.9	9.0	1.166886
6.9	9.5	1.146954
6.9	10.0	1.163135

TABLE 3.3

By Value Measured in Forward Approach

Initial Values – for GA
Variable boundaries: Rigid
Population size: 90
Total no. of generations: 50
Cross over probability: 0.9000
Mutation probability (real): 0.0500
Number of real-coded variables: 9
Total runs to be performed: **10**
Exponent (n for SBX): 2.00
Exponent (n for mutation): 100.00
Lower and upper bounds: **4.0000** <= x_real[i] <= **6.0000**

The shape we can design using the result of Figure 3.36a,b is not smooth although it fits the required object of 1 T in the air-gap. It cannot be practicably realized.

Thereafter, the constraint $h_i \geq h_{i+1}$ was set in order to get a smooth shape and we obtained a good result (see Tables 3.6 and 3.7 and Figures 3.37 and 3.38).

3.6 A Test Problem from Alternator Rotor Design

3.6.1 Problem Definition

We will briefly describe here the design of the rotor of an alternator so as to produce a sinusoidal flux density in the air-gap. This example is about determining the pole-face contour of a salient pole synchronous generator to

TABLE 3.4

Result Sets without Any Constraints Set

Ten Different Sets of Values for h
Best ever fitness: 0.020956 (from generation: 49) Variable vector: 5.012679 4.871028 5.265669 4.210631 4.054786 4.282740 4.276830 4.236632 4.182740
Best ever fitness: 0.026815 (from generation: 13) Variable vector: 5.326403 5.112372 4.150428 4.433471 4.547620 4.289636 4.644031 4.300652 4.242340
Best ever fitness: 0.028357 (from generation: 5) Variable vector: 4.912639 4.870801 4.742357 4.105071 4.114241 4.211418 4.743663 4.158798 5.002710
Best ever fitness: 0.023955 (from generation: 26) Variable vector: 4.712271 4.949989 5.127415 4.448242 4.164439 4.347501 4.147932 4.158028 4.982341
Best ever fitness: 0.022125 (from generation: 9) Variable vector: 5.012679 4.866797 5.196280 4.240957 4.165248 4.093611 4.154891 4.081246 4.372730
Best ever fitness: 0.028347 (from generation: 35) Variable vector: 5.444831 5.03279 4.484883 4.053361 4.306067 4.375677 4.414737 4.364856 4.412441
Best ever fitness: 0.027625 (from generation: 16) Variable vector: 5.197021 5.012679 4.084607 4.033482 4.312485 4.228772 4.482273 4.472361 5.00120
Best ever fitness: 0.023478 (from generation: 49) Variable vector: 5.312679 4.943095 4.974926 4.168272 4.113802 4.399413 4.063702 4.277921 4.983340
Best ever fitness: 0.031048 (from generation: 7) Variable vector: 4.872959 5.012679 4.763752 4.354236 4.242119 4.068927 4.019170 4.216181 4.581122
Best ever fitness: 0.027831 (from generation: 46) Variable vector: 5.113639 5.454329 4.339498 4.080595 4.037304 4.105972 4.704814 4.294497 4.222741

TABLE 3.5

By Value Measured Using GA without Setting Any Constraints

y	x	By
6.9	5.5	0.863251
6.9	6.0	0.966588
6.9	6.5	0.996262
6.9	7.0	0.996262
6.9	7.5	1.027163
6.9	8.0	0.985977
6.9	8.5	0.997835
6.9	9.0	1.005534
6.9	9.5	0.991576
6.9	10.0	0.991576

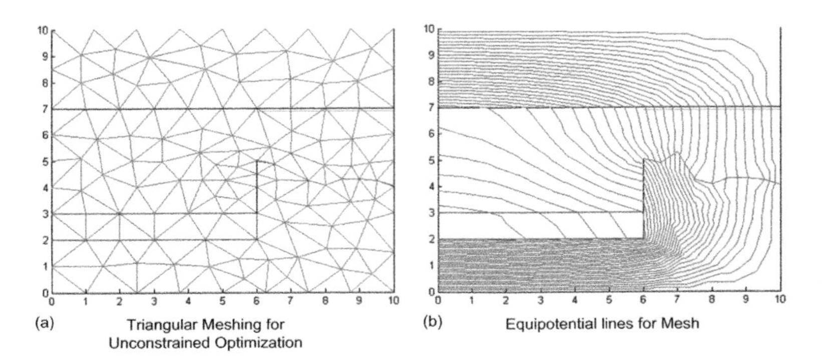

FIGURE 3.36
The pole face shaped without constraints. (a) Triangular meshing for unconstrained optimization, (b) Equipotential lines for mesh.

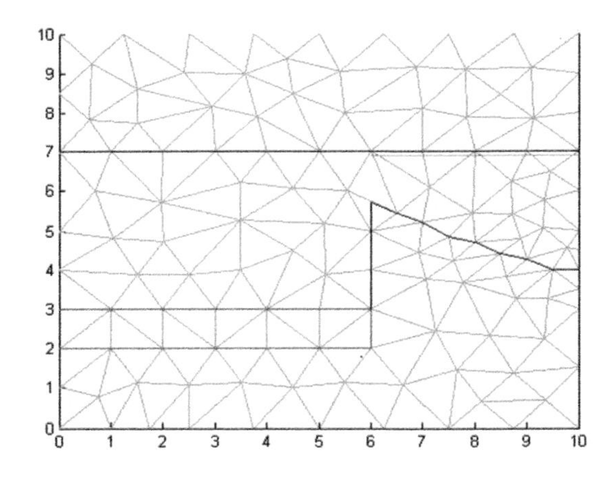

FIGURE 3.37
Triangular meshing using constraints.

TABLE 3.6
By Value Measured Using GA without Set Constraints

y	x	By
6.9	5.5	0.897568
6.9	6.0	1.090641
6.9	6.5	1.101291
6.9	7.0	1.101291
6.9	7.5	1.087689
6.9	8.0	1.007916
6.9	8.5	1.047785
6.9	9.0	0.994623
6.9	9.5	0.993329
6.9	10.0	0.993161

TABLE 3.7

Result Set with Minimum Fitness Core out of Ten Results Using Constraints

Set of Values for h with Best Fitness
Best ever fitness: 0.049387 (from generation: 50) Variable vector: **5.522679 5.454090 5.207238 4.834951 4.683326 4.414775 4.267622 4.005445 4.002250**

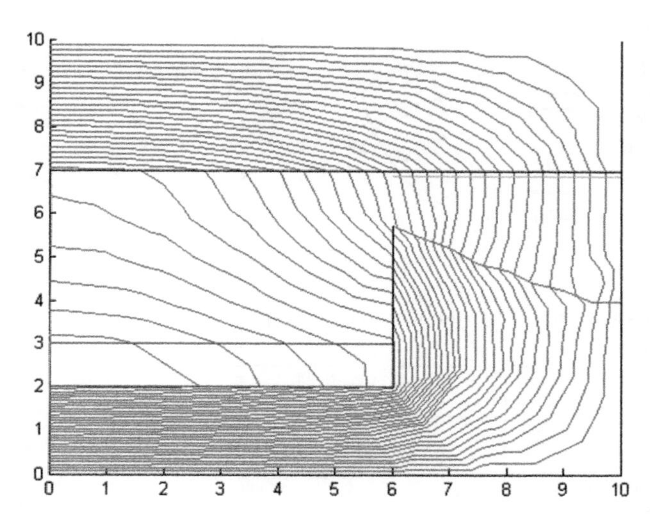

FIGURE 3.38
View of the equipotential lines with constraints.

demonstrate the parametrized mesh generator and matrix-solution software as applied to constrained optimization. In addition to geometry, the current density in the excitation coil and the geometric parameters that define the shape of the pole piece have to be predicted in order to achieve a sinusoidal distribution of the air-gap flux with a peak value of 1.0 T and reduce the flux leakage while the air-gap is constrained to a minimum to prevent the motor from hitting the stator (see Figure 3.39).

Figure 3.40 shows the related dimensions, material properties, and field excitation values used. The symmetry of the magnetic fields with respect to the pole axis enables the modeling of just half of the pole pitch, where the pole axis, which is the line of symmetry, is located at the right boundary to the finite element solution domain. The stator is idealized as a solid steel region without slots, and both the stator and the rotor are made of linear steel with a relative permeability of 2000. We will optimize the device with constraints of current density $J \leq 2.0\ A/mm^2$ which is the limit for copper windings, air-gap between stator and rotor $x < 2$ cm and flux go through the points A_1 and $A_2 < 0.3 \times$ flux go through the points A_3 and A_1 which means allowable leakage flux is 30%.

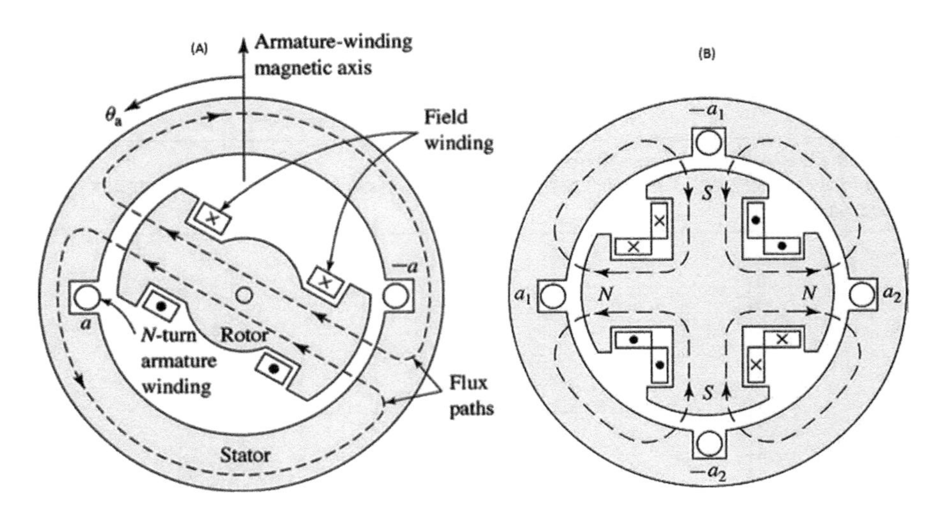

FIGURE 3.39
A synchronous generator (A) Two-pole and (B) Four-pole.

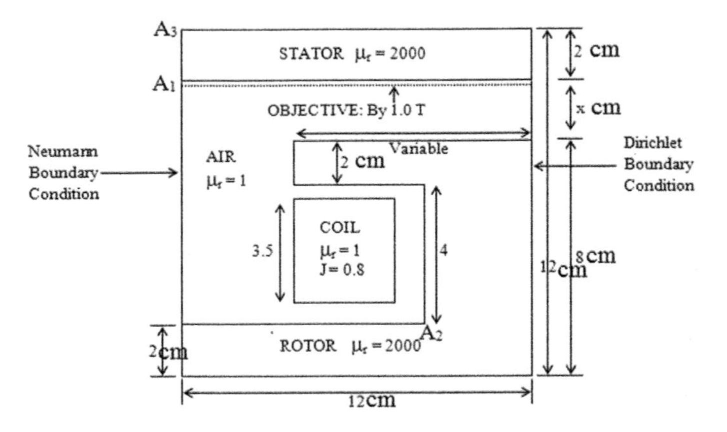

FIGURE 3.40
Parametrized geometry of salient pole.

3.6.2 The Alternator Rotor: Problem Model

Figure 3.41 explains the parameters of the problem and how to model this problem. There are 14 fixed points, 16 variable heights (h1, …, h16), eight measuring points, and four materials in this problem (see Figure 3.41). Figure 3.42 shows the corresponding starting mesh for this problem. Figure 3.43 shows the flux lines of this salient pole synchronous generator at starting.

When we optimize this problem directly by defining an independent parameter for the displacement of each point in the rotor, like in the previous example, the shape we get is given in Figure 3.43.

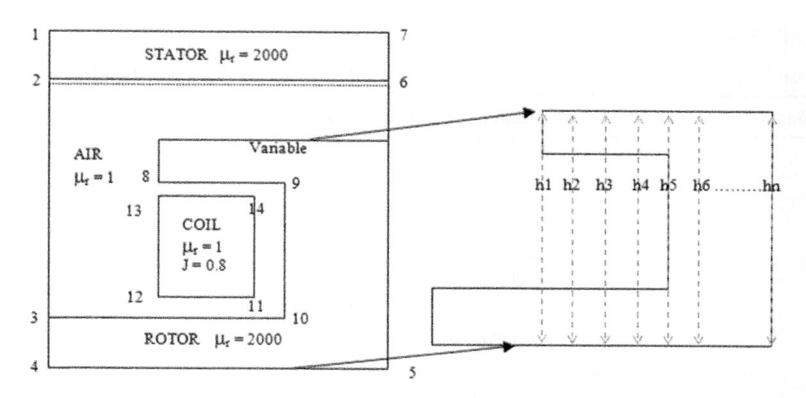

FIGURE 3.41
Defining the problem.

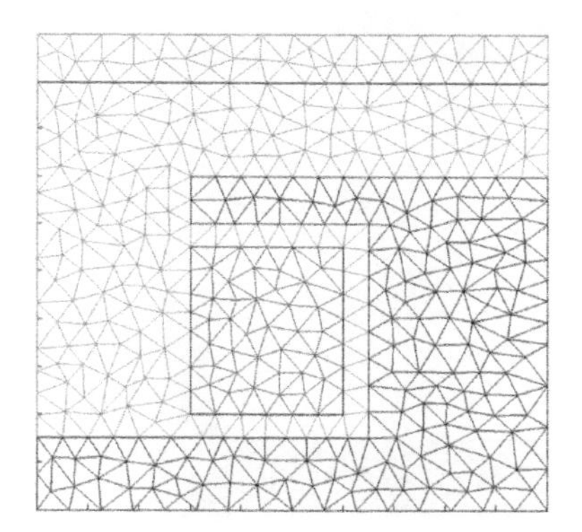

FIGURE 3.42
Initial mesh.

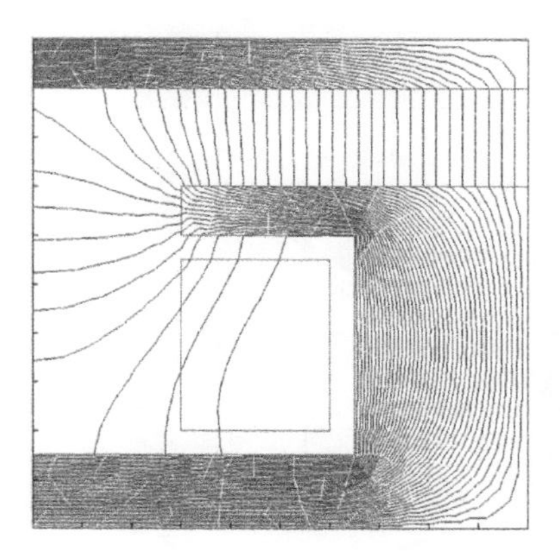

FIGURE 3.43
Flux lines from initial mesh.

TABLE 3.8
The Flux Distribution as the Direct Result of Constrained Optimization

Measuring Points	Flux Density (in Tesla)	Target Flux Density (in Tesla)
1	0.0202	0.0000
2	0.1869	0.1736
3	0.3721	0.3420
4	0.4987	0.5000
5	0.6321	0.6428
6	0.7453	0.7660
7	0.8769	0.8660
8	0.9215	0.9397
9	0.9709	0.9848
10	1.0091	1.0000

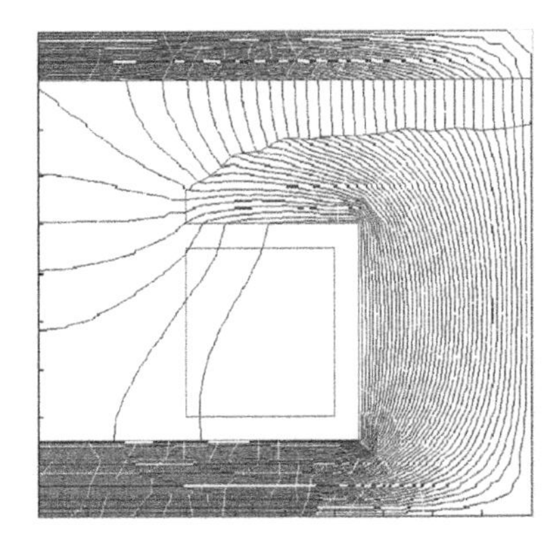

FIGURE 3.44
Optimized shape.

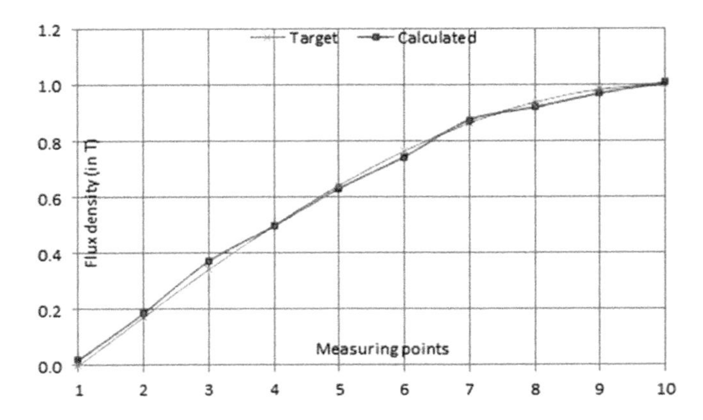

FIGURE 3.45
Slight mismatch between target and its realization.

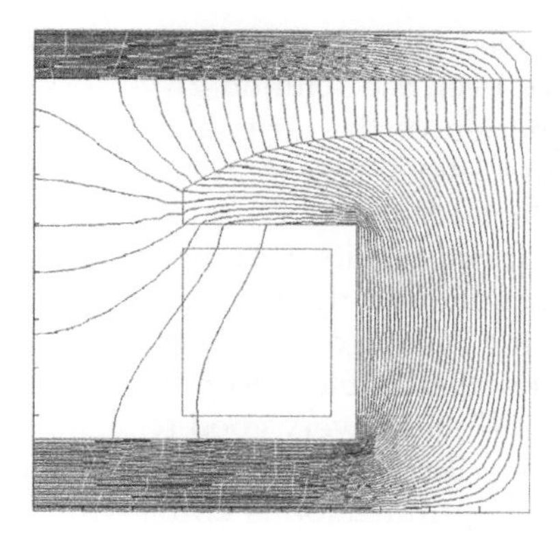

FIGURE 3.46
Flux lines from smoothened shape of Figure 3.43.

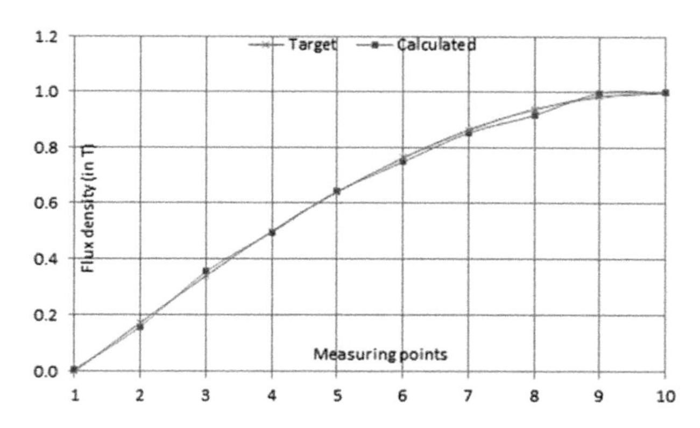

FIGURE 3.47
Excellent match between target and its realization.

TABLE 3.9
The Flux Distribution after Optimization with Smoothened Shape

Measuring Points	Flux Density (in Tesla)	Target Flux Density (in Tesla)
1	0.0083	0.0000
2	0.1596	0.1736
3	0.3560	0.3420
4	0.4964	0.5000
5	0.6434	0.6428
6	0.7517	0.7660
7	0.8549	0.8660
8	0.9183	0.9397
9	0.9986	0.9848
10	1.0035	1.0000

This was then optimized with the constraints. This shape though manufacturable is rough; however, the solution gives very good results in terms of a sinusoidal distribution of the air-gap flux with a peak value of 1 T (see Table 3.8 and Figure 3.44). It is seen in Figure 3.45 that the realized flux density is slightly off the target.

Table 3.8 gives the realized results in numerical form.

We then used an averaging technique to remove sharp bends. To this end, we took five neighboring values of a height and calculated the mean for every variable solution with suitable modification for boundary variables to get Figure 3.46. Figure 3.47 shows very good realization of the target.

The results are given in Table 3.9 and Figures 3.45 and 3.46. The average error of un-smoothened optimization is 2.96% while the smoothened optimization method has an error of 2.24%.

4

Some Basic Matrix Solution Schemes

4.1 Matrix Solution

In all our numerical methods in field computation, we will solve for the potential or some such quantity at discrete points such as the nodes of a mesh or the middle of an element. This process yields a matrix equation. The matrix-solution methods available are either iterative (where we progressively improve the solution from iteration to iteration), or direct methods where we obtain the solution once and for all. The conjugate-gradients method is the most popular of the iterative methods and the Cholesky-factorization method is the most used of the direct methods. Nonetheless, in this chapter we will briefly review the commonly used matrix solution schemes.

4.2 Matrix Solution by Gaussian Elimination

Once we form the matrix equation from the relevant method of approximation, we have, say,

$$\mathbf{Ax} = \mathbf{b}, \tag{4.1}$$

Where we have used $\mathbf{b}$ in place of $\mathbf{B}$ to indicate that the right-hand side is a column vector as happens in finite element field computation.

From our approximations, we need to solve the equation to obtain the solution $\mathbf{x}$ explicitly. For the moment we shall content ourselves with just getting the solution, without devoting much thought to the speed of solution and storage requirements (or efficiency), which we shall take up in detail in the next chapter.

In this section we shall briefly describe the Gaussian elimination scheme for direct matrix solution – one of the most general methods of matrix solution available (Stewart, 1973). Suffice it to say for now that the implementation presented here does not exploit any possible symmetry or sparsity of the coefficient matrix **A**. In the algorithm presented here, we only use the fact that the matrix is diagonally dominant, so that in the process of elimination, we shall never encounter a zero along the diagonal. In general, matrix solution by Gaussian elimination, however, zeros may occur on the diagonal, in which case we may resort to the swapping of some rows. However, in finite element field computation, the need never arises because the matrix of coefficients is always diagonally dominant.

In solving Equation (4.1) by Gaussian elimination, we make the diagonal term of row 1 unity by dividing the whole equation by the diagonal coefficient of row 1, and then subtract an appropriate multiple of that row from every subsequent row, to make all entries of column 1 that lie below row 1 vanish. Then we proceed to row 2, make the diagonal term 1, and make all entries of columns 2 that lie below row 2 vanish. Proceeding in this manner we will arrive at a matrix equation with all terms below the diagonal zero. The matrix equation so altered is in upper triangular form. The last row of the equation now only involves the last unknown x_n of the unknown column vector, so that it may be found. The penultimate row involving x_{n-1} and x_n may then be used to find x_{n-1}. Proceeding thus, all variables down to x_1 may be found. This process is known as back substitution. This procedure is algorithmically described in Algorithm 4.1. We shall appreciate the algorithm more fully through the hand-worked example for solving the equation:

$$\begin{bmatrix} 1.2500 & -0.2500 & 0.0000 & -1.0000 \\ -0.2500 & 2.5000 & -2.0000 & 0.0000 \\ 0.0000 & -2.0000 & 5.0000 & -0.5000 \\ -1.0000 & 0.0000 & -0.5000 & 2.5000 \end{bmatrix} \begin{bmatrix} A_1 \\ A_2 \\ A_3 \\ A_4 \end{bmatrix} = \begin{bmatrix} 2.0000 \\ 1.0000 \\ 2.0000 \\ 1.0000 \end{bmatrix}. \tag{4.2}$$

We shall solve by Gaussian elimination, discussed earlier in Section 2.3. We first divide the first equation by a factor of 1.25 to make the first diagonal term unity, then subtract an appropriate multiple of the resulting first row from the subsequent rows to make the first column below the diagonal term zero to get:

$$\begin{bmatrix} 1.0000 & -0.2000 & 0.0000 & -0.8000 \\ 0.0000 & 2.4500 & -2.0000 & -0.2000 \\ 0.0000 & -2.0000 & 5.0000 & -0.5000 \\ 0.0000 & 0.2000 & -0.5000 & 1.7000 \end{bmatrix} \begin{bmatrix} A_1 \\ A_2 \\ A_3 \\ A_4 \end{bmatrix} = \begin{bmatrix} 1.6000 \\ 1.4000 \\ 2.0000 \\ 2.6000 \end{bmatrix}. \tag{4.3}$$

Proceeding thus, we get the upper triangle matrix equation

$$\begin{bmatrix} 1.0000 & -0.2000 & 0.0000 & -0.8000 \\ 0.0000 & 1.0000 & -0.81633 & 0.81633 \\ 0.0000 & 0.0000 & 1.0000 & -0.19697 \\ 0.0000 & 0.0000 & 0.0000 & 1.55303 \end{bmatrix} \begin{bmatrix} A_1 \\ A_2 \\ A_3 \\ A_4 \end{bmatrix} = \begin{bmatrix} 1.6000 \\ 0.5714 \\ 0.93333 \\ 3.33333 \end{bmatrix}. \tag{4.4}$$

Now from the last equation, we have

$$A_4 = \frac{3.33333}{1.55303} = 2.14634. \quad (4.5)$$

From the second to last equation, the penultimate row:

$$A_3 = 0.93333 + 0.1969 A_4 = 1.35610. \quad (4.6)$$

By such back substitution

$$A_2 = 0.571429 + 0.81633 A_3 + 0.08163 A_4 = 1.85366, \quad (4.7)$$

and

$$A_1 = 0.2 A_2 + 0.8 A_4 + 1.6 = 3.6878. \quad (4.8)$$

Algorithm 4.1: Gaussian Elimination

```
Procedure Gauss (x, A, b, n)
{Function: Solves the n×n matrix equation Ax=b by
Gaussian elimination
Output: x = the solution, an n vector
Inputs: A = a positive definite n×n matrix; b = the
right-hand side of the equation; n = the
size of the matrix}
Begin
   {Begin: Make the matrix upper triangular}
   For i←1 To n-1 Do
     {Begin: scale row i}
     For j←i+1 To n Do
       A[i,j]←A[i,j]/A[i,i]
     b[i]←b[i]/A[i,i]
     A[i,i]←1.0
     {End: Scale row i}
     {Begin: Make column i below row i zero
      That is, subtract A[j,i] times row i from row j
      for j > i}
      For j←i+1 To n Do
        b[j]←b[j]-A[j,i]*b[i]
```

```
      For k←i+1 To n Do
        A[j, k] ←A[j, k] -A[j, i] *A[i, k]
      A[j, i]←0.0
     {End: Making column i below row i zero}
   {End: Making the matrix upper triangular}
   {Begin: Back substitution}
   x[n] ←b[n]/A[n, n]
   For i←1 To n-1 Do
     j←n-i
     x[j] ←b[j]
     For k←j+1 To n Do
      x[j] ←x[j] -A[j,k] *x[k]
End
```

4.3 The SOR Method

The sucessive over-relaxation (SOR) algorithm, was the algorithm of choice in the early years of field computation when memory was limited, and still is in some large matrix systems. In this algorithm, an initial guess is made of the solution, and in each iteration we go through every row and compute the variable corresponding to that row, assuming of course that all the other variables are known. To algorithmically describe this:

$$x[i] := \frac{B[i] - \sum_{k \neq i; k=1}^{n} A[i,k]^{*} x[k]}{A[i,i]} \qquad i = 1, \ldots, n \tag{4.9}$$

must be repeatedly implemented until the elements $x[i]$ imperceptibly change. This is described in Algorithm 4.2. Note that the checking for $k \neq i$ required in Equation (4.9) is avoided by adding $A[i,i]^{*} x[k]$ to the quantity Sum of Algorithm 4.2 and then taking it out. Acceleration is also used, with a factor of 1.25, exploiting the consistent convergence of the approximate solution from one side of the true solution. That is, if $x[i]$ moves, say, from 100 in an iteration to 140 in the next, then the true solution is taken to require a movement in the same direction from 100 to 140 and beyond. Acceleration, therefore, nudges the answer further along the way by multiplying the change by the acceleration factor 1.25, so that the answer is modified to $100 + 1.25 * 40 = 150$. Schemes exist for computing the ideal acceleration factor that lies in the range 1 to 2 and may be employed (Zienkiewicz and Lohner, 1985). A factor that is too large will push the approximate solution beyond the true solution, and therefore a safe value of 1.25 is used here.

Algorithm 4.2: Liebmann's (or SOR) Matrix-Solution Scheme

```
Procedure Liebmann(x, A, b, n, Precis)
{Function: Solves the matrix equation Ax=b
Output: x = the solution, an n vector
Inputs: A = the n×n matrix of coefficients; b = the
right-hand side, an n vector; n = the size of the
equation; Precis = the percentage precision to which
convergence is required}
Begin
 Repeat
   MaxX;← 0
   MaxChange ← 0
   For i ←1 To n Do
    Old ← x[i]
    Sum ← 0
    For k ←1 To n Do
      Sum ← Sum+A[i,k]*x[k]
    x[i]← (b[i]-Sum+A[i,i]*x[i])/b[i,i]
    Change ← x[i] - Old
    x[i] ← Old + 1.25* Change {Acceleration}
    AbsChange ← Abs(Change)
    AbsX ← Abs(x[i])
    If Abs Change > MaxChange
      Then MaxChange ← AbsChange
    If AbsX > MaxX
      Then MaxX ← AbsX
   Error ← 100*MaxChange/MaxX
 Until (Error < Precis)
End
```

It is easily shown that in a uniform finite difference mesh, since the diagonal terms are 4 and the off-diagonal terms are –1, Algorithm 4.2 reduces to iteratively setting the value at a node to the average of the values at the four surrounding nodes. As a result, no storage needs to be allocated for a matrix. Moreover, a lot of time is saved by avoiding the several unnecessary multiplications $A[i,k]*x[k]$ inherent to Algorithm 4.2, even when the coefficient $A[i,k]$ is zero.

As an example, consider the solution of the equation

$$\begin{bmatrix} 2 & 1 \\ 1 & 2 \end{bmatrix} \begin{Bmatrix} x_1 \\ x_2 \end{Bmatrix} = \begin{Bmatrix} 4 \\ 5 \end{Bmatrix} \tag{4.10}$$

In general, it is common to start with our initial guess at {0,0}. So, from the first equation we get

$$2x_1 + 1 \times 0 = 4;\ \text{or}\ x_1 = 2.0 \tag{4.11}$$

From the second equation, using the latest value of x_2

$$1 \times 2.0 + 2 \times x_2 = 5;\ \text{or}\ x_2 = 1.5 \tag{4.12}$$

That finishes iteration 1 with values ($x_1 = 2.0$ and $x_2 = 1.5$). Now in iteration 2,

$$2x_1 + 1 \times 1.5 = 4;\ \text{or}\ x_1 = 1.25 \tag{4.13}$$

$$1 \times 1.25 + 2 \times x_2 = 5;\ \text{or}\ x_1 = 1.875 \tag{4.14}$$

At the end of this iteration 2, we have values $\left(x_1 = 1.25 \text{ and } x_2 = 1.875\right)$. Proceeding with the iterating, we get as close as we want to the solution ($x_1 = 1.0$ and $x_2 = 2.0$).

Note that x_1 has gone from 0.0 to 2.0 to 1.25. That is, in early iterations our assumption of acceleration, that convergence is consistently toward the solution, does not hold. But it remains true after the first few iterations. If we had acceleration, say with acceleration factor $\alpha = 1.25$, as soon as we got $x_1 = 2$, we would have changed this to

$$x_1 = 0.0 + 1.25 \times \left(2.0 - 0.0\right) = 2.5. \tag{4.15}$$

Now the computation of x_2 has to be modified to use the latest value:

$$1 \times 2.5 + 2 \times x_2 = 5;\ \text{or}\ x_2 = 1.25 \tag{4.16}$$

Accelerating

$$x_2 = 0.0 + 1.25 \times \left(1.25 - 0.0\right) = 1.5625 \tag{4.17}$$

etc.

It is also easy to do these iterations on a spreadsheet. For example, the above Gauss algorithm on an Excel spreadsheet may be represented by the formulae of column F, A to B, where columns 1 to 2 represent the matrix A and column D represents the right-hand side. Column F gives the formula.

And the results after iterations are shown in the next screenshot

	A	B	C	D	E	F
1	2	1		4		=(D1 - B1*F2)/A1
2	1	2		5		=(D2 - A2*F1)/B2

	A	B	C	D	E	F
1	2	1		4		0.99999
2	1	2		5		2.000005

4.4 The Cholesky-Factorization Scheme

The Cholesky-factorization technique is a direct scheme for solving the matrix equation

$$\mathbf{Ax} = \mathbf{b}, \tag{4.1}$$

and is particularly suited to symmetric matrices **A**. This relies on the factorization

$$\mathbf{A} = \mathbf{LL}^t, \tag{4.18}$$

where **L** is a lower triangular matrix. The solution method relies on splitting Equation (4.1) into two:

$$\mathbf{Lz} = \mathbf{b} \tag{4.19}$$

and

$$\mathbf{L}^t\mathbf{x} = \mathbf{z}, \tag{4.20}$$

which is the same as

$$\mathbf{LL}^t\mathbf{x} = \mathbf{b}. \tag{4.21}$$

Thus, if we can find **L**, from Equation (4.19), **z** may be computed by forward elimination and then from Equation (4.20) in which **z** is now known, **x** may be computed by back substitution. For example, if we had

$$\begin{bmatrix} 4 & 2 \\ 2 & 5 \end{bmatrix}\begin{bmatrix} x_1 \\ x_2 \end{bmatrix} = \begin{bmatrix} 1 \\ 1 \end{bmatrix}, \tag{4.22}$$

we may use the factorization

$$\begin{bmatrix} 4 & 2 \\ 2 & 5 \end{bmatrix} = \begin{bmatrix} 2 & 0 \\ 1 & 2 \end{bmatrix}\begin{bmatrix} 2 & 1 \\ 0 & 2 \end{bmatrix}, \tag{4.23}$$

so that corresponding to Equation (4.19), we have

$$\begin{bmatrix} 2 & 0 \\ 1 & 2 \end{bmatrix}\begin{bmatrix} z_1 \\ z_2 \end{bmatrix} = \begin{bmatrix} 1 \\ 1 \end{bmatrix}. \tag{4.24}$$

Using the first row we have $2z_1 = 1$ or $z_1 = 1/2$, and from the second row we have $z_1 + 2z_2 = 1$, which gives us $z_2 = 1/4$. Now corresponding to Equation (4.20):

$$\begin{bmatrix} 2 & 1 \\ 0 & 2 \end{bmatrix} \begin{bmatrix} x_1 \\ x_2 \end{bmatrix} = \begin{bmatrix} \frac{1}{2} \\ \frac{1}{4} \end{bmatrix}. \tag{4.25}$$

From the last row we now have $2x_2 = 1/4$ or $x_2 = 1/8$, and from the first row we have $2x_1 + x_2 = 1/2$, using the now known value of $x_1 = 3/16$.

It is seen that once the Cholesky factorization of the matrix is made, the computation of **x** is trivial.

To determine the rule for computing **L**, consider:

$$\begin{bmatrix} L_{11} & 0 & 0 & \cdots \\ L_{21} & L_{22} & 0 & \cdots \\ L_{31} & L_{32} & L_{33} & \cdots \\ \vdots & \vdots & \vdots & \end{bmatrix} \begin{bmatrix} L_{11} & L_{21} & L_{31} & \cdots \\ 0 & L_{22} & L_{32} & \cdots \\ 0 & 0 & L_{33} & \cdots \\ \vdots & \vdots & \vdots & \end{bmatrix} = \begin{bmatrix} A_{11} & A_{21} & A_{31} & \cdots \\ A_{21} & A_{22} & A_{23} & \cdots \\ A_{31} & A_{32} & A_{33} & \cdots \\ \vdots & \vdots & \vdots & \end{bmatrix}. \tag{4.26}$$

To obtain an element $\mathbf{A}\left[i, j\right]$ of **A**, we multiply the *i*th row $\mathbf{L}^i$ of **L**, with the *j*th column of $\mathbf{L}^t$, which is $\mathbf{L}^j$, the *j*th row of **L**, both of *n* terms. Now since **A** is symmetric, the equation of A_{12} will be the same as that for A_{21}, etc. So, confining ourselves to the lower half of **A** where $i \geq j$:

$$\mathbf{L}^i = \left[L_{i1} \quad L_{i2} \quad L_{i3} \quad \ldots \quad L_{ij} \quad \ldots \quad L_{ii} \quad 0 \quad 0 \quad \ldots\right], \tag{4.27}$$

$$\mathbf{L}^j = \left[L_{j1} \quad L_{j2} \quad L_{j3} \quad \ldots \quad L_{jj} \quad 0 \quad 0 \quad \ldots\right], \tag{4.28}$$

where the diagonal or last nonzero term of $\mathbf{L}^j$ appears not later than its counterpart in $\mathbf{L}^i$ and both vectors go to n terms. Performing the multiplication, we need to consider two cases. First $i = j$:

$$A_{ii} = \mathbf{L}^i\mathbf{L}^i = L_{i1}^2 + L_{i2}^2 + L_{i3}^2 + \ldots + L_{ii}^2 = \sum_{k=1}^{i} L_{ik}^2, \tag{4.29}$$

and second $i > j$:

$$A_{ij} = \mathbf{L}^i\mathbf{L}^j = L_{i1}L_{j1} + L_{i2}L_{j2} + L_{i3}L_{j3} + \ldots + L_{ij}L_{jj} = \sum_{k=1}^{i} L_{ik}L_{jk}. \tag{4.30}$$

Rearranging, we have

$$L_{ij} = \begin{cases} \sqrt{\left[A_{ij} - \sum_{k=1}^{i-1} L_{ik}^2 \right]} & i = j > 1 \\ \dfrac{A_{ij} - \sum_{k=1}^{j-1} L_{ik} L_{jk}}{L_{jj}} & i > j \end{cases} \tag{4.31}$$

Observe also that, by multiplying $\mathbf{L}^i$ and $\mathbf{L}^{t1}$ in Equation (4.26),

$$A_{11} = L_{11}^2. \tag{4.32}$$

Knowing L_{11}, then, we have from Equation (4.31):

$$L_{21} = \frac{A_{21}}{L_{11}}, \tag{4.33}$$

$$L_{22} = \sqrt{\left[A_{22} - L_{21}^2 \right]}, \tag{4.34}$$

$$L_{31} = \frac{A_{31}}{L_{11}}, \tag{4.35}$$

$$L_{32} = \frac{A_{32} - L_{31} L_{21}}{L_{22}}, \tag{4.36}$$

$$L_{33} = \sqrt{\left[A_{33} - L_{31}^2 - L_{32}^2 \right]}, \tag{4.37}$$

and so on. It should be noted that once L_{ij} is computed, A_{ij} is never required. This allows us to write L_{ij} in the same storage location as assigned to A_{ij} at great savings in memory. The chief disadvantage in the method is that as the matrix size becomes large, the memory requirements become excessive. Moreover, on account of large round-off errors in the solution for large matrices, so as to quantify and check the round-off error, the need to store **A** becomes greater with increasing matrix size, so that the storage savings realized by overwriting **L** on **A** vanish.

4.5 The Conjugate-Gradients Algorithm

The conjugate-gradients scheme (Kershaw, 1978; Meijerink and van der Vost, 1977) is used for the solution of sparse positive definite symmetric matrix equations of the type seen in Equation (4.1). In this method, the solution is found as a series of vectors $\mathbf{p}_i$

$$\mathbf{x} = \alpha_1\mathbf{p}_1 + \alpha_2\mathbf{p}_2 + \ldots + \alpha_m\mathbf{p}_m, \tag{4.38}$$

where m is no larger than the matrix size n and depends on the starting direction $\mathbf{p}_1$ and the scatter between the eigenvalues of $\mathbf{A}$, or the spectral radius of $\mathbf{A}$. Hence, although this is an iterative scheme, it is guaranteed to converge in n or fewer iterations; the term *semi-iterative* is therefore commonly employed.

The underlying principle of the conjugate gradients scheme is the existence of n vectors $\mathbf{p}_i$ orthogonal to the matrix of coefficients:

$$\mathbf{p}_i^t\mathbf{A}\mathbf{p}_j \begin{cases} = 0 & \text{if } i \neq j \\ \neq 0 & \text{if } i = j \end{cases} \tag{4.39}$$

Thus, by substituting Equation (4.38) in Equation (4.18) and pre-multiplying by $\mathbf{p}_i^t$:

$$\mathbf{p}_i^t\mathbf{A}\left[\alpha_1\mathbf{p}_1 + \alpha_2\mathbf{p}_2 + \ldots + \alpha_m\mathbf{p}_m\right] = \mathbf{p}_i^t\mathbf{b}, \tag{4.40}$$

which gives us from Equation (4.39)

$$\alpha_i = \frac{\mathbf{p}_i^t\mathbf{b}}{\mathbf{p}_i^t\mathbf{A}\mathbf{p}_i}. \tag{4.41}$$

Thus, the problem of finding the unknown $\mathbf{x}$ reduces to one of finding the orthogonal vectors $\mathbf{p}_i$, having which, using Equation (4.41), we may compute the terms α_i and then $\mathbf{x}$ from Equation (4.38). Since the set of vectors $\mathbf{p}$ is not unique, we are free to assume any value for $\mathbf{p}_1$. In the algorithm for finding the other $\mathbf{p}$s, there is no need to store all of them, so that the demands on computer memory are not heavy.

$$\mathbf{p}_1 = \mathbf{b} - \mathbf{A}x_1 \tag{4.42}$$

is a good value to take for the starting vector $\mathbf{p}_i$, because what we now need to add to the starting value x_1, namely a multiple α of $\mathbf{p}_1$, must have components parallel to the residual. In general, at a step i in the iterative process,

$$\mathbf{x}_{i+1} = \mathbf{x}_i + \alpha_i\mathbf{p}_i. \tag{4.43}$$

We need at this point to find α_i. If we regard the operation at this iterative step as solving the matrix equation

$$\mathbf{A}\mathbf{e}_i = \mathbf{b} - \mathbf{A}\mathbf{x}_i = \mathbf{r}_i \tag{4.44}$$

for the vector residual

$$\mathbf{e}_i = \mathbf{x} - \mathbf{x}_i = \alpha_i\mathbf{p}_i + \alpha_{i+1}\mathbf{p}_{i+1} + \ldots \tag{4.45}$$

in the current solution $\mathbf{x}_i$, then according to Equation (4.41):

$$\alpha_i = \frac{\mathbf{p}_i^t \mathbf{r}_i}{\mathbf{p}_i^t \mathbf{A} \mathbf{p}_i}. \tag{4.46}$$

After computing α, we may update the solution to $\mathbf{x}_{i+1}$, using Equation (4.45), corresponding to which we will have a new residual $\mathbf{r}_{i+1}$. Now we need to find $\mathbf{p}_{i+1}$, to make another improvement in the solution, and, as before, we may ideally take $\mathbf{p}_{i+1}$ as the new residual $\mathbf{r}_{i+1}$. However, having accounted for all components in the solution parallel to $\mathbf{p}_i$, we do not wish $\mathbf{p}_{i+1}$ to have any components parallel to $\mathbf{p}_i$, with which no more gains in accuracy may be made in the solution. Therefore, we need to take off from the residual an appropriate multiple β of $\mathbf{p}_i$:

$$\mathbf{p}_{i+1} = \mathbf{r}_{i+1} - \beta_i \mathbf{p}_i. \tag{4.47}$$

The value of β_i is found by requiring the orthogonality relationship Equation (4.39):

$$\mathbf{p}_i^t \mathbf{A} \mathbf{p}_{i+1} = \mathbf{p}_i^t \mathbf{A} \mathbf{r}_{i+1} - \beta_i \mathbf{p}_i^t \mathbf{A} \mathbf{p}_i = 0 \tag{4.48}$$

whence

$$\beta_i = \frac{\mathbf{p}_i^t \mathbf{A} \mathbf{r}_{i+1}}{\mathbf{p}_i^t \mathbf{A} \mathbf{p}_i}. \tag{4.49}$$

Algorithm 4.3: The Conjugate-Gradients Algorithm

```
Guess x
r ← b - Ax
p ← r
ε ← Permitted Error
e ← Norm(r)
While e > ε Do
  α ← p^t p/p^t Ap
  x ← x + αp
  r ← b - Ax
  e ← Norm(r)
  β ← -r^t Ap/p^t Ap
  p ← r - βp
```

Now we may compute $\mathbf{p}_{i+1}$ and we are ready for the next iterative improvement in $\mathbf{x}$. This procedure is algorithmically stated in Algorithm 4.3.

In its initial form, as proposed by Hestenes and Stiefel (1952), the conjugate-gradients method was not competitive with other schemes of solution. Thus, it did not come into its own until the introduction of the preconditioning schemes of recent years, relying upon solving an equivalent matrix equation with improved convergence properties. It may be shown that the number of conjugate gradient steps required for convergence is the same as the number of independent eigenvalues in the matrix of coefficients $\mathbf{A}$. Preconditioning, therefore, is an attempt to solve an equivalent matrix equation with clustered eigenvalues. The ideal matrix equation has the identity matrix as the coefficient matrix, since it has all n eigenvalues the same, given by the solution of $[\lambda - 1]^n = 0$. The method becomes competitive with direct methods in its preconditioned form, and is today the preferred method for many. In general, the modified equation takes the form

$$\mathbf{B}\mathbf{A}\mathbf{B}^t\left[\mathbf{B}^{-t}\mathbf{x}\right] = \mathbf{B}\mathbf{b}, \tag{4.50}$$

where the preconditioning matrix $\mathbf{B}$ is any matrix that clusters the eigenvalues of the new coefficient matrix $\mathbf{B}\mathbf{A}\mathbf{B}^t$, or makes it as close to the unit matrix as possible. Attention is drawn to the use of the term $\mathbf{B}^t\mathbf{B}^{-t}$ and the change of unknowns from $\mathbf{x}$ to $\mathbf{B}^{-t}\mathbf{x}$ so as to keep the coefficient matrix symmetric in keeping with the requirements of the conjugate gradients algorithm. Thus, in the incomplete Cholesky preconditioning, $\mathbf{B}$ is the inverse of an approximate Cholesky factor $\mathbf{L}$ defined by Equation (4.18). If $\mathbf{B}$ were made exactly equal to $\mathbf{L}^{-1}$, then the coefficient matrix is $\mathbf{L}^{-1}\mathbf{L}\mathbf{L}^t\mathbf{L}^{-t}$ or the identity matrix, but the cost of computing $\mathbf{L}$ will be high; and besides, there will be no need to use the conjugate gradients algorithm, since we may easily compute the solution from $\mathbf{L}$. The approximate Cholesky factor places no large memory load and is easy to compute. It is arrived at by ignoring terms at locations where $\mathbf{A}$ has zeros. Since the exact factor will have nonzero terms only to the right of the leading nonzero of the profile of $\mathbf{A}$, any approximation will be within the profile. In Evans's preconditioning, a similar splitting of the matrix $\mathbf{A}$ is employed, after scaling Equation (5.1) to have unit terms on the diagonal, using the factorization

$$\mathbf{A} = \mathbf{D}\mathbf{S}\mathbf{D}, \tag{4.51}$$

where the diagonal matrix $\mathbf{D}$ is

$$\mathbf{D}_{ij} = \begin{cases} \sqrt{A_{ij}} & i = j \\ 0 & i \neq j \end{cases}, \tag{4.52}$$

and

$$S_{ij} = \frac{A_{ij}}{\left(D_{ii}D_{jj}\right)}. \tag{4.53}$$

For example, the splitting of the following coefficient matrix **A** is

$$A = \begin{bmatrix} 4 & 1 \\ 1 & 9 \end{bmatrix} = \begin{bmatrix} \sqrt{4} & 0 \\ 0 & \sqrt{9} \end{bmatrix} \begin{bmatrix} \frac{4}{\sqrt{4}\sqrt{4}} & \frac{1}{\sqrt{4}\sqrt{9}} \\ \frac{1}{\sqrt{9}\sqrt{4}} & \frac{9}{\sqrt{9}\sqrt{9}} \end{bmatrix} \begin{bmatrix} \sqrt{4} & 0 \\ 0 & \sqrt{9} \end{bmatrix}. \tag{4.54}$$

Note that only the diagonal terms of **D** need to be stored. And since we know that the diagonal terms of the new coefficient matrix **S** are 1, we may store them here and overwrite the rest of **S** upon **A**. Whenever necessary, **A** may be recovered from **S**, so that no extra memory is taken up by **S** or **D**. Equation (5.1), therefore reduces by this scaling, using $\mathbf{DSDx} = \mathbf{b}$ to

$$\mathbf{S}(\mathbf{Dx}) = \left(\mathbf{D}^{-1}\mathbf{b}\right) \tag{4.55}$$

or

$$\mathbf{Sz} = \mathbf{c}, \tag{4.56}$$

where, using the fact that **D** is diagonal, the new variable **z** is defined by

$$z_i = D_{ii}x_i = x_i\sqrt{A_{ii}}, \tag{4.57}$$

and the new right-hand side by

$$c_i = \frac{b_i}{D_{ii}} = \frac{c_i}{\sqrt{A_{ii}}}. \tag{4.58}$$

c, too, may be overwritten on **b**. Once the solution **z** is found by applying the conjugate gradients algorithm to Equation (4.56), **x** may be found through the application of Equation (4.57).

Interestingly, with this scaling, even if we do not use any preconditioning and employ the simple conjugate gradients iterations, convergence will be found to have been improved. Apparently, when the diagonal terms are all of the same size, the eigenvalues must be getting clustered. In Evans's or SOR preconditioning for matrices with 1 everywhere along the diagonal, **L** is assumed to be

$$L_{ij} = \begin{cases} A_{ij} & \text{if } i > k \\ A_{ii} = 1 & \text{if } i = j \\ 0 & \text{if } i < j \end{cases} \tag{4.59}$$

and as a result, no extra storage is required for the matrix **L**. It is elementary to verify that this approximately fits Equation (4.18) if the diagonal dominance of **A** is accounted for; since off diagonal terms are smaller than unity, their squares may be assumed negligible. For example, for the diagonally dominant symmetric matrix **A** scaled to have 1s along the diagonal:

$$\begin{bmatrix} 1.0 & 0.1 & 0.2 \\ 0.1 & 1.0 & 0.3 \\ 0.2 & 0.3 & 1.0 \end{bmatrix},$$

taking the lower triangular part as its Cholesky factor **L**, we have:

$$\begin{bmatrix} 1.0 & 0.0 & 0.0 \\ 0.1 & 1.0 & 0.0 \\ 0.2 & 0.3 & 1.0 \end{bmatrix}\begin{bmatrix} 1.0 & 0.1 & 0.2 \\ 0.0 & 1.0 & 0.3 \\ 0.0 & 0.0 & 1.0 \end{bmatrix} = \begin{bmatrix} 1.0 & 0.1 & 0.2 \\ 0.1 & 1.01 & 0.32 \\ 0.2 & 0.32 & 1.13 \end{bmatrix} \tag{4.60}$$

or approximately

$$\mathbf{L}\mathbf{L}^t \approx \mathbf{A}, \tag{4.61}$$

Thus, it is seen that both methods of preconditioning assume some approximate Cholesky splitting of **A**; while the incomplete Cholesky scheme has the advantage of computing **L** more accurately, Evans's method uses a splitting whose coefficients are already found in **A** and so does not require any extra storage or computation. Significantly, as we shall see, both methods assume the same sparsity pattern for the Cholesky splits as the original matrix **A**, and so ignore all elements of the decomposition factor which fall within the profile.

With a preconditioning matrix **B**, the conjugate-gradients method has been shown to reduce to Algorithm 4.4 (Kershaw, 1978). Observe that when **B** is lower triangular as with Evans's or Cholesky preconditioning, the computation of $\mathbf{B}^{-1}\mathbf{r}$ and $\mathbf{B}^{-t}\mathbf{r}$ is, respectively, by forward elimination and back substitution, so that the inverses never need to be computed explicitly. That is, we in turn solve the equations $\mathbf{Bx}=\mathbf{r}$ (which gives us $\mathbf{B}^{-1}\mathbf{r}$) and $\mathbf{B}^t\mathbf{y} = \mathbf{x}$ (which gives us $\mathbf{B}^{-t}\mathbf{B}^{-1}\mathbf{r}$). In the algorithm the vector **p** is made to contain $\mathbf{B}^{-t}\mathbf{B}^{-1}\mathbf{p}$ where **p** is the vector of the theory above. The algorithm is therefore considerably simplified by our never having to compute the coefficient matrix $\mathbf{BAB}^t$.

Algorithm 4.4: The Preconditioned Conjugate-Gradients Algorithm

```
p ← 0
γ ← 1
x ← B⁻¹b {Guess x}
x ← B⁻ᵗx
ε ← Permitted Error
Repeat
  r ← b - Ax
  r ← B⁻¹r
  e ← Norm(r)
  θ ← rᵗr
  β ← -θ/γ
  γ ← θ
  r ← B⁻¹r
  p ← r - βp
  δ ← pᵗAp
  α ← γ/δ
  x ← x + αp
Until e < ε
```

Recognizing that many graduate students have difficulties implementing this algorithm quickly, a working MATLAB program, consisting of Code Box 4.1 together with Code Box 4.2, is given which the reader may find a useful exercise to transfer to Java or c.

CODE BOX 4.1 FOR THE CONJUGATE-GRADIENTS ALGORITHM

```
% The purpose of this program is to semi-iteratively
  solve a sparse positive
% definite symmetric matrix equation of the type 'Ax=b'
% Input Variables:
% 1. 'AV' -- an 'n x n' sparse, positive definite matrix
stored as
%a one dimensional array of lower triangular elements
% 2. 'Col' -- column indices corresponding to AV elements
% 3. 'b' -- the right-hand side of the equation
% 4. 'x' -- an initial guess of the solution
% 5. 'err' -- permitted error to the solution
% Output Variables:
% 1. 'x' -- the solution vector to the matrix equation
```

```
% Related Functions:
% 1. 'SparseMatMultiply.m'-- solves the equation b=Sx,
where S is a sparse,
%symmetric matrix stored as a 1-dimensnl array of lower
triangular elements
% Format:
% [x] = Conj_Grad(AV,Col,x,b,err)
function [x] = Conj_Grad(AV,Col,x,b,err)
[guess] = SparseMatMultiply(AV,x,Col); % calculate b=Ax
r = b - guess; % calculate residual based on initial
guess for 'x'
p = r; % set initial search direction 'p' equal to 'r'
mag_r = norm(r); % calculate the magnitude of the
residual to determine necessity
             % for loop iteration
while mag_r > err
  [Sp] = SparseMatMultiply(AV,p,Col); % calculate b=Sx
  alpha = (p'*p)/(p'*Ap); % compute the scalar coefficient
  of the
             % search direction vector 'p' used to locate
             % the point at which f(x) (scalar, quadratic
             % function of matrix A and vectors b and x)
             % is minimized
  x = x + alpha*p; % calculate point at which f(x) is
  minimized along the
          % search line defined by 'p'
  [new_guess] = SparseMatMultiply(AV,x,Col); % calculate
  b=Ax
  r = b - new_guess; % calculate new residual based on
  updated solution 'x'
  mag_r = norm(r) % calculate the magnitude of new
  residual
  beta = (-r'*Ap)/(p'*Ap); % calculate the search
  direction coefficient
      % used to subtract from the updated residual any
      % components that are not
      % S-orthogonal (i.e., d(i)'*A*d(j) not equal to
      zero)
      % to previous search vectors
  p = r - beta*p; % calculate a new search vector that is
  S-orthogonal to
                 % all previous search vectors
end
x = real(x); % eliminate any imaginary numbers from
solution
```

CODE BOX 4.2 EVALUATING B = AX

```
% The purpose of the program is to evaluate b = Ax, where
A is a sparse symmetric
% matrix stored as a vector of lower diagonal elements
and x is a column vector
% Input Variables:
% 1. 'AV' -- lower diagonal non-zero elements of a
sparse, symmetric n x n
     %matrix
% 2. 'x' -- n x 1 column vector
% 3. 'Col' -- vector of indices identifying the column
location of
     %non-zero elements in a sparse, symmetric matrix
% Output Variables:
% 1. 'b' -- the solution vector to the matrix
multiplication
% Format:
% [b] = SparseMatMultiply(AV,x,Col)
function [b] = SparseMatMultiply(AV,x,Col)
NElems = length(AV); % determine number of non-zero
elements in S
b = zeros(size(x,1),1); % initiate b for use in for loop
Row = 1; % set row index equal to one
for ij = 1:NElems % for all matrix elements
  Column = Col(ij); % set Column index equal to matrix
element column value
  b(Row) = b(Row) + AV(ij)*x(Column);
    % sum products of corresponding elements
    % matrix A rows and vector x rows
  if Row == Column
    Row = Row + 1; % advance row index
  else
    b(Column) = b(Column) + AV(ij)*x(Row);
% sum products of corresponding matrix A column elements
% and vector x rows
  end
end
```

5

Matrix Computation with Sparse Matrices

5.1 The Importance of Efficiency

5.1.1 Seeking Efficiency

In the computer exercises we have come across in Chapter 2, one notices that the matrix equation for n unknowns involves dimensioning an $n \times n$ real matrix with n^2 storage locations. Obviously, with this kind of storage even the most powerful machine (let alone small machines) would be on its knees for a simple problem with 1000 unknowns. This situation is unacceptable, and we should take into account the architectural details of the computer, the symmetry of the matrix, and whether the matrix is sparse, in the sense that most of the numbers of such a matrix are zero.

Such considerations result in an efficient use of the computing power at our disposal and will allow us to solve problems faster and, indeed, allow us to attempt ever larger problems. The fact that many elements in a sparse matrix are zero means that the basis of these efforts is to avoid multiplications with a zero since on a computer whether we multiply a number with a zero or nonzero, the effort is the same.

In the course of our finite difference or finite element analysis of electromagnetic fields, among a myriad of other sciences, sparse, positive definite, diagonally dominant, and (often) symmetric matrices must be solved. The matrix solution often requires the most amount of computing in field analysis, and any attempt at making our programs efficient should first be directed to this arena. Two kinds of matrix storage are commonly used in solving the equation, sparse storage where only the nonzero elements of half of the matrix are stored, and profile storage where all elements of a matrix from the first nonzero of each row are stored.

5.1.2 Sparse Matrices

We would note that generally when two nodes, say i and j, in a finite element mesh belong to the same element, the energy contributions from the element would fill up column i of row j and column j of row i, in addition to the diagonal terms, of the global matrix which becomes the matrix of coefficients.

For explanatory purposes, consider the problem of a square cable in a square sheath as shown in Figure 5.1. The minimal boundary-value problem consists of an eighth of the geometry with a Dirichlet boundary and two Neumann boundaries. This has been divided into a mesh of eight first order linear elements with ten nodes of which six are unknown. From the element 2-1-3, we may say that row 1 will have columns 2 and 3 filled, row 2 columns 1 and 3, and row 3 columns 1 and 2. Similarly element 5-2-3 informs us that row 2 would have columns 5 and 3 occupied, row 3 columns 2 and 5, etc.

Proceeding thus we arrive at the matrix as shown in Figure 5.2, the mark X standing for occupied locations with unmarked locations being zero. Because the field values at nodes 7 to 10 are known, they do not figure in the equation.

Thus, out of the 36 locations of the coefficient matrix only 15 are occupied. We have used the symmetry of the matrix to store only the diagonal and lower half. There are two schemes of storing sparse matrices, profile storage

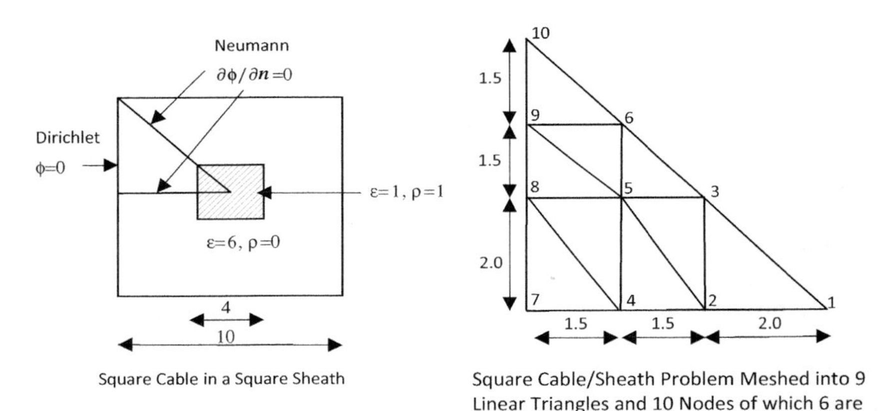

FIGURE 5.1
A square cable and an eighth prepared for finite element analysis

Row\Col	1	2	3	4	5	6
1	X					
2	X	X				
3	X	X	X			
4		X		X		
5		X	X	X	X	
6			X		X	X

FIGURE 5.2
A symmetric lower half of a sparse matrix for mesh of Figure 5.1.

and sparse storage. These will be discussed separately. Once we decide not to allocate storage for all elements of the matrix, we need some means of identifying where in the matrix – which row and which column – a stored number belongs. This involves appropriate data structures.

It seems that we have achieved about 50% storage by eliminating the storage of the zeros. In this case we need 15 number locations instead of 36. However, this reasoning merely from one example involving a small matrix is misleading. In actual fact, sparse storage needs to grow as the matrix size n whereas full storage needs to grow as n^2. For an ideal mesh with isosceles triangles – see Figure 5.3 – a node would be involved in six triangles, that is connected to six other nodes, giving us $6+1$ elements per matrix row where the plus 1 is for the diagonal term. With the exception of the first few rows of the matrix, this generally means four nodes per row in the lower triangle of the matrix including the diagonal. That is, the storage is $4n$ as opposed to n^2 for full storage where n is the matrix size. For large matrices, therefore, the savings are very large – significantly we save more for large matrices, which present the bigger problem of memory shortage. If we used symmetry alone to diminish our storage requirements, the first row would have the number 1, the second row the number 2, and so on so that the need is $1+2+3+\ldots n=n(n+1)/2$ real number locations. That, however, is not good enough since the memory need still grows as n^2.

5.1.3 Computational Time Savings

We resort to sparse storage not only to save memory but also to reduce computational time as we will note in this chapter. Consider the equation

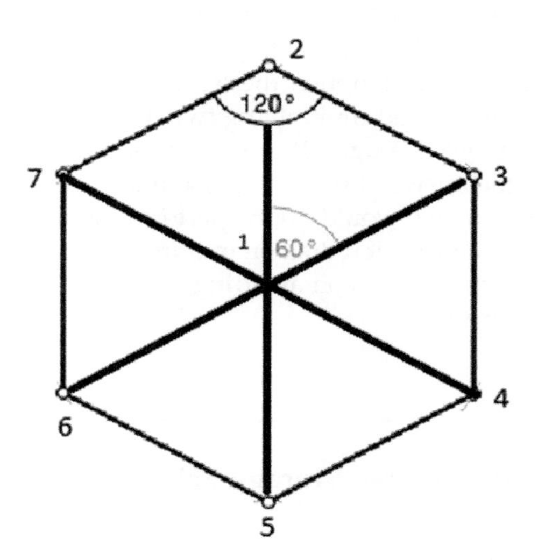

FIGURE 5.3
In a finite element mesh of equilateral triangles, a node 1 is connected to 6 others.

$[A^G]\{x\} = \{b^G\}$ where the matrix $[A^G]$ is a sparse matrix as shown in Figure 5.2. If, for example, we were using an iterative scheme of solution such as SOR, the essence of the algorithm at iteration k would be the modifying Equation (4.9)

$$x_i^{k+1} = \frac{1}{A_{ii}^G}\left\{b_i^G - \sum_{j=1}^{i-1} A_{i,j}^G x_j^{k+1} - \sum_{j=i+1}^{n} A_{i,j}^G x_j^k\right\} \tag{5.1}$$

where improving the $k+1$-th iterate of x_i we use the i-th row of the matrix equation using the latest values of all other values x_j, $j=1$ to n except $j=i$. Since the operation of a particular iteration $k+1$ goes on row i at a time, i ranging from the first to the last row, all j values before i have already been improved and therefore we use the latest value from iteration $k+1$ and all j values after i use the value from the previous iteration k. To over-relax we would accelerate with factor $1 < \alpha < 2$ according to

$$x_i^{k+1} = x_i^k + \alpha\left(x_i^{k+1} - x_i^k\right) \tag{5.2}$$

We note that each operation involves $n-1$ multiplications $A_{i,j}^G x_j$ per row. Since there are n rows to be dealt with per iteration there are $n(n-1)$ multiplications per iteration. Since the number of iterations for convergence is at least of order n, there are at least n^3 multiplications to perform.

However, in a finite element or finite difference mesh (see Section 2.3), for example, of the ideal isosceles triangles (see Figure 5.3), each node will be surrounded by six triangles giving us six connected nodes. That is, per row equation, we will be making n-1 multiplications $A_{i,j}^G x_j$ (mostly with $A_{i,j}^G = 0$) whereas only about six of these multiplications will be with non-zero numbers. Although in our heads we would quickly say that the multiplication $A_{i,j}^G x_j$ is zero when one of the numbers is zero, on a computer there is no difference between multiplying two nonzero numbers and multiplying two numbers of which one is zero; for the computer will go through all the operations with the registers before arriving at zero as the answer. So, with sparse matrices if we can identify the zeros or eliminate them, we would not be performing any multiplication with them. This leads to a significant reduction in computational time.

Sparse storage schemes focus on avoiding these multiplications with zero and not storing these zeros.

5.2 Symmetric and Sparse Storage Schemes – Suitable Data Structures

The exploitation of the symmetry of a matrix is very straightforward. Since, for a symmetric matrix **A**, $A[i,j] = A[j,i]$, we need to store only the diagonal

```
1
2    3
4    5    6
7    8    9    10
...  ...  ...  ...  ...
```

Element (i,j) at 1 + 2 + 3 + …+i – 1 + j = i*(i–1) div 2 + j

E.g., Element (4,2) at 1 + 2 + 3 + 2 = (4*3)/2 + 2 = 8

FIGURE 5.4
Symmetric matrix storage.

and the upper or lower half. The storage of the diagonal and the lower half is as shown in Figure 5.4.

Although by using the property of symmetry we may cut down storage by approximately a half, storage requirements are still of the order of n^2. For a 1000*1000 matrix, with symmetry, we may bring storage from 10^6 numbers to $(1000+1)*1000/2=500{,}500$, still a considerable number, despite the near 50% savings.

It is seen that $A[1,1]$ is stored as $AV[1]$ of a vector **AV**, $A[2,1]$ and $A[1,2]$ as $AV[2]$, $A[2,2]$ as $AV[3]$, and so on. Next, on row 1, we have one number; on row 2, two numbers; on row 3, three numbers; and on row $i-1$, $i-1$ numbers. $A[i,j]$ and $A[j,i]$ for $i \geq j$ will then be at a location given by all the numbers up to row $i-1$ plus j numbers. Since $1+2+\ldots+i-1$ is $(i-1)^* i/2$, we have $A[i,j]$ stored at $AV[(i-1)^* i/2+j]$.

To describe this algorithmically, we may write a procedure or subroutine to determine the location Loc where $A[i,j]$ is stored, as described in Algorithm 5.1 in FORTRAN. Thus, in the program, we would store the $n \times n$ matrix **A** as a vector **AV** of length $(n+1)*n/2$ and save approximately half the space originally used. Thereafter, wherever we find $A[i,j]$ in the program, we would replace this by $AV[\text{Loc}(i,j)]$; here $\text{Loc}(i,j)$ will call the function Loc, which in turn will compute and return in Loc the integer location *ij* of $A[i,j]$ in **AV**.

Algorithm 5.1: Locating Coefficients in Symmetric Storage

```
Integer Function Loc(i,j)
Integer i, j, ij
If(i . LT . j) Then  !Number in Upper Half, Stored at
(j,i)
```

```
    ij=(j-1)*j/2+i
  Else  !Number in Lower Half, Stored at (i,j)
    ij=(i-1)*i/2+j
  EndIf
  Loc = ij
  Return
  End
```

Although we cut storage almost by a factor of 2, the storage requirement still is of order n^3 and is therefore enormous. Real gains are attained by exploiting sparsity. A sparse matrix is one in which most of the numbers are zero, and it is the exploitation of this that allows us to solve large problems by the finite element or finite difference method.

The use of sparsity brings storage down to the order of n – for a first-order finite element mesh, for example, an equation for the potential at a node will be in terms of that potential itself and the potentials at those nodes that are linked to that node through triangle edges. For a regular mesh, triangles will tend to be near-equilateral (see Figure 5.3), so that each node will be surrounded by approximately six others. Probabilistically, of these six, three will have smaller numbers than that node, and therefore the corresponding coefficients in the lower half that need to be stored are three in number. In total then, including the diagonal, we would have four real numbers per row to store, giving us a total of $4n$ real numbers to store. But, because the nonzero coefficients appear in arbitrary locations, we also need to store the row and column corresponding to each element of the matrix so that we might know where it belongs in the matrix. That is, total storage is $4n$ real numbers, $4n$ integers for row locations, and $4n$ integers for the column locations – a storage requirement growing only as n to our advantage. For the example of 1000 nodes, we then need to store 4000 real numbers and 8000 integers.

How is data accessed when we have this storage? Let the nonzero coefficients of a matrix **A** be stored as a vector **AV**, such that item $\mathrm{A}[i,j]$ of **A** is [*ij*] , $i=\mathrm{Row}[ij]$, and $j=\mathrm{Col}[ij]$. As an illustration, consider the 10×10 symmetric matrix **A** of Figure 5.5A, coming from a finite element discretization, with nodes 1, 9, and 10 being on Dirichlet boundaries and the others in the interior of the region of solution. This matrix is stored as the real vector **AV** and integer vectors **Row** and **Col** and the number of nonzero elements NElems, as shown in Figure 5.5B. The 10×10 matrix has been stored with 28 real numbers and 56 integers. In terms of savings this example does not show the large gain usually attendant upon this storage method, because it is for large n that the difference is apparent – since in one scheme the requirements increase as n and in the other as n^2.

A =

4	–1	0	0	–1	–1	0	0	0	0
–1	4	–1	0	–1	0	0	–1	0	0
0	–1	4	0	0	–1	–1	0	0	–1
0	0	0	4	–1	–1	–1	0	–1	0
–1	–1	0	–1	4	–1	0	0	0	0
–1	0	–1	–1	0	4	0	–1	0	0
0	0	–1	–1	0	0	4	0	–1	–1
0	–1	0	0	0	–1	0	4	–1	–1
0	0	0	–1	0	0	–1	–1	4	0
0	0	–1	0	0	0	–1	–1	0	4

A

AV =	**Col** =	**Row** =	**NElems** =
[4,	[1,	[1,	28
-1, 4,	1, 2,	2, 2	
-1, 4,	2, 3,	3, 3	
4,	4,	4,	
-1, -1, -1, 4,	1, 2, 4, 5,	5, 5, 5, 5,	
-1, -1, -1, 4,	1, 3, 4, 6,	6, 6, 6, 6,	
-1, -1, 4,	3, 4, 7,	7, 7, 7,	
-1, -1, 4,	2, 6, 8,	8, 8, 8,	
-1, -1, -1, 4,	4, 7, 8, 9,	9, 9, 9, 9,	
-1, -1, -1, 4]	3, 7, 8, 10]	10, 10, 10, 10]	

B

Diag = [1, 3, 5, 6, 10, 14, 17, 20, 24, 28]

C

FIGURE 5.5
A sparse symmetric matrix and storage: A, The matrix **A;** B, Possible storage scheme as vector **AV;** and C, Replacement of vector **Row** by vector **Diag.**

Algorithm 5.2: Efficient Sparse Symmetric Storage

```
Subroutine Locate(ij, Found, Diag, Col, NElems, i, j)
   Integer Row(NPoints), Col(NElems), ij, NElems, i,
   j, p, q
   Logical Found
  Cmmnt:  Is (i, j) in lower or upper half of A?
   If (i .Ge. j) Then
```

```
    Cmmnt:  We are in the lower half
       p = i
       q = j
     Else
    Cmmnt:  We are in the unstored upper half. Pick
equal number at (j, i)
       p = j
       q = i
     EndIf
     Cmmnt:  Pick location of Row p, Column q
     If (p.Eq.1) Then
       ij = 1
       Found = .True.
       Go To 2
     Else
       ij = Diag(p-1)+1
  1    If(Col(ij).Eq.q) Then
        Found =. True.
        Else
        If((Col(ij).Gt.q).Or.(ij.Eq.Diag(p)) Then
  Cmmnt:  We have passed A(p, q) without finding it
  Cmmnt:  so that A(p, q) is zero.
          Found = .False.
        Else
          ij = ij+1
          Go To 1
        EndIf
       EndIf
     EndIf
  2  Return
     End
```

A more efficient storage scheme naturally results from our noticing that the vector **Row** does not change for each row. That is, it is sufficient to denote where the changes take place, since for a given row the value of **Row** is that row number. We therefore change our data structure and replace the vector **Row** by the vector of integers **Diag** of the same length as the matrix size n. Item i of **Diag** tells us where in **AV** the last nonzero item of row i of **A** (the diagonal term, commonly) is stored. For the example Algorithm 5.2, **Diag** is of length 10 and is given in Figure 5.5C. This states that the last item of row 1 is **AV**[1], of row 2 is **AV**[3], of row 3 is **AV**[5], and so on. Another example of this storage is given in Figure 5.4. We have, by these means, for the example of the first-order regular finite element mesh, cut down storage requirements from $4n$ real number + $8n$ integers, to $4n$ real numbers for **AV** + $4n$ integers for **Col** + n integers for **Diag**. And how do we identify $A[i,j]$ from this data structure? We use the fact that all numbers of row i are stored between locations **Diag**$[i-1]+1$ and **Diag**$[i]$ in **AV**; for example, in the matrix in Algorithm 5.2, all items of row 5 are between the 7th (**Diag**[5 – 1]) and 10th (**Diag**[5]) locations of **AV**. The algorithm for this operation of identifying the location is given in Algorithm 5.2, where a procedure Locate identifies the location ij in the vector **AV** at which the nonzero coefficient of the matrix $\mathbf{A}[i,j]$ is stored. A Boolean variable Found is used to flag true when the location row i, column j exists in the un-stored matrix **A**.

With this procedure called Locate, one may replace those parts of a program with full storage where the statement $\mathbf{A}[i,j]$ appears with a call to Locate followed by using for $\mathbf{A}[i,j]$ the number $\mathbf{AV}[ij]$ when Found is true and 0.0 when Found is false. However, this is wasteful of computer power since in a large matrix most coefficients are zero and Locate will be called several times unnecessarily, returning Found = False. We would then not be maximizing the second advantage of sparse storage, besides memory savings – that is, time saving through the avoidance of multiplications with zeros. By this scheme, although we may avoid a multiplication, much time will be expended in searching for zeros.

However, under certain circumstances, it is possible to enhance efficiency by running through every element of **AV** sequentially, thereby skipping all the zeros. Conventionally, on the other hand, the sequence of operations will be through the rows, as represented by the first Do loop of Algorithm 5.2, for example. Naturally, this entails some modification of the algorithm. In the typical Liebmann (or Gauss–Seidel) Algorithm 5.2 for iterative matrix equation solution, we improve the unknown $\mathbf{x}[i]$ of every row i as we run down the rows of the matrix. In computing $\mathbf{x}[i]$, we step through the column numbers k and perform operations using $\mathbf{A}[i,j]$. But because we store only the lower triangular part of a symmetric matrix, the coefficients $\mathbf{A}[i,j]$ will be stored sequentially up to column $j=i$; thereafter, since the coefficients belong to the upper triangular part, their reflections along the diagonal in the lower triangle will need to be used. When we move sequentially along a row in the upper triangle, however, we will be moving down a column in the lower triangle. That is, these are stored in sequential rows and not columns so that they are not stored sequentially in the array **AV**. This will cause some searching as well as page faults. It is therefore more efficient to convert to the original and slower Gaussian iteration where we improve the unknown vector **x** only at the end of an iteration. This is presented in Algorithm 5.3. Observe that if the matrix of coefficients is preconditioned as in Equation (4.50), convergence will be faster and there will be no need to divide the residual in the array **Temp** by the diagonal coefficients, which are 1.

Algorithm 5.3: Gauss Iterations for Sparse Matrices

```
Procedure Liebmann(x, A, B, n, Precis, Col, Diag,
NElems)
{Function: solves the matrix equation Ax=b
Output x = The solution, an n vector
Inputs: A = The n×n matrix of coefficients stored as
a vector without zeros; b=The right-
hand side, an n vector; n=The size of the
equation; Precis=The percentage precision to
which convergence is required; Col,
Diag, NElems=sparse matrix data structures}
Begin
   Repeat
     For k ← 1 To n Do
       Temp[i] ← b[i]
     Row ← 2
     For ij ← 2 To NElems Do
         Column ← Col[ij]
         If Column=Row
         Then
           Row ← Row + 1
         Else
          Temp[Row] ← Temp[Row] - A[ij]*x[Column]
          Temp[Column] ← Temp[Column] - A[ij]*x[Row]
      MaxX ← 0
      MaxChange ← 0
      For k ← 1 To n Do
        Old ← x[i]
        x[i] ← Temp[i]/A[Diag[i]] {Division
        unnecessary if diagonals of A are scaled to 1}
        Change ← x[i] - Old
        x[i] ← Old + 1.25*Change {Acceleration}
        AbsChange ← Abs(Change)
        AbsX ← Abs(x[i])
    If AbsChange > MaxChange
      Then MaxChange ← AbsChange
    If AbsX > MaxX
      Then MaxX ← AbsX
      Error ← 100*MaxChange/MaxX
   Until (Error < Precis)
End
```

5.3 Profile Storage and Fill-in: The Cholesky Scheme

5.3.1 Data Structures for Profile Storage

Profile storage is used whenever the sparse matrix of coefficients encountered in finite element analysis is decomposed to alter it. A good such decomposition method, the Cholesky scheme, has been presented in Section 2.10. There we developed the key **LU**-Decomposition equations for the matrix of coefficients **A** of the equation to be solved **Ax** = **b**:

$$L_{ij} = \begin{cases} \text{When } j = 1: A_{ij} = L_{i1}L_{11} \text{ or } L_{i1} = \dfrac{1}{L_{11}} A_{i1} \\ \text{When } j \neq 1: L_{i1}L_{j1} + L_{i2}L_{j2} + \ldots + L_{ij}L_{jj} \text{ or } L_{ij} = \dfrac{1}{L_{jj}}\left(A_{ij} - \displaystyle\sum_{k=1}^{j-1} L_{ik}L_{jk} \right) \end{cases} \tag{5.3}$$

$$L_{ii}\left(i \neq 1\right) = \sqrt{L_{11}^2 + L_{21}^2 + \ldots + L_{ii}^2} \text{ or } L_{ii} = \sqrt{\left(\sum_{k=1}^{i-1} L_{ik}^2 \right)} \tag{5.4}$$

In this process we have what is called a fill-in: a zero that is to the right of the first nonzero element on a row becomes a nonzero during decomposition. But for sparse matrices we make an important observation. Consider row 4 of the particular matrix shown in Figure 5.2:

$$L_{41} = \frac{A_{41}}{L_{11}} = 0 \tag{5.5}$$

Similarly,

$$L_{42} = \frac{1}{L_{22}}\left(A_{42} - \sum_{k=1}^{1} L_{4k}L_{2k} \right) \tag{5.6}$$

In general for any A_{4j} if all the numbers on row 4 before A_{4j} are zero and A_{4j} is also zero, then L_{4j} too would be zero. So, in place of all the leading zeros of **A**, the corresponding **L** terms would be zero. The first nonzero of **A** on a row would yield a nonzero **L**. Let us now assume that there is a nonzero **L** as at L_{42} followed by a zero in **A** as at A_{43}. Using the previous formula, although A_{43} is zero, L_{43} might well become nonzero because of L_{42} in the formula.

$$L_{43} = \frac{1}{L_{33}}\left(A_{43} - \sum_{k=1}^{2} L_{4k}L_{4k} \right) \tag{5.7}$$

This is what we call fill-in – a nonzero L_{43} appears where we had a zero A_{43}. Recalling that we write **L** in the space allocated for **A** because we will not need **A** after computing **L**, this means that we need to assign storage for all A_{ij}s, whether zero or not, to the right of the leading nonzero on a row as shown shaded in Figure 5.6. As the left border seems like the profile of a face, this is called profile storage. Sometimes for no particular reason scientists and engineers (in particular civil engineers) choose to store the upper half of the matrix. In this case it is called skyline storage as in place of the profile we would have what looks like the skyscraper buildings of a big modern city.

In profile storage, as with sparse storage, the elements of **A** are stored in a vector **AV** row by row. For the example shown

$$\{\mathbf{AV}\} = [A_{11}, A_{21}, A_{22}, A_{31}, A_{32}, A_{33}, A_{42}, 0.0, A_{44}, A_{52}, A_{53}, A_{54}, A_{55}, A_{63}, 0.0, A_{65}, A_{66}] \quad (5.8)$$

Observe that although A_{43} and A_{64} are zero we have assigned space for them since these numbers will be occupied by L_{43} and L_{64} as a result of fill-in.

Now the question arises as to which row and which column these numbers of {**AV**} belong to in [A]. For without that information we cannot compute the solution by reconstructing the matrix [A] in the matrix equation solution process. This can also be done by storing the row and column location of each number:

$$\mathbf{Row} = [1,2,2,3,3,3,4,4,4,5,5,5,5,6,6,6,6] \quad (5.9)$$

$$\mathbf{Col} = [1,1,2,1,2,3,2,3,4,2,3,4,5,3,4,5,6] \quad (5.10)$$

For **Col**, this is seen to be predictable in a way and therefore a waste of memory; many of the numbers are redundant. Only transitions are required. While many data structures exist, a commonly useful one uses two vectors of integers. One of these stores the first column of each row in FC (the other column numbers will increase one by one up to the diagonal term and therefore do not need to be stored) and the location in **AV** of each diagonal term A_{11}, A_{22}, A_{33}, etc. in **Diag**:

Row\Col	1	2	3	4	5	6
1	X					
2	X	X				
3	X	X	X			
4		X		X		
5		X	X	X	X	
6			X		X	X

FIGURE 5.6
An example of profile storage for a symmetric matrix.

$$\mathbf{FC} = [1,1,1,2,2,3] \tag{5.11}$$

$$\mathbf{Diag} = [1,3,6,9,13,17] \tag{5.12}$$

For example, this data structure says that the first nonzero entry in row 4 of **A** is at column number FC(4) = 2 and that the element A(4,4) is located at Diag(4) = 9 of **AV**. That is, AV(Diag(4)) = A(4,4).

In our original classical algorithm we may get the term A(i,j). But now having switched to profile storage we no longer have [A] in memory and instead we have only the vector {**AV**}. So wherever we have A(i,j) in the original code, we replace it by FAProfile(i,j,**AV**,Diag, FC). Here FAProfile is the function, using the ever-ubiquitous MATLAB, as shown in Code Box 5.1.

CODE BOX 5.1 Extracting A(i,j) from AV in Profile Storage

```
function FAProfile(i,j, AV, Diag, FC)
  if j<FC(i) % Then location (i,j) is out of the profile
     FAProfile=0.0;
  else
     if i==1
        Location = 1
     else
        Location = Diag(i-1)+1+(j-FC(i))
  FAProfile=AV(Location);
  end;%if
end;
```

The data structures **Diag** and **FC** are constructed by examination of elements for connectivity, *ne* number of elements *e*, each with three nodes v(1), v(2), and v(3) as shown in Code Box 5.2.

CODE BOX 5.2 Making the Data Structures Diag and FC from the Finite ElementMesh Data

```
for e=1:ne %ne is the number of elements in the mesh
    Read v(1), v(2), v(3)
    for i=1:2
      for j=i+1:3; %Compare for i=1 v(1) with v(2) and v(3)
                   %and then for i=2, v(2) with v(3)
    V1=v(i);
    V2=v(j);
```

```
    if V1>V2 %Swap to ensure that (V1,V2) is in lower
      triangle of A
         Temp=V1
         V1=V2;
         V2=Temp;
      end;%if
      if FC(V2)>V1 %Then no space for (V1,V2)and must be
      allocated
         FC(V2)=V1;
      end;%if
     end;%for j
   end; %for i
  end; %for e
  %Now we are ready to make Diag
  Diag(1)=1
  for i=2:n %nxn is the matrix size of A
   Diag(i)=Diag(i-1)+1+i-FC(i); %Allow for FC(i)-i numbers
   on row i
  end; %for i
```

5.3.2 Cholesky's Method with Profile Storage

When direct methods of solution such as Cholesky factorization are employed to solve a sparse symmetric matrix of the type of Figure 5.7A, in the process of decomposition, subtracting a row from subsequent ones, the sparse structure of the matrix is disturbed. For example, in Figure 5.7A, to make all terms of the first column beyond row 1 zero, we would subtract an appropriate multiple of row 1 from all subsequent ones – it is seen that doing this for row 6 will result in some factor of the coefficient (1, 2) appearing in location (6, 2), that was originally zero. This process whereby a zero coefficient becomes nonzero is called a fill-in and occurs only at zero locations to the right of the first nonzero term of a row. Therefore, in direct schemes, storage is allocated for the whole profile as in Figure 5.7B, thus requiring 29 locations in place of the 20, as shown in the example. For larger problems, this difference is much greater.

For the matrix of Figure 5.7A, profile storage would replace the vector **Col**, giving the column numbers of the nonzeros, by the vector **FC**, which shows the first column of a row occupied by a nonzero, as shown in Figure 5.7E. Observe that, as opposed to the sparse storage of Figure 5.7C, the vector **AV** containing the coefficients of the matrix **S** now has several zeros packed into it, shown in bold lettering in Figure 5.7E. In using Cholesky factorization for the solution, the Cholesky factor **L** of the matrix **A** is usually overwritten on **A** to save storage. These additional zeros then are the places where fill-in may be accommodated.

1							
2	3						
0	4	5					
0	0	0	6				
0	7	8	0	9			
10	0	0	11	0	12		
0	0	13	0	14	0	15	
16	17	0	18	0	19	0	20

A

1							
2	3						
4	5	6					
0	7	8	9				
0	0	10	11	12			
0	0	0	0	13	14		
0	0	0	0	15	16	17	
0	0	0	0	0	18	19	20

D

1							
2	3						
0	4	5					
0	0	0	6				
0	7	8	9	10			
11	12	13	14	15	16		
0	0	17	18	19	20	21	
22	23	24	25	26	27	28	29

B

FC = [1, 1, 2, 4, 2, 1, 3, 1]
Dia = [1, 3, 5, 6, 9, 12, 15, 20]
AV = [1.0,
.0.5, 1.0,
0.2, 1.0,
1.0,
–0.1, 0.2, 0.0, 1.0,
–0.15, 0.0, 0.0, 0.4, 0.0, 1.0,
–0.32, 0.0, 0.15, 0.0, 1.0,
0.32, –0.24, 0.0, 0.16,0.0, –0.16, 0.0, 1.0]

E

Col = [1, 1,2, 2,3, 4, 2,3,5, 1,4,6, 3,5,7, 1,2,3,4,6, 8]
Diag = [1, 3, 5, 6 , 9, 12, 15, 20]
AV = [1.0,
0.5, 1.0,
0.2, 1.0,
1.0,
–0.1, 0.2, 1.0,
–0.15, 0.4, 1.0,
–0.32, 0.15, 1.0,
0.32, –0.24, 0.16, –0.16, 1.0]

C

FIGURE 5.7
Storage in sparse matrices: A, Sparse symmetric matrix and storage scheme, B, Dense nature of inverse fill-in within band, C, Data structures for sparse symmetric storage corresponding to **A,** D, Renumbered matrix of **A**, and E, Data structures for symmetric profile storage corresponding to **A**.

Algorithm 5.4: Cholesky Factorization with Profile Storage

```
Procedure CholeskyFactor(L, S, FC, Diag, MatSize)
{Function: Computes the lower Cholesky factor L of
the matrix S of size MatSize by
MatSize. Profile storage is used for S, with FC
giving the first column of each row and
Diag the location of the diagonal entry of each row.
In calling, L and S may be made
the same}
Begin
   L[1] ← Sqrt(S[1])
   Row ← 2
   ij ← 2
   Repeat
     For column ← FC[Row] To Row Do
         Sum ←S[ij]
         If Column > 1
           Then
            If FC[Row] > FC[Column]
              Then kStart ← FC[Row]
              Else kStart ← FC[Column]
           rk ← Diag[Row - 1] + 1 + kStart - FC[Row]
           ck ← Diag[Column - 1] + 1 + kStart
           - FC[Column]
           For k ← kStart To Column - 1 Do
             Sum ← Sum - L[rk]*L[ck]
             rk ← rk + 1
             ck ← ck + 1
         If Row = Column
           Then
            L[ij] ← Sqrt(Sum)
            Row ← Row + 1
           Else
            L[ij] ← Sum/L[Diag[Column]]
         ij ← ij + 1
   Until Row > MatSize
End
```

Algorithm 5.4 shows how the Cholesky factor **L** is computed using the data structures **FC** and **Diag**. Note how the term at row i, column j is identified. Row i begins at location **Diag**[i – 1] + 1, occupied by column **FC**[i]. Since all columns to the right of this are occupied up to the diagonal, column j is $j-$**FC**$[i]$ numbers away. So, the term i, j of **S** is at location **Diag**$[i-1]+1+j-$**FC**$[i]$ in the vector **AV**. Algorithms 5.5 and 5.6, respectively, give forward elimination and back substitution schemes with profile storage.

Observe from the foregoing, therefore, that the more zeros there are within the profile, the more costly matrix solution will be, both in terms of storage requirements and computing time. For this reason, in direct schemes, a renumbering of the variables is done so as to minimize the bandwidth of the matrix and reduce the number of zeros in the band so that storage space for the inverse may be saved. The renumbered matrix of Figure 5.7A is shown in Figure 5.7D, where the variables [1, 2, 3, 4, 5, 6, 7, 8] have now been renumbered as [4, 5, 6, 1, 7, 2, 8, 3]. It is now seen that there is no zero within the profile, so that the inverse will take only 20 locations, like the original, as opposed to 29 before renumbering. While for this small matrix we have achieved the ideal of no extra memory for the inverse, it has been shown by Silvester et al. (1973) that for large first-order finite element schemes, the inverse takes about four times as much space as the original matrix and, partly as a result of the relevance of this fact to a microcomputing milieu, direct schemes are tending to be used less, although they are still very popular.

Algorithm 5.5: Forward Elimination for Profile Stored Matrices

```
Procedure Forward(L, x, b, FC, Diag, MatSize)
{Function: Solves the equation Lx = b by forward
elimination. L is a lower triangular matrix
stored as a 1-D array using profile data structures
FC and Diag}
Begin
   x[1] ← b[1]/L[1]
   ij ← 2
   For Row ← 2 To MatSize Do
     Sto ← b[Row]
     Column ← FC[Row]
     While Column < Row Do
        Sto ← Sto - L[ij*x[Column]
        ij ← ij + 1
        Column ← Column + 1
     x[Row] ← Sto/L[ij]
End
```

Algorithm 5.6: Back Substitution for Profile Stored Matrices

```
Procedure Forward(L, x, b, FC, Diag, MatSize)
{Function: Solves the equation Lt x=b by back
substitution. L is a lower triangular matrix
stored as a 1-D array using FC and Diag}
Begin
   For i ← 1 To MatSize Do
     x[i] ← b[i]
   ij ← Diag[MatSize + 1
   Row ← MatSize + 1
   For i ← 2 To MatSize Do
     Row ← Row - 1
 {or For Row ← MatSize Down To 2 Do in place of the
  last three lines}
     ij ← ij - 1
     x[Row] ← x[Row]/L[j]
     Stop ← Diag[Row - 1] + 1
     Column ← Row
     While ij > Stop Do
        ij ← ij - 1
        Column ← Column - 1
        x[Column] ← x[Column] - L[ij]*x[Row]
   x[1] ← x[1]/L[1]
End
```

5.4 Sparse Storage for SOR

Let us be clear in our terminology. Although profile storage is used where the matrix of coefficients is sparse, that is with zeros in the middle, we do not call profile storage sparse storage because as seen in the previous section, all holes in the middle are also allocated storage. With sparse storage, those holes do not have storage assigned.

In iterative schemes of solution, the matrix of coefficients [A] remains unchanged. Therefore, fill-in does not need to be provided for. However, we do not have column values increasing by 1 after the first nonzero. This means that we need to track both the row number and column number of an element of **A**, exploiting the fact that if we store row by row, the row number would change only occasionally. So, examining the storage of row and column indices for the example matrix above,

$$\mathbf{Row} = [1,2,2,3,3,3,4,4,5,5,5,5,6,6,6] \tag{5.13}$$

$$\mathbf{Col} = [1,1,2,1,2,3,2,4,2,3,4,5,3,5,6] \tag{5.14}$$

for the 15 nonzero numbers of **A**. We see the redundancy in **Row**, so instead we allow only for **Col** and **Diag** which is the same as before:

$$\mathbf{Diag} = [1,3,6,8,12,15] \tag{5.15}$$

$$\mathbf{Col} = [1,1,2,1,2,3,2,4,2,3,4,5,3,5,6] \tag{5.16}$$

Here Diag(i) contains the location in **AV** of the number A(i,i) and Col(i) is the column location of the number AV(i).

How now do we generate **Diag** and **Col**? As before with profile storage, it is by examining the connectivity between nodes in the mesh where the three vertices v(1), v(2), and v(3) of each element e (numbering ne) are stored in a file. This is shown in Code Box 5.3.

CODE BOX 5.3 Generating Diag and Col from Mesh Data

```
%Start with only the diagonal terms of the matrix.
%The rest we fill up as we find the need.
for i=1:n %n is the matrix size
  Col(i)=i;
  Diag(i)=i;
end;
for e=1:ne %ne is the number of finite elements;
  Read v(1), v(2), v(3) % First order triangular
                         %elements are assumed with three
                         vertices
  for i=1:2
   for j=i+1:3; %Compare for i=1 v(1) with v(2) and v(3)
                %and then for i=2, v(2) with v(3)
       V1=v(i);
       V2=v(j);
       if V1>V2 %Ensure that (V1,V2) is in lower triangle of P
        Temp=V1;
        V1=V2;%V1 is the row
        V2=Temp;%V2 is the Column
     end;%if
     %Search for column V2 on row V1
     NotFound=true;
```

```
      Passed=false;
      Location = Diag(V1-1)+1;
      while (NotFound && Location<Diag(V1) &&
      V1<Col(Location))
        if Col(Location)= V1
           NotFound=false;
         end;%if
         if V1>Col(Location)
            Passed=true;
         end; %if
       end;%while
       if passed %then create space for Column V2 on row V1
         for i=Diag(n):-1:Location %Shift all elements of
         Col right
           Col(i+1)=Col(i);
         end;
         Col(Location)=V2; %Insert space for matrix term
         (V1,V2)
         for i=V1:n
           Diag(i)=Diag(i)+1; %one extra element from row V1
         end;
       end;%if
    end; %for j
   end; %for i
end;%for e
```

In implementing the SOR algorithm with this data structure we should not deal with every Aij, but only the nonzero Aij values. This is a situation where the SOR algorithm needs modification. Now the function FAProfile will take the form FASparse, as shown in Code Box 5.4.

CODE BOX 5.4 Finding A(i,j) in AV with Sparse Storage

```
function FASparse(i,j, AV, Diag, Col)
  if i==1
     FASparse=AV(1);
  else
    if i<j %if no. in upper triangle look for it in lower
    triangle
       temp = i;
       i = j;
```

```
      j = temp;%Swapped i and j
   end;
   if j<Col(Diag(i-1)+1) % Then location (i,j) is out of
   the profile
     FA=0.0;
   else
     k=Diag(i-1)+1;
     while j<Col(k) && j<Diag(i)
      k=k+1; %run through elements of Col in row i;
             % till we find col i (j=Col(k)), pass col i
             (j>Col(k)
             %or reach the diagonal term j=Diag(i)
     if j=Col(k)
      FASparse=AV(k);
     else
      FASparse=0.0;
     end;
   end;
  end;%if
end;%function
```

However, although this procedure works, there are two objections to it:

1. A lot of computing time will be expended in searching for numbers that are zero and therefore not stored.
2. The algorithm is associated with page faults. The computer stores a few numbers in RAM (fast memory) and most numbers in slow memory. If in a computational process we suddenly have need of a number resident in slow memory, the computer would write (i.e., save) all numbers or information in fast memory into slow memory and then read a new batch of numbers including the number we need to process from slow memory into fast memory. This delay is called a page fault. Typically a computer stores a matrix column by column in sequential memory locations – that is, for the matrix A, the SOR algorithm needs A(1,1), A(2,1), …, A(n,1), A(1,2), A(2,2), A(3,2), …, A(n,2), A(3,1) …, etc. For the SOR algorithm, we need a row of a matrix at a time and this is all right for small matrices. But, for example, when dealing with a matrix of coefficients **A** of size 1000×1000, in determining, for example, the 300th unknown, we would need the numbers A(330,1), A(330,2) etc., up to A(300,300), each a distance 1000 numbers after the other. This surely would yield page faults. Instead, using symmetry if we sought these numbers from their reflections A(1,300), A(2,300), … A(300,300), few page faults would occur since these numbers are in a column. After reaching the

diagonal however, we need A(300,301), A(300, 302), …, A(300,1000). To get them in column sequence we use A(301,300), A(302,300), …, A(1000,300). That is, we reorder the SOR algorithm of Equation (4.9) to read as

$$x_i^{k+1} = \frac{1}{A_{ii}^G}\left\{b_i^G - \sum_{j=1}^{i-1} A_{i,j}^G x_j^{k+1} - \sum_{j=i+1}^{n} A_{i,j}^G x_j^k\right\} \tag{5.1}$$

That is, we use numbers of **A** in column order. This has been found to yield immense savings in computation time for large matrices. Although the argument has been made for fully stored matrices, it holds even for sparse stored matrices in part because the numbers of a matrix are stored a row at a time as a vector up to the point that we reach the diagonal. Therefore, we will be using the numbers sequentially as stored, but after the diagonal the numbers would not be in sequential storage. For the numbers of a row after the diagonal we would call for numbers below the diagonal term and these are not stored in sequential order. We must also note that in sparse storage because we do not store the zeros, the matrix is much more compact and therefore there would be less need for page faults which are associated with the need to store large blocks of memory. Note further that these arguments will not apply to small matrices which can be stored all at once in fast memory.

For these two reasons, it is better to abandon the SOR algorithm and revert to the Gauss algorithm which was replaced by SOR. In the revised Gauss algorithm, we do not go row by row but rather go by an element of a matrix after an element of the matrix **A** as shown in Code Box 5.5.

CODE BOX 5.5 Efficient Reversion to Gauss from SOR

```
Error = 1000.00;
Iter=0;
%n is matrix size. We have Diag, AV, and Col. X is the
solution;
IterLimit=n*n*n;
while (Iter < IterLimit && Errror>0.0001)
  maxX=0.0;
  MaxChange=0.0;
  for i=1:n
     sum(i,1)=0.0;
  end;
  row=2;
```

```
    for e=2:Diag(n)
        col=Col(e);
        if row ~=col
          sum(row)=sum(row)-AV(e)*x(col);
          sum(col)=sum(col)-AV(e)*x(row);
        else
          row=row+1;
        end; %if
     end; %for e;
     for row=1:n
        old=x(row);
        x(row)=sum(row)/PV(Diag(row));
        x(row)=old+1.5*(x(row)-old);
        Change=abs(x(row)-old);
        AbsX=abs(x(row));
        If Change>MaxChange
         MaxChange=Change;
        end;
        if AbsX>MaxX
         MaxX=AbsX;
        end;
    end;%for row
    Error=100*Maxchange/MaxX;
end; %while
```

Several simple schemes like those described previously in this section exist for gaining efficiency in matrix computation. Algorithms 5.7 and 5.8 give simple schemes for gaining efficiency in matrix operations, for matrix addition and matrix multiplication, respectively. Most computers, when we dimension an $n \times n$ matrix, store the matrix column by column. Thus, although we may designate, for example, **A** as an $n \times n$ matrix, the machine stores it as a vector; that is, column 1 of the matrix **A** occupies the first n locations of a vector entity **AV** in the machine, column 2 occupies the next n numbers, and so on. In mathematical terms, $S[i, j]$ is stored in a vector **AV** within the computer, at location $(i-1)^* n + j$. Therefore, whenever a matrix is dimensioned and we call one of its elements $A[i, j]$, the computer identifies **A** as a matrix through the dimension statement and picks out the coefficient of the vector in which it is stored, by evaluating its location $(i-1)^* n + j$. We may exploit this fact to save some cpu time on the machine by immediately dimensioning all matrices as vectors, instead of letting the machine do it for us; thus, we often spare the machine the burden of computing the locations of terms. In other words, an $n \times n$ matrix **A** is dimensioned directly as a vector **AV** of length n^2 and whenever we wish to refer to $A[i, j]$, we devise special algorithms to locate an element without performing the operation $(i-1)^* n + j$. When this cannot be avoided, we refer to $\mathbf{AV}[(i-1)^* n + j]$ after computing $(i-1)^* n + j$.

Algorithm 5.7: Efficient Matrix Addition

```
Procedure MatAdd(S, A, B, n, m)
{Function: Adds the n×m column stored matrices A and
B and returns the sum is S
Output: S
Inputs: A, B, n, m}
Begin
  For i←1 To n*m Do
    S[i]←A[i]+B[i]
End
```

Algorithm 5.8: Efficient Matrix Multiplication

```
Procedure MatMultiply(S, A, B, l, m, n)
{Function: Multiplies the l×m matrix A and the m×n
matrix B and returns the result in
the l×n matrix S. A, B, and S are column stored
Output: S
Inputs: A, B, l, m, n}
Begin
  ij←0
  kjStart←1-m
  For j←1 To n Do
    kjStart←kjStart+m
    For i←1 To l Do
      ij←ij+1
      kj←kjStart
      ik←i
      x←0.0
      For k←1 To m Do
        x←x+A[ik]*B[kj]
        ik←ik+1
        kj←kj+1
      S[ij]←x
End
```

For example, as shown in Algorithm 5.7, matrix addition

$$S[i,j] = A[i,j] + B[i,j] \quad i = 1,\dots,n; \ j = 1,\dots,n \tag{5.17}$$

becomes straightforward without ever having to perform a multiplication for locating a coefficient. Algorithm 5.8 shows a possible scheme for doing a matrix multiplication

$$S[i,j] = \sum_{k=1}^{m} \mathrm{A}[i,k]^{*} \mathrm{B}[k,j] \quad i = 1,\dots,l; \ j = 1,\dots,n, \tag{5.18}$$

exploiting the way the computer stores numbers. In this algorithm, *ij* represents the location of $\mathbf{S}[i,j]$ in the one-dimensional array **S**; similarly, *ik* and *kj* correspond to Equation (5.18). We have used the fact that when we move along a column as in $B[k,j]$, *kj* will increment by 1, and when along a row as in $A[i,k]$, the indices *ik* will increment by the number of rows of **A**, which is l. A dummy store x is used so as not to cause any page faults (described in the next paragraph) in repeatedly calling the matrix **S**, until the summation in Equation (5.10) is complete.

Another means of saving time exploits the algorithms we are using. Most machines are allowed only a limited memory in the processor. Thus, while a program may deal with several kilobytes, only a small portion of it remains in core – and the moment we refer through the program to a number that is in secondary storage, what is in core is sent out to secondary storage, and what must be operated on by the computer is called into core. This swapping is called a page fault, because it involves the slow process of reading from and writing on secondary storage, but no cpu time is expended. In matrix operations, for example in a machine that allows 100 words in internal memory, the moment we refer to $S[1, 1]$ of a 100×100 matrix, the computer will call in the first column of the matrix, since 100 real numbers and no more may be kept in core. After operating on $S[1, 1]$, if we now refer to $S[2, 1]$, since $S[2, 1]$ is already in core, the machine will quickly operate on it. However, should we operate on $S[1, 2]$ after operating on $S[1, 1]$, since $S[2, 1]$ is at location 101, which is outside the machine, the machine will have to send out all or a part of the first column and bring in the second column. This is slow and will increase both the page faults on the machine and total running time.

To avoid the disadvantages of this system of paging in a computer, if we look at our SOR Algorithm 5.2.3, (also known as Gauss iteration) we will notice that we always refer to the matrix row by row. Thus, running a large matrix solution program with regular matrix or column-by-column storage will yield a large number of page faults. This problem is easily overcome by resorting to row-by-row storage of the matrix. That is, we would dimension the matrix as a vector **AV** of length n^2 and store item A[i, j] in SV$[(i-1)^{*} n + j]$. This writer recalls with some amusement (and, needless to say, embarrassment), his days as a graduate student nearly 40 years ago when he tried solving a 600×600 matrix by the SOR scheme with full storage on a multiuser VAX machine and had to crash the program after two days! Finally, the answer was obtained in two hours with row storage. Memory is now less of an issue. It is important to stress, however, that this may not be the best scheme when we are using another algorithm in which we do not refer to numbers consecutively along a row.

5.5 Sparse Storage and the Conjugate Gradients Algorithm

The previous procedures described for enhancing efficiency relate to getting the best efficiency from a given method. However, advances in efficiency may also be made in changing over completely to a different method of equation solution. For finding the solution for nonsingular matrices, two approaches are commonly taken; these are through the direct and iterative schemes. In Section 2.3, we were introduced to the direct Gaussian elimination scheme. In Section 3.2, we gained some experience with the SOR scheme, which is an iterative scheme. It is seen that in the direct scheme we have a solution at the end of the computational process. The error in the solution will depend on the accuracy of the computer used and is caused by round-off error arising in the arithmetic operations. Thus, as the size of the problem grows, so will the error. In the iterative schemes, on the other hand, we have an approximation to the solution, at any point in the solution process, which is successively improved with the iterations. Thus, it is we, the users, who determine when to stop computation. The matrix coefficients are always preserved, so that the error in the solution will not increase with the size of the problem. Iterative schemes have a special advantage when we are able to make a good guess at the solution, as in solving a nonlinear problem or in solving a refined mesh for which a solution corresponding to its cruder predecessor exists; for the approximate solution may be used as a starting solution and the convergence time may be cut down. Moreover, it will be explained in the sections to come that iterative schemes are more amenable to having the sparsity of the matrix exploited. It may be stated in general that as the problem size grows, iterative schemes become more and more preferable. When the transition in economy occurs is a very subjective issue (Hoole et al., 1989).

Currently, the two most commonly used methods of matrix solution in field computation are the frontal solution algorithm (Irons, 1970), resorting to a direct decomposition of the matrix, and the semi-iterative preconditioned conjugate gradient algorithm (Jennings, 1977; Kershaw, 1978; Hoole et al., 1986), relying upon finding, for an $n \times n$ matrix, the solution as a sum of n or fewer vectors, each of which is found during an iteration.

Sparse $n \times n$ matrices using full storage require memory growing as n^2, although the nonzero coefficients usually grow only as n. For example, in a finite difference mesh, the equation at each node is expressed in terms of the four surrounding nodes, so that only five coefficients appear in a row, giving us $5n$

entries in the matrix. Where the problem is surrounded by a Dirichlet boundary, the matrix equation is symmetric in addition to being sparse, so that only the lower or upper half of the matrix needs to be stored, and numbers from the half that is not stored may be obtained from that which is stored. If we decide to store the lower half on a particular row, the stored coefficients will correspond to columns numbered lower than the row. So out of the four nodes surrounding a node, probabilistically taking two to have lower numbers, we need to store only three coefficients, counting the diagonal terms, giving us $3n$ entries. As already seen in Chapter 2, the finite element matrices are always symmetric. In an ideal, first-order, optimally triangular finite element mesh, all triangles will have similar error, and therefore triangles will be either equilateral or close to it. Therefore, as pointed out, each node will be approximately connected to six others; that is, an equation corresponding to a particular node will have seven coefficients relating the potential at that node to those at the other six nodes.

Thus, following the previous argument, if the symmetry and sparsity of the system are exploited, storing the diagonal coefficient and three off-diagonal ones in the lower half, only $4n$ coefficients need be stored. More generally, the storage requirement for the coefficients is cn where c is the approximate connectivity of a node, or, with symmetric matrices, half of the average connectivity of a node. However, storing only the nonzero coefficients also means storing the rows and columns of the coefficients. Special schemes exist (Silvester et al., 1973) for storing all the information of the matrix with nc real numbers and $(n+1)c$ integers. This is shown in Figure 5.7, where Figure 5.7D shows the data structure corresponding to the matrix of Figure 5.7A. The 20 nonzero coefficients of the 8×8 matrix are stored in a vector **AV** of real numbers, and **Col**, an array of 20 integers, shows the corresponding column numbers of the coefficients. **Diag**, an array of eight integers, shows the positions in **AV** of the diagonal terms.

Therefore, in solving sparse matrices a fair deal of time must first be expended in generating the data structure as in Figure 5.7C. We go through the matrix equation, or alternatively the physical system that produces the equation (such as a finite element grid), and, using the connectivities, compute the arrays **Col** and **Diag** or similar arrays coming from some variant of the storage scheme. This process can take as much as 30% of total solution time, depending on the efficiencies of the algorithms used for sparsity computation and matrix solution.

Although computing the data structures for sparse storage takes up time, we undertake it for two reasons. First, by reducing storage requirements to the order of the matrix size from the previous square, the solution of large problems is made practicable. Second, by not storing the nonzero coefficients, several multiplications by zero (which on a computer take the same time as multiplications with nonzero numbers) are avoided, thus more than offsetting the time lost in computing the sparsity.

Algorithms 5.9, 5.10, 5.11, and 5.12 give the relevant sparse algorithms for the Conjugate Gradients method for matrix solution.

Algorithm 5.9: Forward Elimination for Sparse Matrices

```
Procedure Forward(L, x, b, Diag, Col, NElems,
MatSize)
{Function: solves the equation Lx = b by forward
elimination. L is a lower triangular sparse
matrix stored as a 1-D array}
Begin
   x[1] ← b[1]/L[1]
   ij ← 2
   For Row ← 2 To MatSize Do
     Sto ← b[Row]
     Column ← Col[ij]
     While Column < Row Do
       Sto ← Sto - L[ij]*x[Column]
       ij ← ij + 1
       Column ← Col[ij]
     x[Row] ← Sto/L[ij]
End
```

Algorithm 5.10: Back Substitution for Sparse Matrices

```
Procedure Backward(L, x, b, Diag, Col, NElems,
MatSize)
{Function: solves the equation Ltx = b by back
substitution. L is a lower triangular sparse
matrix stored as a 1-D array}
Begin
  For i ← 1 To MatSize Do
    x[i] ← b[i]
  ij ← NElems + 1
  Row ← MatSize + 1
  For i ← 2 To MatSize Do
    Row ← Row - 1
{or For Row ← MatSize DownTo 2 Do in place of the
last three lines}
    ij ← ij - 1
    x[Row] ← x[Row]/L[ij]
    Stop ← Diag[Row - 1] + 1
```

```
    While ij > Stop Do
      ij ← ij - 1
      Column ← Col[ij]
      x[Column] ← x[Column] - L[ij]*x[Row]
  x[1] ← x[1]/L[1]
End
```

Algorithm 5.11: Performing the Multiplication P^tSP for Sparse Matrices

```
Function TransPeeSPee(S, p, Diag, Col, NElems,
MatSize): Real
{Function: Computes ptSP where S is a sparse matrix
and p is a column vector}
Begin
  Answer ← 0.0
  RowSum ← 0.0
  Row ← 1
  For ij ← 1 To NElems Do
     Column ← Col[ij]
     RowSum ← RowSum + S[ij]*p[Column]
  If Column = Row
      Then
        Answer ← Answer + p[Row]*(2*RowSum
        - S[ij]*p[Row])
        Row ← Row + 1
        RowSum ← 0.0
  TransPeeSPee ← Answer
 End
```

Algorithm 5.12: Procedures for ICCG Algorithm with Sparse Matrices

A. Sparse Matrix Multiplication

```
Procedure SparseMatMultiply(S, x, b, Diag, Col,
NElems, MatSize)
```

```
{Function: Evaluates b = Sx, where S is a sparse
matrix and b is a vector}
   {Since the matrix S is stored as a vector, we run
   through the elements of S, identify }
   {the row and column values of the element and
   perform the multiplication}
Begin
   For Row ← 1 To MatSize Do
     b[Row] ← 0.0
     Row ← 1
     For ij ← 1 To NElems Do
        Column ← Col[ij]
        b[Row] ← b[Row] + S[ij]*x[Column]
        If Row = Column
          Then Row ← Row + 1
          Else b[Column] ← b[Column] + S[ij]*x[Row]
End
```

B. The Infinity Norm of a Vector

```
Function Norm(r, n)
   {Function: Finds the largest component of the
   vector r of length n}
   {Required for checking convergence}
Begin
   Answer ← 0.0
   For i ← 1 To n Do
     Temp ← Abs(r[i])
     If Temp > Answer
        Then Answer ← Temp
   Norm ← Answer
End
```

Algorithm 5.13 The ICCG Algorithm with Sparse Matrices

```
p ← 0
γ ← 1
Forward(A, x, b, Diag, Col, NElems, MatSize)
Backward(A, x, b, Diag, Col, NElems, MatSize)
ε ← Permitted Error
Repeat
   SparseMatMultiply(A, x, r, Diag, Col, NElems,
   MatSize)
   For i ← 1 To MatSize Do
```

```
    r[i] ← b[i] - r[i]
  Forward(A, r, r, Diag, Col, NElems, MatSize)
  e ← Norm(r, MatSize)
  θ ← 0.0
  For i ← 1 To MatSize Do
      θ ← θ + r[i]*r[i]
  β ← θ/γ
  γ ← θ
  Backward(A, r, r, Diag, Col, NElems, MatSize)
  For i ← 1 To MatSize Do
    p[i] ← r[i] - β*p[i]
  δ ← TransPeeSPee(A, p, Diag, Col, NElems, MatSize)
  α ← γ/δ
  For i ← 1 To MatSize Do
    x[i] ← x[i] - α*p[i]
Until e < ε
```

Similarly, the procedures of forward elimination, back substitution, and computation of the product $\delta = \mathbf{p}^t\,\mathbf{S}\mathbf{p}$ in the preconditioned conjugate gradients algorithm (Algorithm 5.13) may be based on sequentially taking up the stored nonzero coefficients of the matrix. The procedures are presented in Algorithms 5.9, 5.10, 5.11, and 5.12, respectively. Observe that in solving by back substitution, after determining a term **x**[*i*], we are free to back substitute in advance using all coefficients of the matrix at column *i* above row *i*. Note also that in Algorithms 5.9 and 5.10, **x** may be overwritten on **b** so that in the calling statement the same array may stand for **x** and **b**. In Algorithm 5.11 we use the fact that a coefficient **S**[*ij*] standing for **S**[Row, Column] appears twice if it is off the diagonal and only once if Row = Column – thus, we subtract the diagonal contribution from the term RowSum since it would have been added once too often.

In Algorithm 5.13 (which uses Algorithm 5.12), the same principles are used in rewriting the Incomplete Cholesky Conjugate Gradients Algorithm of Algorithm 5.2 for sparse matrices. Because Evans's preconditioning is used, it is assumed that the diagonal coefficients of the matrix **S** are 1 and that the lower triangular part of **S** (i.e., the stored part of **S**) is the incomplete Cholesky factor **B**.

5.6 Renumbering of Variables: The Cuthill–Mckee Algorithm

Renumbering of the variables is normally done with direct schemes so as to reduce the band of the matrix and thereby save on storage. That is, we minimize

$$K = \sum_i D_i \quad \text{where } D_i = \min\left[i - j\right] \text{ for every } j \text{ connected to } i. \qquad (5.19)$$

This reduces the spread between elements of a matrix and thereby reduces zeros in the matrix.

In the foregoing, D_i is clearly the largest distance between the diagonal term of row i and any other term j on that row. Of the many renumbering schemes available, the most commonly used is the Cuthill–McKee algorithm (Cuthill and McKee, 1969). This is demonstrated in the matrix of Figure 5.7A and yields the matrix of Figure 5.7D. The scheme is highly systematic and lends itself well to computer implementation.

We start the renumbering scheme by searching for the connectivities between variables and making the second and third columns of Figure 5.8. The second column says, for example, that variable 1 is connected to three others, and these three are given as variables 2, 6, and 8 in the third column.

The rule for renumbering is that we start with a least connected variable, 4 or 7 in this case, and call it 1. Had we chosen to call variable 4 our new variable 1 as in the example, we would enter it under column 4; and variable 6, the least connected of the connections of 4 (i.e., connections 6 and 8 of column 3) becomes 2; and variable 8, the other variable connected to 4, becomes 3. Now we do the same thing to the new node 2 (old 6). Of its connections 1, 4, and 8, 4 and 8 are already renumbered, and therefore 1 takes the new number 4. Proceeding thus, all the variables are renumbered as in column 4. It has been shown (George, 1973) that reversing the algorithm results in more efficient renumbering. It is also known that as a result of the arbitrariness in numbering equally connected variables, different strategies result in different storage requirements. But we will not delve into this aspect, which is not of direct relevance here.

Original Node Number	No. of Connections	Connection to	New Number	New Connections to
1	3	2, 6, 8	4	5, 2, 3
2	4	1, 3, 5, 8	5	4, 6, 7, 3
3	3	2, 5, 7	6	5, 7, 8
4	2	6, 8	1	2, 3
5	3	2, 3, 7	7	5, 6, 8
6	3	1, 4, 8	2	4, 1, 3
7	2	3, 5	8	6, 7
8	4	1, 2, 4, 6	3	4, 5, 1, 2

FIGURE 5.8
Cuthill–McKee renumbering algorithm implementation connecting matrix of Figure 5.7A to that of Figure 5.7D.

After renumbering, we have generated all the connectivities in Figure 5.7 and these may be profitably used to compute the sparsity data structure of Figure 5.7C, which otherwise would take a large portion of the total computing time. That is, we have found that the time for sparsity computation is only marginally smaller than the time for renumbering plus the time to compute sparsity using the tables generated in Figure 5.7.

To compute the sparsity of the renumbered matrix using Figure 5.7, we generate column 5 of the tables, which is in fact an overwritten column 3. For example, looking at the new variable 4 of the first row, which was the old variable 1 and connected to the old variables 2, 6, and 8 of column 3, we would say that it is now connected to 5, 2, and 3 of the last column, which is overwritten on the third; this is obtained by saying that 2 is now 5, 6 is now 2, and so on. Having generated this column, we then reorder it to go in increasing integers 2, 3, 5 (which in fact is the array **Col** of Figure 5.7C) once we drop 5 belonging to the upper triangle of the matrix, which is not stored in view of symmetry. Thus, including the diagonal term, columns 2, 3, 4 will belong to that part of **Col** corresponding to row 4 in the renumbered matrix, that is nonzero elements 7, 8, and 9 or row 4 in Figure 5.7D.

5.7 Renumbering and Preconditioning

In this section, we show that although there is nothing to be gained by renumbering in terms of memory, in the preconditioned conjugate gradients algorithm, there is much to be gained in terms of time, when preconditioning methods are employed using approximate inverses with the same profile as the matrix solved. We will also show that the data on the connectivity of the matrix to be computed for renumbering may be used with profit to allocate storage for the sparse matrix in the manner of Figure 5.7C, so that the time expended in renumbering may be compensated for by savings in time for allocation of storage.

We have already seen that in the two commonly used methods of preconditioning, Evans's and Incomplete Cholesky, the approximate inverse is assumed to have the same sparsity structure as the original; that is, we ignore those terms of the inverse that take the positions of the zero coefficients within the profile.

Therefore, the more zero terms there are within the profile of the matrix, the more terms of the inverse we will be ignoring. As seen in the example of Figure 5.7, the inverse of the matrix of Figure 5.7A with 20 coefficients requires 29 locations, and therefore the approximate inverse will be neglecting nine terms. However, had we renumbered the matrix of Figure 5.7A to get that of Figure 5.7D, before solving, the inverse would take only 20 locations. Thus, using the same sparsity structure as the original neglects no number. Although, in this example, renumbering results in a storage requirement of only 20, commonly this will be slightly above the number of nonzero coefficients, since for large problems renumbering rarely results in no zeros within the profile as here. This error coming from neglecting terms builds up as we compute the incomplete Cholesky factor, since we start from the top of the matrix (Silvester and Ferrari, 1983) and keep working down. Thus, as we go down we will be using less and less accurate values of the decomposed factor to compute those at current locations.

What happens, therefore, in renumbering, is that the approximate inverse is improved so that the matrix equation we solve has a better approximation to the unit matrix. That is, the radius of the matrix has been reduced by clustering the eigenvalues, and therefore fewer search directions may be expected in the conjugate gradients process.

Test runs of the renumbered and un-renumbered conjugate gradients algorithm were made using Evans's method of preconditioning. When the same problem is taken and run with increasing refinement, the total time for solution against matrix size on a log–log graph was obtained. The total time for solution is by itself not of significance, since it is clearly a function of the computer and operating system used and the efficiency of the algorithms.

6

Other Formulations, Equations and Elements

6.1 Introduction to the Galerkin Method and Function Spaces

So far we have relied on the variational approach to finite elements. That is, to solve a differential equation, we need a variational functional that at its minimum would satisfy the differential equation whose solution we seek.

We have already used and seen from energy-related arguments in Chapter 2 that the satisfaction of the Poisson equation (Equation 2.12) in a region with Neumann or Dirichlet boundary conditions, is the equivalent of extremizing the functional

$$\mathcal{L}(\varphi) = \iiint_R \left[\frac{1}{2} \epsilon (\nabla \varphi)^2 - \rho \varphi \right] dR \tag{6.1}$$

In the variational approach to finite elements, we must always identify a functional whose minimization corresponds to the satisfaction of the differential equation being solved.

The variational approach, in fact, allows us to obtain energy-related values of one higher order and, by the conversion of the $\nabla^2 \varphi$ term to the term $[\nabla \varphi]^2$ in Equation (6.1), allows us to use the simple first-order trial functions for φ, which will not be allowed when dealing with $\nabla^2 \varphi$, which will reduce the second derivative to zero. Moreover, the $[\nabla \varphi]^2$ term yields the symmetry behind the finite element matrices, as an examination of Chapter 2 will show.

Be that as it may, the variational approach has one strong disadvantage – the need to identify a functional with the differential equation being solved. This is not always possible, unless one is ready to use such frequently inconvenient methods as a least-square minimization of the residual of the differential equation. The Galerkin method is a powerful scheme suited to many a differential equation especially when there is no functional identified with that equation.

Prior to describing the Galerkin method, we will first say something of function spaces on which the Galerkin method is based. Therefore, before getting into the Galerkin method, some parallels between vector spaces and function spaces need to be pointed out. These parallels in concepts are summarized in Table 6.1. Those who are familiar with vector spaces will understand from the table those operations that will apply to function spaces.

The concept of a Hilbert space generalizes the ideas underlying the Euclidean space, extending concepts of vector algebra from two and three dimensions to any number of dimensions employing geometric intuition. The idea of a perpendicular projection on to a subspace helps with the idea of minimum distance allowing us to use approximation theory. Thus, if we are solving a partial differential equation for $\varphi \in U$, the universal space, while the solution is confined to trial functions φ^* belonging to a subspace S of U, then the idea of the distance between φ and φ^*, or the projection of φ in U onto φ^* in S brings about the concept of the norm $\|\cdot\|$ relating

TABLE 6.1

Parallels between Vector Spaces and Function Spaces

	Vector Space U	Function Space U
Spanning Set {S}	Every $\underline{V} \in U, \underline{V} = V_{s1} + \ldots + V_{sm}$ $=\{S\} = \{c_1 V_{s1}, \ldots, c_m V_{sm}\}$	Every $f \in U$ $f = c_1 f_{s1} + \ldots + c_m f_{sm}$ $\{S\} = \{f_{s1}, \ldots, f_{sm}\}$
Basis Set {B} is the smallest spanning Set. Dimension n = size of smallest spanning set	Any vector $\underline{V} = d_1 \underline{V}_{b1} + \ldots + d_n \underline{V}_{bn}$ $\{B\} = \{\underline{V}_{b1}, \ldots, \underline{V}\}$ $n \leq m$	Any function $f = f_{b1} + \ldots + f_{bn}$ $\{S\} = \{d_1 f_{b1}, \ldots, d_{m1} f_{bm}\}$ $n \leq m$
Closure	$\underline{V} = \underline{V}_1 + \underline{V}_2 \in U$ if $\underline{V}_1, \underline{V}_2 \in U$	$f = f_1 + f_2 \in U$ if $f_1, f_2 \in U$
Additive Commutativity	$\underline{V}_1 + \underline{V}_2 = \underline{V}_2 + \underline{V}_1$ if $\underline{V}_1, \underline{V}_2 \in U$	$f_1 + f_2 = f_2 + f_1$ if $f_1, f_2 \in U$
Associativity of Addition	$\underline{V}_1 + (\underline{V}_2 + \underline{V}_3) = (\underline{V}_1 + \underline{V}_2) + \underline{V}_3$	$f_1 + (f_2 + f_3) = (f_1 + f_2) + f_3$
Associativity of Multiplication	$\alpha(\beta \underline{V}) = (\alpha\beta)\underline{V}$	$\alpha(\beta f) = (\alpha\beta) f$
The Additive Identity Element	$\exists a$ unique element $\underline{0} \in U$ s.t $\underline{V} + \underline{0}$ $= \underline{V}$ for every $\underline{V} \in U$	$\exists a$ unique element $0 \in U$ s.t $f + 0$ $= f$ for every $f \in U$
Multiplicative Identity	$1\underline{V} = \underline{V}$	Similarly $\exists a$ unique identity element $1 \in U$ s.t $1f =$ for every $f \in U$
Zero Vector/ Function	A vector $\underline{V} = 0$ in a space with basis vectors $\underline{V}_{bi}$ if and only if $\underline{V}_{bi} \cdot \underline{V} = 0$ for every i	A function $f = 0$ if and only if $\langle f_{bi} \mid f \rangle = 0$ for every i where the inner product $\langle f \mid g \rangle = \iint_R fg dR$

every element f to a non-negative number meeting the following axiomatic conditions:

$$\|f(x)\| \geq 0 \tag{6.2}$$

$$\|f(x)\| = 0 \text{ if and only if } f(x) = 0. \tag{6.3}$$

$$\|\alpha f(x)\| = |\alpha| \cdot f(x) \tag{6.4}$$

$$\|f_1(x) + f_2(x)\| \leq \|f_1(x)\| + \|f_2(x)\| \tag{6.5}$$

A common definition of a norm is the L-2 norm which satisfies the above properties:

$$\|f(x)\|_2 = \sqrt{\int |f(x)|^2 \, dx} \tag{6.6}$$

This may be used to get a least-square error between the actual φ and its approximate trial function φ^*. A more common norm in general space $\underline{x}$ of dimension n is

$$\|f(x) \cdot g(\underline{x})\| = \int_R f(\underline{x}) g(\underline{x}) dR \tag{6.7}$$

This introduction suffices for now to rather simplistically describe the Galerkin method.

6.2 The Generalized Galerkin Approach to Finite Elements

The Galerkin method of approximately solving differential equations is a powerful and general one, the details of which are much beyond the scope of this book. Indeed, it may be shown that the above variational formulation may also be arrived at by the Galerkin method. In fact, while the variational method cannot be applied unless we are able to identify a functional, no such restriction applies to the Galerkin method.

In solving the operator equation

$$A\varphi(x, y) = \rho(x, y), \tag{6.8}$$

the actual solution φ in general has no restrictions on it other than the boundary conditions. But φ^*, the approximate solution we obtain, belongs to a restricted subspace of the universal space – the approximation space. For instance, in Equation (2.42) we have restricted the solution to belong to the subspace of linear polynomials in x and y. The Galerkin method provides us with the best of the various solutions within the approximation space. Thus, the approximate solution is limited by our trial function and gives us the best-fit of the governing equation from the trial space.

Supposing we seek an approximation in a space of order n, with basis functions $\alpha_1,\ldots,\alpha_n$, and the right-hand side ρ, or an approximation of it ρ^*, belongs to a space of order m, with basis $\beta_1,\ldots,\beta_m$; then

$$\varphi^* = \sum_{i=1}^{n} \alpha_i \varphi_i, \tag{6.9}$$

$$\rho^* = \sum_{i=1}^{m} \beta_i \rho_i, \tag{6.10}$$

At this point it is in order to give a rather simplistic definition of basis functions, drawing upon an analogy with basis vectors. A one-dimensional vector space, we know, has the unit vector $\mathbf{u}_x$ in the direction of the dimension x as basis, so that any one-dimensional vector is expressed as $a\mathbf{u}_x$. Similarly in two dimensions, we have a basis consisting of $\mathbf{u}_x$, $\mathbf{u}_y$, so that any vector from a two-dimensional subspace is defined by $a\mathbf{u}_x + b\mathbf{u}_y$ for unique values of a and b. Analogously, function spaces are generated by basis functions. For example, the subspace of constants is generated by the basis function 1, so that any constant is c. 1; the subspace of first-order polynomials in two-dimensional space is generated by the three basis functions 1, x, and y, so that first-order polynomials in x and y belong to a subspace of dimension 3. Second-order polynomials in x and y are generated by 1, x, y, xy, x^2, and y^2, so that they belong to a subspace of dimension 6. Please note that just as basis vectors are not unique, basis functions too are nonunique. For example, $\mathbf{u}_x$ and $(1/2)\mathbf{u}_y$ also form a basis for two-dimensional vectors, just as the set 1, x, and $x+y$ is a basis for first-order polynomials in x and y. However, since the dimension of the subspace cannot vary, any two basis sets must of necessity have the same number of elements, this number being the dimension of the space. Moreover, it is easily shown that the elements of a basis set must be independent for the expansion of an element from the subspace to be unique. That is, an element of a basis set cannot be expressed by a combination of the other elements.

With this definition, we may define the generalized Galerkin method, which, once we have decided on a trial function φ^* lying in a subspace of dimension n, allows us to get that of the various functions in that subspace that has least error. It states that, for any basis $\gamma_1,\ldots,\gamma_n$ of the space of φ^* (not necessarily the same as the basis set $(\alpha_1, \alpha_2,\ldots,\alpha_n)$, the best approximation to the equation being solved, Equation (6.8), is when

$$\sum_{i=1}^{n} \varphi_i \langle \gamma_k \mid A\alpha_l \rangle = \sum_{i=1}^{m} \rho_i \langle \gamma_k \mid \beta_i \rangle \quad k = 1,\ldots,n \tag{6.11}$$

where $\langle | \rangle$ denotes what is called an inner product, which is commonly defined as:

$$\langle f(x,y) | g(x,y) \rangle = \int fg dR. \tag{6.12}$$

Be it noted that Equation (6.11) is a matrix equation, since the inner products are known once we decide upon the approximation spaces and, therefore, the basis sets to employ. We have n such equations for every γ_k, for $k=1,\ldots, n$, so that Equation (6.11) is in reality an $n \times n$ matrix equation for the n unknowns φ_k. Moreover, since the function elements of a basis are independent, the coefficient matrix will be nonsingular and therefore invertible. The matrix form of the equation is

$$\mathbf{P\Phi} = \mathbf{q}, \tag{6.13}$$

where the matrix $\mathbf{P}$ and the vector $\mathbf{q}$, respectively of size $n \times n$ and $n \times 1$, and the column vector $\mathbf{\Phi}$ of length n are defined by and computed from

$$\mathbf{P}(i,j) = \int \gamma_i A \alpha_j \, dR \quad i=1,\ldots,n;\ j=1,\ldots,n, \tag{6.14}$$

$$A^r \mathbf{T}(i,j) = \int \gamma_i \beta_j \, dR \quad i=1,\ldots,n;\ j=1,\ldots,m, \tag{6.15}$$

$$\mathbf{q} = A^r \mathbf{Tr}, \tag{6.16}$$

$$A^r = \int dR = \text{Area}. \tag{6.17}$$

Previously we symbolized Equation (6.13) in the form $\mathbf{Ax}=\mathbf{b}$ but now we use $\mathbf{P}$ and $\mathbf{q}$ since, as we shall see, they do not always correspond to $\mathbf{A}$ and $\mathbf{b}$. The elements of $\mathbf{\Phi}$ and $\mathbf{r}$ are, respectively, φ and ρ; that is, $\Phi(i) = \varphi_i$ and $q(i) = \rho_i$. In finite element analysis, the region of solution is often divided into several smaller subregions over each of which a different trial function with interelement relationships is defined. Thus, the integrations are performed over the different elements and then summed to get the total integral over the solution region. This summation is similar to the building of the global matrix from the local matrix demonstrated with the variational method in Section 2.5.

The generalized Galerkin method in West European and North American literature is also known as the method of weighted residuals or the method of moments, because the functions γ serve as weighting functions. While the method may be justified in elaborately-rigid mathematical terms, a simplistic explanation of how it works again makes recourse to the analogy with vector spaces; here we regard the Galerkin method as setting every component of the residual to zero, in the approximation space. For the approximations of Equations (6.9) and (6.10), the residual from (6.8) is

$$\Re(\varphi) = \sum_{i=1}^{n} \varphi_i A\alpha_i - \sum_{i=1}^{m} \rho_i \beta_i, \tag{6.18}$$

which is still being generated by the basis vectors $\alpha_1,\ldots,\alpha_n$. This residual lies in the approximation space having the independent functions $\zeta_1,\ldots,\zeta_n$ as basis. Thus, if we multiply the residual by each ζ_i and set the integral over the solution region R to zero, we would ensure the orthonormality of the residual to each element of a basis of the approximation space; that is, if the residual in the approximation space is expanded in terms of the ζ's, the coefficient corresponding to each ζ is made zero. To draw on our analogy with vector spaces again, if, corresponding to the residual $\mathfrak{R}$, we had the vector $\mathfrak{R}$ given by $f_1\mathbf{u}_1 + f_2\mathbf{u}_2 + \ldots + f_i\mathbf{u}_i + \ldots + f_n\mathbf{u}_n$, then to set the component f_i of $\mathfrak{R}$ in the direction $\mathbf{u}_i$ to zero, we would, using the dot (or scalar) product, set $\mathbf{u}_i \cdot \Re = f_i = 0$. Thus, to set $\mathfrak{R}$ to zero in every direction, we would do the same thing with each of the basis elements, $\mathbf{u}_1,\ldots,\mathbf{u}_n$. In dealing with function spaces as opposed to vector spaces, however, we must set $\Re \cdot \zeta_i$ to zero for every i. However, the resulting terms being functions of space, an overall optimal reduction of the residual is achieved by the integration of the product over the solution region, as in Equation (6.11).

Since the approximation space has different bases, the choice of basis would affect the final approximation of φ, and hence the individual components ζ_i weight the solution in the different directions. In this sense, the basis set ζ employed gives the generalized Galerkin method different names. For example, if all three bases α, β, and γ are the same, the method is known as the Bubnov–Galerkin method. On the other hand, if we take $\gamma_i = \partial\Re/\partial\varphi_i = A\alpha_i,\ \ i = 1,\ldots,n,$ as the basis γ, we have the *least-squares* method. In the least-squares method, we say that the nonnegative quantity $\Re^2$ should be a minimum. When we impose this condition on Equation (6.13), after squaring and differentiating with respect to each φ_i, we obtain

$$\sum_{i=1}^{n} \varphi_i \langle A\alpha_k \mid A\alpha_i \rangle = \sum_{i=1}^{m} \rho_i A \langle \alpha_k \mid \beta_i \rangle \quad k = 1,\ldots,n. \tag{6.19}$$

The same Galerkin method becomes the so-called *collocation* method when we set the residual $\mathfrak{R}$ to zero at n preselected points called *collocation points*, and thereby obtain n equations for solution. Here the choice of the location of the points will determine the accuracy of the solution, and it may be shown that collocation falls under the generalized Galerkin method, with γ_i taken as the Dirac-Delta function $\delta(x - x_i)$ at the collocation point $x = x_i$.

A further generalization of the Galerkin method may be made for a function governed by two or more equations. For example, if φ is governed by Equation (6.8) and the equation

$$P\varphi(x,y)=\sigma(x,y), \tag{6.21}$$

then the corresponding generalized Galerkin approximation for the equation pair, by applying the same principle to the sum of the residuals from the two governing equations when the trial functions are put in, is

$$\sum_{i=1}^{n}\varphi_i\langle\gamma_k\,|(A+\lambda P)\alpha_i\rangle=\sum_{i=1}^{m}(\rho_i+\lambda\sigma_i)\langle\gamma_k\,|\,\beta_i\rangle\quad k=1,\ldots,n, \tag{6.21}$$

where λ is a weighting function, often called a *Lagrangian Multiplier.* For large λ, the solution would tend to satisfy Equation (6.20) more (in the sense of the norm of the residual), and for small λ, the answer would match Equation (6.8) better.

Another way in which the Galerkin method is described in the literature is, corresponding to Equation (6.11),

$$\int\psi[A\varphi-\rho]dR=0, \tag{6.22}$$

where φ is the unknown in the approximation space and ψ too is a complete function in the same approximation space. By *complete* is meant a function that may be described only by all the basis functions of the subspace. Thus, ψ becomes a sum of the ζ functions and, equating each of the n components of the expression Equation (6.22), we obtain Equation (6.11). Similarly, corresponding to Equation (6.21), we have

$$\int\psi[A\varphi-\rho]dR+\lambda\int\psi[P\varphi-\sigma]dR=0. \tag{6.23}$$

6.3 Normal Gradient Boundary Conditions in Finite Elements – The Neumann Condition

6.3.1 Forced and Natural Boundary Conditions

In solving a differential equation, we may impose the boundary conditions through trial functions. For example, if we are using a first-order trial function such as

$$\varphi=a+bx+cy=\varphi_1\zeta_1+\varphi_2\zeta_2+\varphi_3\zeta_3 \tag{6.24}$$

in a triangle where we have node 1 at a Dirichlet boundary with potential 100, then we merely set φ_1 to 100 in Equation (6.24), and automatically, the Dirichlet boundary condition is satisfied at that node. A Dirichlet boundary condition is referred to as an *essential* boundary condition, because at least one point on the boundary is required to be a Dirichlet boundary for uniqueness – see Section 2.6 of this book (Hoole, 1988).

Similarly, Neumann boundaries also may be imposed through trial functions. For the example of Equation (6.24), if a particular side of the triangle is a Neumann boundary, then we would explicitly impose a relationship between the three vertex potentials so as to satisfy the Neumann condition. This would result in two potentials, instead of the three independently variable potentials φ_1, φ_2, and φ_3 as before in the triangle to be solved. Correspondingly, the approximation space within the triangle will no longer be spanned by the functions ζ_1, ζ_2, and ζ_3, but rather by two appropriate functions, which explicitly satisfy the Neumann condition. This way of explicitly imposing Neumann conditions through the trial functions is known as a *strong imposition*. Since our solution space is already approximate and of limited dimension, strongly asserted conditions further reduce the freedom of the solution to fit the governing equation and are generally considered undesirable; for we have a situation where a quantity must obey two laws (the governing equation and the boundary conditions), and our approximation does not allow that quantity to obey them both completely, since the actual answer will not necessarily take the form of our trial function. The finite element solution can fit the governing equation only in an optimal sense and exactly only when, by chance, we happen to take a trial function having the same form as the exact solution. Therefore, we will distort the solution (already distorted by the trial functions) further, by forcing the solution to obey one of the laws completely. What is preferred is to impose both laws in the same way so that the approximation may obey them both as much as possible, and, within the limits imposed by the approximation space, arrive at a solution that satisfies both conditions as best as it can, or optimally.

When we do not explicitly impose the Neumann conditions exactly, the procedure is called a *weak formulation*. Equations (6.18) and (6.22), when generalized to more than one equation, give us a natural way of imposing the Neumann condition. For the Poisson equation, the residual of the first equation is $[-\varepsilon\nabla^2\varphi - \rho]$ operating within the region of solution R, and the residual of the Neumann-boundary condition, specifying that the normal gradient of the field is σ along a part $\boldsymbol{\Gamma}$ of the total boundary S, is $[\partial\varphi/\partial n - \sigma]$, so that the Galerkin formulation, corresponding to Equation (6.23), is

$$\iint_R \psi\left[-\varepsilon\nabla^2\varphi - \rho\right]dR + \int_\Gamma \lambda\psi\left[\frac{\partial}{\partial n}\varphi - \sigma\right]dS = 0. \tag{6.25}$$

The choice of λ would determine toward which of the two equations the solution is to be weighted, and $\lambda = \varepsilon$ would be the most natural choice without biasing the solution toward either equation, since the operator $-\nabla^2$ is already scaled by the factor ε. Using the well-known vector identity

$$\nabla \cdot (\varphi A) = \varphi \nabla \cdot A + A \cdot \nabla \varphi \tag{6.26}$$

with $\varphi = \psi$ and $\mathbf{A} = \nabla \varphi$, integrating over the solution region R and applying Gauss's theorem

$$\iint A \cdot dS = \iiint \nabla \cdot A dR \tag{6.27}$$

where the surface S bounds the volume R, we obtain

$$\iint_R [\varepsilon \nabla \psi \cdot \nabla \varphi - \psi \rho] dR - \int_S \varepsilon \psi \frac{\partial}{\partial n} \varphi \, dS + \int_\Gamma \varepsilon \psi \left[\frac{\partial}{\partial n} \varphi - \sigma \right] dS = 0. \tag{6.28}$$

Now ψ is a sum of the components of a basis spanning the approximation space, and is such as to strongly satisfy the boundary conditions along the Dirichlet portions of S; that is, along portions of S excluding $\mathbf{\Gamma}$. On these Dirichlet portions of S, since the value of φ is imposed through the trial functions, the approximation space, having no freedom there, will have no contribution to make to ψ. Therefore, the first surface integral over S will be zero over the Dirichlet boundary ($S - \mathbf{\Gamma}$), the nonzero portion of the integral being only over $\mathbf{\Gamma}$. The first surface integral will therefore cancel the first part of the second surface integral over $\mathbf{\Gamma}$, giving:

$$\iint_R [\varepsilon \nabla \psi \cdot \nabla \varphi - \psi \rho] dR - \int_\Gamma \varepsilon \psi \sigma \, dS = 0. \tag{6.29}$$

If we expand φ as a sum of the n terms $\zeta_i \varphi_i$, where the φ_i are to be determined, and ρ as a sum of the m terms $\beta_i \rho_i$; and equate every component γ_i making up ψ, to zero, we obtain:

$$\sum_{i=1}^{n} \varphi_i \langle \nabla \gamma_k \mid \nabla \zeta_i \rangle = \sum_{i=1}^{m} \rho_i \langle \gamma_k \mid \beta_i \rangle + \varepsilon \sigma \langle \gamma_k \mid 1 \rangle \quad k = 1, \ldots, m. \tag{6.30}$$

A very important relationship between the variational derivation of Equation (6.1) and the generalized Galerkin expression of Equation (6.30) may be seen here. If we express φ as a complete unknown in the approximation space in Equation (6.1), we obtain:

$$\mathcal{L}[\varphi] = \frac{1}{2} \sum_{i=1}^{n} \sum_{k=1}^{n} \varepsilon \varphi_i \langle \nabla \zeta_k \mid \nabla \zeta_i \rangle \varphi_k - \sum_{i=1}^{m} \sum_{k=1}^{n} \varphi_k \langle \beta_i \mid \zeta_k \rangle \rho_i, \tag{6.31}$$

$$\frac{\partial}{\partial \varphi_k} \mathcal{L}[\varphi] = \sum_{i=1}^{n} \varepsilon \varphi_i \langle \nabla \zeta_k \mid \nabla \zeta_i \rangle - \sum_{i=1}^{m} \rho_i \langle \zeta_k \mid \beta_i \rangle = 0 \quad k = 1, \ldots, n, \tag{6.32}$$

which is exactly the same as Equation (6.30) but for $\gamma_i = \zeta_i$ and σ, the specified normal gradient at Neumann boundaries, being zero. Therefore, the variational formulation presented above is the same as the Bubnov–Galerkin method for zero Neumann derivative.

This similarity has an important implication for the variational formulation. What this equivalence states is that whenever the normal gradient of the field is zero, we simply do not bother about it when we employ the variational formulation. It will automatically turn out to be so. This is, therefore, called a *natural* boundary condition. In Equation (6.6) we saw that when the Neumann condition is imposed and the energy $\mathcal{L}$ is minimized, the Poisson equation is satisfied. What we have just learned is that we need not impose simple Neumann conditions at all. We merely extremize the energy and we will simultaneously satisfy the simple Neumann condition on non-Dirichlet boundaries and the Poisson equation everywhere within the region of solution. That this condition turns out automatically in one-dimensional problems from symmetry conditions has already been shown in Section 2.7 using the energy functional. Here we have used the Galerkin method to prove the same thing in two dimensions as well. An extension of the proof for two dimensions using the energy functional is given by Hoole (1988).

Needless to say, since we have by choosing an approximation space limited the ability of our solution to satisfy these equations exactly, the solution we obtain will be that function of our approximation space that satisfies them best; that is, while our solution will neither fit the Poisson equation nor the simple Neumann boundary exactly, it will satisfy both in an optimal sense. This optimality will be in terms of the energy $\mathcal{L}$, which we are minimizing; and that is why, as was discussed in Equation (2.52), the positive term $O^2(\kappa\eta)$ shows that the energy computed from the solution for φ will be of an order of accuracy higher than that of φ. This means that energy based on finite element computations, a function of $(\nabla\varphi)^2$, will be more accurate than any φ we may compute.

We have already seen from Chapter 2 how all the terms but $\int \varepsilon\psi\sigma\, dS$ in Equation (6.29) are handled. To determine the handling of this term, consider a triangle 123 with edge 23 forming a part of the boundary on which σ is specified. Since ζ_1 is zero along this edge, the unknown φ is $\zeta_2\varphi_2 + \zeta_3\varphi_3$ for first-order interpolation, so that γ_k will assume the values ζ_2 and ζ_3 in Equation (6.30). The relevant terms of Equation (6.30), then, corresponding to the three weighting functions ζ_1, ζ_2, and ζ_3 are, respectively, zero. For these,

$$\int \varepsilon\sigma\zeta_2\, dS = \varepsilon\sigma \int_0^1 \zeta_2 L_{23}\, d\zeta_2 = \frac{1}{2}\varepsilon\sigma L_{23}, \tag{6.33}$$

and

$$\int \varepsilon\sigma\zeta_3\, dS = \varepsilon\sigma \int_0^1 \zeta_3 L_{23}\, d\zeta_3 = \frac{1}{2}\varepsilon\sigma L_{23}, \tag{6.34}$$

where L_{23} is the length of the edge 23. This means that zero, $(1/2)\varepsilon\sigma L_{23}$, and $(1/2)\varepsilon\sigma L_{23}$, respectively, must be added to the right-hand-side terms of q of Equation (6.13). This additional term is

$$\mathbf{v} = \frac{1}{2}\varepsilon\sigma L_{23}\begin{bmatrix}0 & 1 & 1\end{bmatrix}^t. \tag{6.35}$$

In variational terms this corresponds to minimizing the functional

$$L(\varphi) = \iint\left\{\frac{1}{2}\varepsilon[\nabla\varphi]^2 - \varphi\rho\right\}dR + \int\varepsilon\sigma\varphi\, dS. \tag{6.36}$$

In magnetic field problems, this specified normal-gradient boundary condition is useful if the input flux into the solution region is specified. In electric field problems, this may correspond to a boundary charge specification. In magnetic field problems, after getting the solution of the magnetization interior to permeable materials, we may use the normal gradient of the magnetic field on the surface of such materials to obtain the field in the exterior free space.

For example, consider the electric fuse of Figure 6.1. This is a typical current flow problem. When current exceeds a predesigned amount, the fuse melts and opens up the circuit, particularly localized to any constriction we may place, on account of the high current density and attendant heating there. To model a current I flowing through the conductor of width w, we know the current density J is I/w.

Supposing we are given, instead of the input and output voltages, the total flow of current I. Then we have at the input, the normal component of current density J as:

$$\sigma\nabla\varphi\cdot\mathbf{u}_n = \frac{I}{w}, \tag{6.37}$$

where σ is the conductivity and w is the width of the fuse. This fuse may be posed as the boundary value problem of Figure 6.1A and solved. The left and right boundaries are at a constant but unknown potential, because they are far away from the constriction. In addition, the left boundary has Equation (6.37). We also need to specify φ at least at one point for uniqueness (see

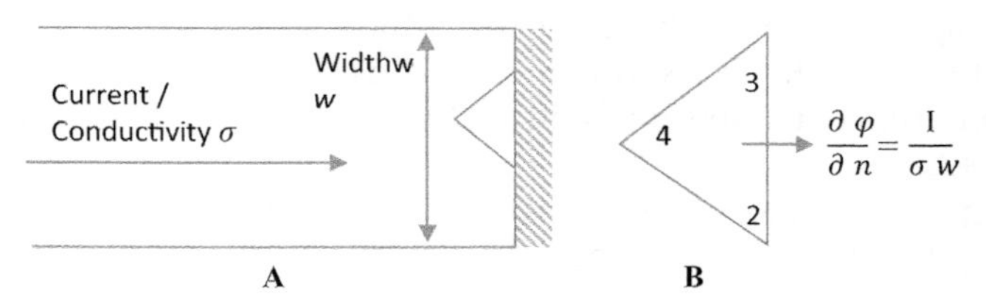

FIGURE 6.1
The electric fuse: A. The fuse and B. An element 234 on the boundary.

Section 2.6). We shall choose this to be on our right boundary and thereby make the entire boundary be at zero potential. As the current I is increased, the potential on the left boundary will rise to accommodate it.

The Neumann conditions imply no current egress along the other boundaries. Moreover, since this is a linear problem in a homogeneous medium, the conductivity may be taken as 1, without affecting our solution for φ.

The formation of the finite element matrices will be as here, except that the boundary elements, such as 234 for example, will have the term

$$\mathbf{v} = \frac{1}{2}\left(\frac{IL_{24}}{\omega}\right)\begin{bmatrix}0 & 1 & 1\end{bmatrix}^t \tag{6.38}$$

(L_{24} being the length of the edge 24) added to the usual right-hand-side vector $\mathbf{q}$ of Equation (6.11), as required by Equation (6.35). However, if by chance the boundary element 234 had been numbered 342, with the first and second nodes being on the boundary, it is easy to verify that what we need to add is

$$\mathbf{v} = \frac{1}{2}\left(\frac{IL_{24}}{\omega}\right)\begin{bmatrix}1 & 1 & 0\end{bmatrix}^t \tag{6.39}$$

6.3.2 Handling Interior Line Charges in Finite Elements

The foregoing considerations are the mathematically parallel to what we need to do to model line charges. Particularly in electron devices, it is common to encounter a specified line charge distribution interior to the solution region. By considering a small cylindrical pillbox with lids on either side of the boundary and applying Gauss' theorem to that pillbox, it can be shown that

$$D_{n1} = -\varepsilon_1 \nabla\varphi \cdot \mathbf{u}_{n1} = -D_{n2} + \mathbf{q} = \varepsilon_2 \nabla\varphi \cdot \mathbf{u}_{n2} - \mathbf{q}, \tag{6.40}$$

where $\mathbf{q}$ is the line charge density. Observe that the difference in signs is on account of the normal $n1$ and $n2$ along S being oppositely directed. We know how to specify the gradient at a boundary using Equation (6.36), but how do we specify the condition interior to the solution region? To discover this, let the line containing the charge be C. We shall extend it along either end with sections C_1 and C_2 without charge Q, to cut the solution region R and boundary S into two parts R_1 and R_2 and S_1 and S_2, as shown in Figure 6.2. Let g be $\nabla\varphi \cdot \mathbf{u}_n$ along this extended line with values g_1 and g_2 on either side of this line $C_1 + C + C_2$.

Applying Equation (6.36) to the two parts separately and imposing these unknown gradients g_1 and g_2 along the common boundary of the two regions,

$$\mathcal{L}_1(\varphi) = \frac{1}{2}\iint_{R1}\left\{\varepsilon_1[\nabla\varphi]^2 - 2\varphi\rho\right\}dR + \int_{C_1+C+C_2} \varepsilon_1 g_1 \varphi \, dC, \tag{6.41}$$

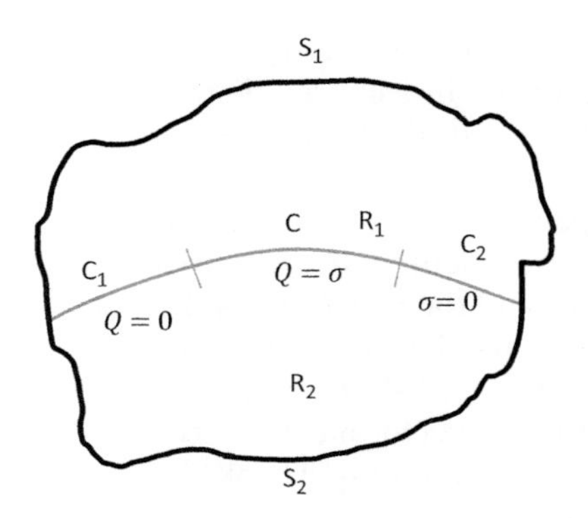

FIGURE 6.2
Modeling interior line charges.

$$\mathcal{L}_2(\varphi)=\frac{1}{2}\iint_{R2}\left\{\varepsilon[\nabla\varphi]^2-2\varphi\rho\right\}dR+\int_{C_1+C+C_2}\varepsilon_2 g_2\varphi\, dC. \tag{6.42}$$

Observe from Equation (6.40) that $\varepsilon_1 g_1+\varepsilon_2 g_2$ is zero along C_1 and C_2 because Q is zero there, and that it is $-Q$ along C. Moreover, note that a simple Neumann or Dirichlet condition has been assumed along S. Adding the above two equations, we have

$$\mathcal{L}(\varphi)=\frac{1}{2}\iint_{R}\left\{\varepsilon[\nabla\varphi]^2-2\varphi\rho\right\}dR-\int_{C}Q\varphi\, dC \tag{6.43}$$

This was to be expected because it corresponds to the integral $\varphi\rho$ over R as in the second term, representing the energy expended in bringing Q to its present location.

To implement the functional of Equation (6.43), consider the addition to the energy $\Delta\mathcal{L}$ from a segment of the line charge between two adjacent nodes 1 and 2. Assuming a linear interpolation, we have

$$\begin{aligned}\Delta\mathcal{L}&=\int_1^2 Q\left\{\varphi_1+\frac{[C-L_1][\varphi_2-\varphi_1]}{L_2-L_1}\right\}dC\\&=\frac{1}{2}QL_{12}[\varphi_1+\varphi_2],\end{aligned} \tag{6.44}$$

where L_{12} is the length between the two nodes, and L_1 and L_2 are the coordinates C of the nodes along some coordinate system down the line charge. When the functional is extremized with respect to the potentials, the φ terms drop off so that the terms $(1/2)QL_{12}$ contribute to the first and second rows of the right-hand side of the global finite element matrix equation $\tilde{\mathbf{q}}$. For example, if 3, 7, and 5 are adjacent nodes along the line charge Q, then node 7 is an interior node and row 7 of $\tilde{\mathbf{q}}$ will have the terms $(1/2)QL_{37}$ and $(1/2)QL_{75}$

added to it, on account of the line charge contribution Q. However, if node 7 had been at an end of the line charge distribution, with, say, line 57 to its right but no charge to its left, then only the term $(1/2)QL_{75}$ will be added to the 7th row of $\underset{\sim}{q}$.

Let us say that a parallel-plate capacitor with a specified plate charge Q C/m was solved by this method to demonstrate the handling of interior line charges. An additional data set defines the line segments 12, 23, 34, and 45, etc., containing the line charges. After forming the usual finite element matrix by going through this list of line charge elements, for every line element ij, the right-hand-side vector has the term $(1/2)QL_{ij}$ added to the rows i and j of the right-hand-side vector $\underset{\sim}{q}$. Think about how you would formulate this problem on a computer.

6.3.3 Natural Impedance Boundary Conditions

The impedance boundary condition (Hoole and Carpenter, 1985) permits us to block off from our solution region those parts experiencing eddy currents, and in which we are not interested. This significantly reduces computational costs because in an area subject to eddy current effects, the variation in φ (the equivalent of the vector potential we solve for) is rapid, in fact exponential. This necessitates a fine mesh for accurate modeling of exponential changes in φ. The impedance boundary condition may in general be expressed as

$$\frac{\partial A}{\partial n} = kA. \tag{6.45}$$

Such a boundary condition is encountered in the escape of heat from a surface in heat-flow studies and can be easily imposed naturally (Visser, 1968). Applying the Galerkin principle as in Equation (6.23)

$$\iint_R \psi\left[-\upsilon\nabla^2 A - J\right]dR + \int_\Gamma \upsilon\psi\left[\frac{\partial A}{\partial n} - kA\right]dS = 0, \tag{6.46}$$

where R is the solution region with boundary S. Γ is that part of S on which the impedance boundary condition applies. Manipulating as before, using the divergence theorem:

$$\iint_R [\upsilon\nabla\psi\cdot\nabla A - \psi J]dR - \int_S \upsilon\psi\frac{\partial A}{\partial n}dS + \int_\Gamma \upsilon\psi\left[\frac{\partial A}{\partial n} - kA\right]dS = 0. \tag{6.47}$$

Since the first surface integral vanishes on account of the trial functions along those parts of the surface where A is fixed, we obtain the local matrix equation from

$$\iint_A [\upsilon\nabla\psi\cdot\nabla A - \psi]dR - \int_S \upsilon k\psi A\, dR = 0. \tag{6.48}$$

When the finite element trial functions are put in, the surface integral term gives the contribution

$$\Delta\mathbf{P} = \frac{1}{6}L\begin{bmatrix} 2 & 1 \\ 1 & 2 \end{bmatrix} \tag{6.49}$$

to the local matrix **P**. L is the length of the element edge along the surface, and the associated finite element 2-vector contains the vector potentials at the two vertices of the edge.

6.4 A Simple Hand-Worked Example

6.4.1 A Test Problem with an Analytical Solution

As a means of explicating the approximation theory of the preceding section, we shall here take up a simple numerical example. We shall try to approximate, by the various methods detailed, the solution to a problem that we know exactly by analytical methods.

Consider the function φ defined by:

$$\frac{d^2}{dx^2}\varphi = x \quad 0 \le x \le 1, \tag{6.50}$$

$$\frac{d}{dx}\varphi = 0 \quad x = 1, \tag{6.51}$$

$$\varphi = 0 \quad x = 0. \tag{6.52}$$

This is basically the Poisson equation in one dimension with $\rho(x) = -x$; the region of solution R is the closed interval [0, 1] with the two points $x=0$ and $x=1$ as boundary. At the left boundary we have a Dirichlet condition, Equation (6.52), and at the right boundary a simple Neumann condition, Equation (6.51).

To obtain the analytical solution, integrating Equation (6.50), we get

$$\frac{d}{dx}\varphi = \frac{1}{2}x^2 + c_1. \tag{6.53}$$

Using the condition of Equation (6.51) we have $c_1 = -(1/2)$. Further integration gives

$$\varphi = \frac{x^3}{6} - \frac{1}{2}x + c_2. \tag{6.54}$$

Finally, using the Dirichlet boundary condition of Equation (6.52) we obtain $c_2 = 0$, so that

$$\varphi = \frac{x^3}{6} - \frac{x}{2}. \tag{6.55}$$

With this analytical solution available for purposes of comparison, let us try to obtain the same solution by approximate methods. Since the exact solution is of order 3 in x, so as not to arrive at the exact solution by approximate methods, we will always restrict our trial functions to order 2 or lower.

6.4.2 Galerkin – Strong Neumann, One Second-Order Element

Let our trial function be

$$\varphi = a_0 + a_1 x + a_2 x^2 \tag{6.56}$$

over the whole region of solution. The Dirichlet boundary condition of Equation (6.52) is always explicitly imposed, so that a_0 must be zero. Moreover, since we are also explicitly imposing the Neumann condition here:

$$\frac{d}{dx}\varphi = a_1 + 2a_2 x. \tag{6.57}$$

Putting in Equation (6.51), we have $a_1 + 2a_2 = 0$, so that

$$\varphi = \left(-2x + x^2\right) a_2. \tag{6.58}$$

We see that by explicitly imposing the Neumann condition, the trial function has been reduced to one free parameter a_2, which may vary to satisfy the governing Equation (6.50).

Therefore, the approximation space is spanned by the function $(-2x + x^2)$, since any solution must be a multiple of it. That is, the dimension of the approximation space, n, is 1 and the only element of this basis, a_1, is $(-2x + x^2)$. What must be determined then is a_2 of Equation (6.57).

To this end, applying the Galerkin formulation of Equation (6.11), with $\gamma = \alpha$, $n = 1$, and the operator $A = d^2/dx^2$, we get

$$\int_0^1 \left(-2x + x^2\right) \frac{d^2}{dx^2}\left(-2x + x^2\right) a_2 \, dx = \int_0^1 \left(2x + x_2\right) x \, dx. \tag{6.59}$$

Solving,

$$a_2 = \frac{5}{16}, \tag{6.60}$$

$$\varphi = \frac{5\left(-2x + x^2\right)}{16}. \tag{6.61}$$

6.4.3 Collocation: Explicit Neumann, One Second-Order Element

When we explicitly impose the Neumann condition, with a second-order element encompassing the whole region of solution, the trial function would be exactly the same as Equation (6.57). The residual for this trial function at any point, in accordance with Equation (6.18), is

$$\mathfrak{R} = \left[\frac{d^2}{dx^2}\left(-2x + x^2\right)a_2 - x\right] \tag{6.62}$$

and should be zero. To determine a_2 through collocation, we should determine one collocation point at which we may set the residual to zero. For the region of solution [0, 1], $x = 0.5$ is a representative point, so that, expanding Equation (6.62), we get

$$\mathfrak{R} = 2a_2 - x = 2a_2 - \frac{1}{2} = 0; \tag{6.63}$$

from which we get

$$a_2 = \frac{1}{4}, \tag{6.64}$$

yielding

$$\varphi = \left(-2x + x^2\right)a_2 = \frac{-2x + x^2}{4}, \tag{6.65}$$

which is reasonably close to the exact solution of Equation (6.52) as seen from Figure 6.3. The importance of the choice of the collocation point to accuracy is seen from the fact that, had we chosen our collocation point as $x = 0$, we would have obtained $a_2 = 0$, and for the point $x = 1$, $a_2 = (1/2)$ – both answers widely differing from the exact.

6.4.4 Least Squares: Strong Neumann, One Second-Order Element

Again we have the same trial function of Equation (6.57), and, according to Equation (6.19), the least-squares technique minimizes the integral of the residual $\mathfrak{R}$ of Equation (6.62) with respect to the free coefficient a_2:

$$\frac{\partial}{\partial a_2}\int_0^1 \mathfrak{R}^2\,dx = \frac{\partial}{\partial a_2}\int_0^1 \left[2a_2 - x\right]^2 dx = \frac{\partial}{\partial a_2}\left[4a_2^2 - 2a_2 + \frac{1}{3}\right] = 0, \tag{6.66}$$

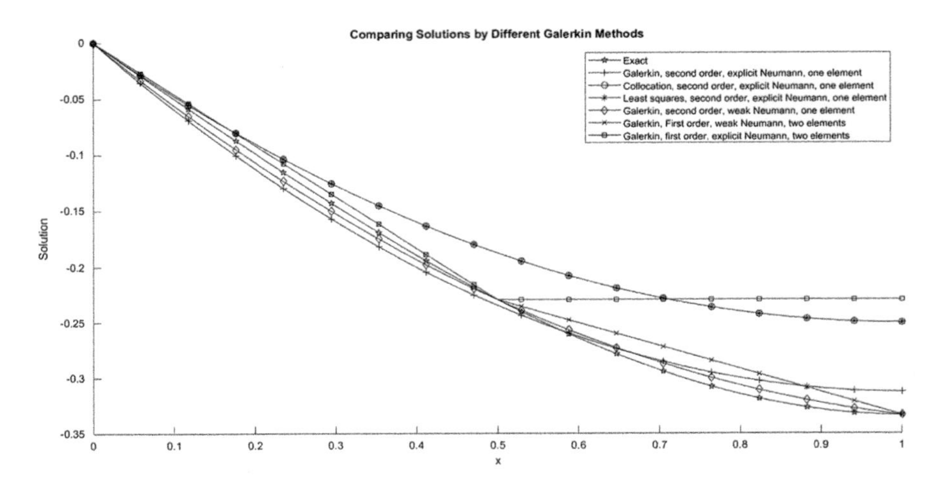

FIGURE 6.3
A plot of various Galerkin solutions of the test problem. (1) Exact. (2) Galerkin, second order, explicit Neumann, one element. (3) Collocation, second order, explicit Neumann, one element. (4) Least squares, second order, explicit Neumann, one element. (5) Galerkin, second order, weak Neumann, one element. (6) Galerkin, first order, weak Neumann, two elements. (7) Galerkin, first order, explicit Neumann, two elements.

giving the same results as Equations (6.63) and (6.64). A way of obtaining the same answer through a direct application of Equation (6.19) is to note first that $A\alpha_1 = (d^2/dx^2)(-2x + x^2) = 2$, so that

$$\int_0^1 2 * 2a_2\,dx = \int_0^1 2 * x\,dx, \tag{6.67}$$

which again is the same result

$$\varphi = \left(-2x + x^2\right)a_2 = \frac{-2x + x^2}{4}, \tag{6.68}$$

6.4.5 Galerkin: Weak Neumann, One Second-Order Element

Here we do not explicitly impose the Neumann condition, but use the weak formulation of Equation (6.32). The trial function is Equation (6.56), with a_0 zero in view of the Dirichlet condition at $x = 0$.

$$\varphi = a_1x + a_2x^2. \tag{6.69}$$

The dimension n of our approximation space has, as a result of our relaxing the Neumann condition, gone up by one, to two. The two functions ζ_1 and

ζ_2, which span the approximation space, are x and x^2, with the coefficients a_1 and a_2 to be determined. The right-hand-side term, $\rho = -x$, may be taken to be spanned by the function $\beta_1 = x$, belonging to the first-order space with $m=1$. The operator ∇ being d/dx, we have $\nabla\zeta_1 = 1$ and $\nabla\zeta_2 = 2x$; applying Equation (6.32) for $k=1$,

$$\int_0^1 1 * 1\, dx a_1 + \int_0^1 1 * 2x\, dx a_2 - \int_0^1 x * (-x)\, dx = 0. \tag{6.70}$$

This reduces to

$$a_1 + a_2 = -\frac{1}{3}. \tag{6.71}$$

Similarly, applying for $k=2$,

$$\int_0^1 2x * 1\, dx a_1 + \int_0^1 2x * 2x\, dx a_2 - \int_0^1 x^2 * (-x)\, dx = 0. \tag{6.72}$$

giving

$$a_1 + \frac{4a_2}{3} = -\frac{1}{4}, \tag{6.73}$$

$$a_1 = -\frac{7}{12}; \quad a_2 = \frac{1}{4}, \tag{6.74}$$

$$\varphi = -\frac{7x}{12} + \frac{x^2}{4}. \tag{6.75}$$

6.4.6 Galerkin: Weak Neumann, Two First-Order Elements

This hand-worked example illustrates again the use of general first-order linear interpolation and the assembly of contributions coming from the various elements of our mesh.

Let us divide our region of solution, [0, 1], into two elements, [0, 0.5] and [0.5, 1], with interpolation nodes 1, 2, and 3, respectively, at $x=0$, $x=0.5$, and $x=1$. Thus, the two elements are described in terms of nodal connections as 1, 2 and 2, 3. In general, consider a first-order element with nodes 1 and 2 at $x = x_1$ and $x = x_2$. The first-order interpolation for this element is, by rearrangement of Equation (2.83),

$$\varphi(x) = \frac{(x_2 - x)}{(x_2 - x_1)}\varphi_1 + \frac{(x - x_2)}{(x_2 - x_1)}\varphi_2 \tag{6.76}$$

This provides us with the line coordinates ζ_1 and ζ_2, corresponding to our triangular coordinates earlier, defined by

$$\zeta_1 = \frac{(x_2 - x)}{(x_2 - x_1)} \tag{6.77}$$

$$\zeta_2 = \frac{(x - x_2)}{(x_2 - x_1)}. \tag{6.78}$$

It may be easily shown that

$$\zeta_1 + \zeta_2 = 1, \tag{6.79}$$

and

$$\int \zeta_1^i \zeta_2^j \, dx = \frac{i!j!1!L}{(i+j+1)!}, \tag{6.80}$$

where $L = x_2 - x_1$, the length of the element, and

$$\frac{\partial \zeta_1}{\partial x} = -\frac{1}{L}; \quad \frac{\partial \zeta_1}{\partial x} = -\frac{1}{L}. \tag{6.81}$$

In general, we have two coefficients φ_1 and φ_2 per element to determine. The first-order Poissonian source $\rho = -x$ may be exactly expanded within the approximation space with the same two basis functions ζ_1 and ζ_2 using

$$x = x_1\zeta_1 + x_2\zeta_2, \tag{6.82}$$

so that $\beta = \alpha = \zeta$. Applying Equation (6.32), with $k=1$,

$$\iint_{x=x_1}^{x_2} \frac{-1}{L}\frac{1}{L} dx \, \varphi_1 + \iint_{x=x_1}^{x_2} \frac{-1}{L}\frac{1}{L} dx \, \varphi_2 - \int_{x=x_1}^{x_2} \zeta_1(-\zeta_1 x_1 - \zeta_2 x_2) dx = 0, \tag{6.83}$$

and for $k=2$,

$$\iint_{x=x_1}^{x_2} \frac{1}{L}\frac{-1}{L} dx \, \varphi_1 + \iint_{x=x_1}^{x_2} \frac{1}{L}\frac{1}{L} dx \, \varphi_2 - \int_{x=x_1}^{x_2} \zeta_2(-\zeta_1 x_1 - \zeta_2 x_2) dx = 0, \tag{6.84}$$

The above two equations reduce, with the help of Equation (6.78), to

$$\frac{\varphi_1}{L} - \frac{\varphi_2}{L} \equiv \frac{-L(2x_1 + x_2)}{6}, \tag{6.85}$$

$$\frac{-\varphi_1}{L} + \frac{\varphi_2}{L} \equiv \frac{-L(x_1 + 2x_2)}{6}. \tag{6.86}$$

The congruence sign instead of the equals sign is used because the integrals are evaluated over an element. Equality will hold only when the integration is over the whole solution region or, equivalently, when the integrals over the elements are added together.

It is to be noted that since the first point at $x=0$ is a Dirichlet point, Equation (6.85) will not apply when considering the first element. We get, for the first element, with $x_1 = 0$, $x_2 = 0.5$, and $\varphi_1 = 0$:

$$2\varphi_2 \equiv -\frac{1}{12}, \tag{6.87}$$

and for the second element, with $x_1 = 0.5$ and $x_2 = 1$:

$$2\varphi_2 - 2\varphi \equiv -\frac{1}{6}, \tag{6.88}$$

$$-2\varphi + 2\varphi_3 \equiv -\frac{5}{24} \tag{6.89}$$

Combining the contributions from the two elements, as given by Equations (6.85) through (6.87),

$$\begin{bmatrix} 2+2 & -2 \\ -2 & 2 \end{bmatrix} \begin{bmatrix} \varphi_2 \\ \varphi_3 \end{bmatrix} = \begin{bmatrix} -\frac{1}{6} - \frac{1}{12} \\ -\frac{4}{24} \end{bmatrix}. \tag{6.90}$$

Solving this matrix equation, we get $\varphi_2 = -(11/48)$ and $\varphi_3 = -(1/3)$, so that according to Equation (6.75),

$$\varphi = \begin{cases} -\dfrac{11x}{24} & \text{in} \quad 0 \le x \le 0.5 \\ -\dfrac{11}{48} + \left(-\dfrac{1}{3} + \dfrac{11}{48}\right)(2x-1) & \text{in} \quad 0.5 \le x \le 1 \end{cases} \tag{6.91}$$

6.4.7 Galerkin: Explicit Neumann, Two First-Order Elements

Here, the interpolation will be exactly as before, but for $\varphi_2 = \varphi_3$ in view of the Neumann condition having to apply in the last element. That is, the field has only one degree of freedom. The equation corresponding to Equation (6.85) will remain the same. But in the second element, since $\varphi_3 = \varphi_2$, the energy contribution will yield

$$2\varphi_2 - 2\varphi_2 \equiv -\frac{1}{6}, \tag{6.92}$$

$$-2\varphi_2 + 2\varphi_2 \equiv -\frac{5}{24}. \tag{6.93}$$

That is, there is no stored energy. Perhaps this is an example that vividly demonstrates why we cannot solve the element equations separately – for then, the above two equations become contradictions. Adding the equations together, we obtain:

$$2\varphi_2 = -\frac{1}{12} - \frac{1}{6} - \frac{5}{24}, \tag{6.94}$$

giving $\varphi_2 = -11/48$, so that the solution is

$$\varphi = \begin{cases} -11x/24 & 0 \le x \le 0.5 \\ -\dfrac{11}{48} & 0.5 \le x \le 1. \end{cases} \tag{6.95}$$

6.4.8 Some Observations

Having obtained by hand these various approximations, which are plotted in Figure 6.3, some observations are in order:

1. The Galerkin method with a weak Neumann solution gives the best overall solution, although the Neumann condition is not satisfied exactly. The explicit Neumann imposition results in exact satisfaction of the Neumann condition, but does not give a good overall fit of the solution.
2. The first-order two-element system and second-order one-element system both have two degrees of freedom when the weak Neumann formulation is used. And yet, the second-order system gives a better answer, because the higher polynomial order of the system allows it to satisfy the equation better.
3. The Galerkin method with the explicit Neumann condition cannot be employed with first-order elements, because the operator d^2/dx^2 will return a zero when differentiating an approximation function of order one.

6.5 Higher-Order Finite Elements

6.5.1 Higher-Order Interpolations

We have so far worked with first-order interpolations – or straight-line variations of the trial function. In general, we have seen that the finite element solution is as good as the trial function we have assumed. An improved solution is achieved in one of two ways: either we resort to a finer subdivision of the finite element mesh keeping the same shape functions, or, as will be described in this section, we resort to improved shape functions while keeping the same elements.

Two approaches are commonly taken in improving the shape function. The first approach is problem dependent, and here we know the kind of solutions to expect, such as exponential variations in eddy current problems (Keran and Lavers, 1986; Weeber et al., 1988) and Bessel functions in cylindrical wave-guides. In these cases, we would allow the unknown to vary as these special functions do and apply the Galerkin method to determine the numerical weights of these shape functions. In this section we shall learn of the more general high-order polynomial shape functions, which may be readily applied without any expectation of the solution. The similarity of the Pascal triangle to various polynomial terms will be used to work with suitably placed interpolation nodes in the high-order triangles shown in Figures 6.4 and 6.5. As demonstrated in Figure 6.4, a first-order triangle takes, from the first two rows, $1+2=3$ nodes (each node corresponding to a degree of freedom), a second-order triangle takes, from the first three rows, $1+2+3=6$ nodes, and in general, an nth-order triangle will contain nodes numbering

$$N(n)=1+2+\ldots+n+1=\frac{1}{2}(n+1)(n+2). \tag{6.96}$$

1

$x \quad y$

$x^2 \quad xy \quad y^2$

$x^3 \quad x^2y \quad xy^2 \quad xy^3$

$x^4 \quad \ldots \quad \ldots \quad \ldots \quad y^4$

FIGURE 6.4
The Pascal triangle.

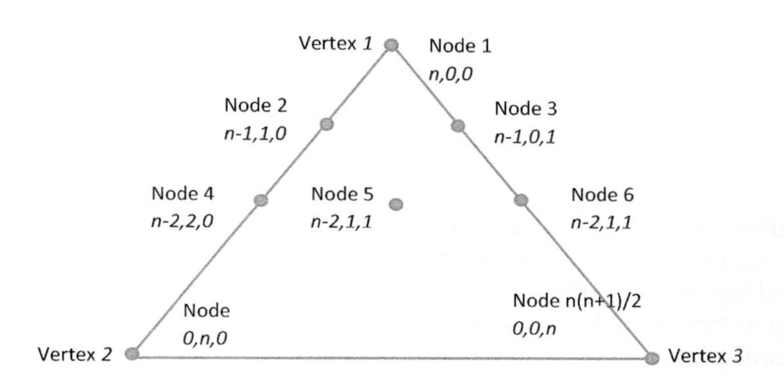

FIGURE 6.5
An nth-order triangle and node-numbering convention. Three-suffix numbering and corresponding regular one-digit numbering.

As defined by Silvester (1969, 1978), the nth-order triangle of Figure 6.5 has the trial function defined by the interpolation

$$\varphi = \sum_{i+j+k=n} \alpha_{ijk}^{n} \varphi = \tilde{\alpha}^{n} \underset{\sim}{\varphi}, \tag{6.97}$$

where each node is uniquely defined by the suffixes i, j, and k totaling n, and Figure 6.5 also gives the three suffix ijk numbering of the nodes and the corresponding regular numbering adopted as convention. The summation in Equation (6.97) is over all possible combinations of i, j, and k. The interpolation row-vector $\tilde{\boldsymbol{\alpha}}^{n}$ has as its elements α_{ijk}^{n}. $\underset{\sim}{\Phi}$ is a column vector with φ_{ijk} as its elements. The triangular coordinates of the nodes are defined by

$$\zeta_1 = \frac{i}{n} \quad \zeta_2 = \frac{j}{n} \quad \zeta_3 = \frac{k}{n} \tag{6.98}$$

and

$$\alpha_{ijk}^{n} = P_i^n(\zeta_1) P_j^n(\zeta_2) P_k^n(\zeta_3), \tag{6.99}$$

where the nth-order polynomials $P_i^n(\zeta)$ are defined by

$$P_i^n(\zeta) = \begin{cases} 1 & \text{if } i = 0 \\ \displaystyle\prod_{j=1}^{i} \frac{n\zeta - j + 1}{j} & \text{Otherwise.} \end{cases} \tag{6.100}$$

To illustrate, for a first-order triangle,

$$\begin{aligned} \varphi &= \alpha_{100}\varphi_{100} + \alpha_{010}\varphi_{010} + \alpha_{001}\varphi_{001} \\ &= P_1^1(\zeta_1) P_0^1(\zeta_2) P_0^1(\zeta_3) \varphi_{100} + P_0^1(\zeta_1) P_1^1(\zeta_2) P_0^1(\zeta_3) \varphi_{010} \\ &\quad + P_0^1(\zeta_1) P_0^1(\zeta_2) P_1^1(\zeta_3) \varphi_{001} = \zeta_1\varphi_1 + \zeta_2\varphi_2 + \zeta_3\varphi_3 \end{aligned} \tag{6.101}$$

since

$$P_1^1(\zeta) = \zeta \quad P_0^1 = 1 \tag{6.102}$$

according to Equation (6.100), and φ_{100}, φ_{010}, and φ_{001} correspond to φ_1, φ_2, and φ_3 at the three vertices as determined from the triangular coordinates defined in Equation (6.98).

As another example, consider the six-noded second-order triangle shown in Figure 6.6, where the trial function becomes

$$\varphi = \alpha_{200}\varphi_{200} + \alpha_{110}\varphi_{110} + \alpha_{101}\varphi_{101} + \alpha_{020}\varphi_{020} + \alpha_{011}\varphi_{011} + \alpha_{002}\varphi_{002} \tag{6.103}$$

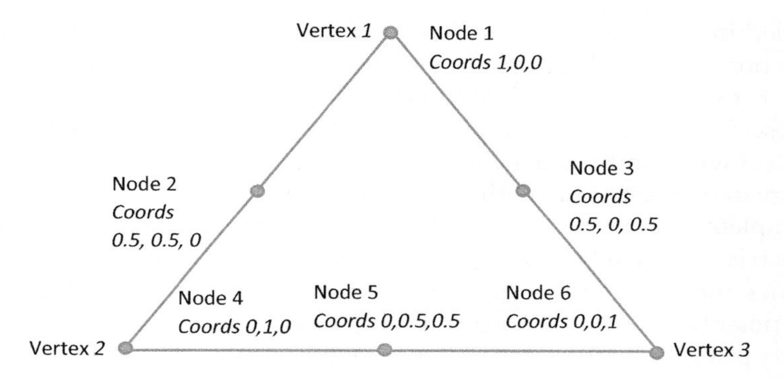

FIGURE 6.6
A second-order triangle (Nodes are interpolation points and vertices are simply the corners of a triangle).

Using Equation (6.100),

$$P_2^2(\zeta) = \left[\frac{2\zeta - 1 + 1}{1}\right]\left[\frac{2\zeta - 2 + 1}{2}\right] = \zeta(2\zeta - 1), \tag{6.104}$$

$$P_1^2(\zeta) = \frac{2\zeta - 1 + 1}{1} = 2\zeta, \tag{6.105}$$

$$P_0^2 = 1. \tag{6.106}$$

Putting Equations (6.99) and (6.104) into Equation (6.103) and simplifying the node numbers as indicated in Figure 6.6, we have

$$\begin{aligned} \varphi = {} & \zeta_1(2\zeta_1 - 1)\varphi_1 + 4\zeta_1\zeta_2\varphi_2 + 4\zeta_1\zeta_3\varphi_3 \\ & + \zeta_2(2\zeta_2 - 1)\varphi_4 + 4\zeta_2\zeta_3\varphi_5 + \zeta_3(2\zeta_3 - 1)\varphi_6 \end{aligned} \tag{6.107}$$

The correctness of these interpolations is easily verified. For example, at vertex 1 of triangular coordinates $(\zeta_1, \zeta_2, \zeta_3) = (1, 0, 0)$, φ becomes φ_1; at vertex 2 with coordinates

$$\left(\frac{1}{2}, \frac{1}{2}, 0\right), \varphi$$

becomes φ_2, and so on. Using the polynomials defined in this section, interpolations of any order may be defined.

6.5.2 Differentiation and Universal Matrices

For purposes of using high-order finite elements, we have already defined the interpolatory trial functions. Now to use them, we need to put them into the Galerkin or variational formulations of Equation (6.1) or Equation (6.11). In this section we shall demonstrate the implementation on the variational formulation, extensions to the Galerkin method being straightforward. We shall also describe pre-computed universal matrices (Silvester, 1978), which allow us to deal with higher-order finite elements using trivial computations that compare in difficulty only with those required for first-order finite elements.

To implement the variational formulation in two dimensions, we need to put the trial function Equation (6.97) into the functional of Equation (6.1) and minimize the functional with respect to the free parameters, which are the nodal potentials that make up φ. Using an approximation for the known quantity ρ similar to Equation (6.1), we have

$$\begin{aligned}\mathcal{L}[\varphi] &= \iint \left\{ \frac{1}{2}[\nabla\varphi]^2 - \rho\varphi \right\} dR \\ &= \iint \left\{ \frac{1}{2}\left[\nabla\tilde{\boldsymbol{\alpha}}^n \underset{\sim}{\Phi}\right]\cdot\left[\nabla\tilde{\boldsymbol{\alpha}}^n \underset{\sim}{\Phi}\right] - \left[\tilde{\boldsymbol{\alpha}}^n \underset{\sim}{\Phi}\right]\left[\tilde{\boldsymbol{\alpha}}^n \tilde{\rho}\right] \right\} dR.\end{aligned} \tag{6.108}$$

The terms $\nabla\tilde{\boldsymbol{\alpha}}$ involve differentiations with respect to x and y, and the results of the differentiations must be integrated over R. Consider for example:

$$\begin{aligned}\frac{\partial\tilde{\boldsymbol{\alpha}}^n}{\partial x} &= \sum_{i=1}^{3} \frac{\partial\tilde{\boldsymbol{\alpha}}^n}{\partial\zeta_i}\frac{\partial\zeta_i}{\partial x} = \sum_{i=1}^{3} \frac{b_i}{\Delta}\frac{\partial\tilde{\boldsymbol{\alpha}}^n}{\partial\zeta_i} \\ &= \tilde{\boldsymbol{\alpha}}^{n-1}\sum_{i=1}^{3}\frac{b_i}{\Delta}\mathbf{G}_i^n = \tilde{\boldsymbol{\alpha}}^{n-1}\mathbf{D}_x^n,\end{aligned} \tag{6.109}$$

which defines the nth-order differentiation matrix $\mathbf{D}_x^n$ and the matrices $\mathbf{G}_i^n$ that make up $\mathbf{D}_x^n$. The derivative of ζ_i with respect to x is obtained using Equation (6.108). Note that $\tilde{\boldsymbol{\alpha}}^n$ being an n-th-order polynomial, its derivative is of order $n-1$ and therefore must be expressible in terms of $\tilde{\boldsymbol{\alpha}}^{n-1}$. And since $\tilde{\boldsymbol{\alpha}}^n$ is a row vector of length $N(n)$ as defined by Equation (6.96), for compatibility $\mathbf{D}_x^n$ must be a matrix of size $N(n-1)\times N(n)$.

As an illustration of the foregoing, consider the second-order interpolation of Equation (6.107):

$$\begin{aligned}\frac{\partial\tilde{\boldsymbol{\alpha}}^2}{\partial\zeta_1} &= \frac{\partial}{\partial\zeta_1}\left[\zeta_1(2\zeta_1-1) \quad 4\zeta_1\zeta_2 \quad 4\zeta_1\zeta_3 \quad \zeta_2(2\zeta_2-1) \quad 4\zeta_2\zeta_3 \quad \zeta_3(2\zeta_3-1)\right] \\ &= \left[4\zeta_1-1 \quad 4\zeta_2 \quad 4\zeta_3 \quad 0 \quad 0 \quad 0\right] \\ &= \left[\zeta_1 \quad \zeta_2 \quad \zeta_3\right]\begin{bmatrix} 3 & 0 & 0 & 0 & 0 & 0 \\ -1 & 4 & 0 & 0 & 0 & 0 \\ -1 & 0 & 4 & 0 & 0 & 0 \end{bmatrix},\end{aligned} \tag{6.110}$$

comparing with Equation (6.109), we have

$$\mathbf{G}_1^2 = \begin{bmatrix} 3 & 0 & 0 & 0 & 0 & 0 \\ -1 & 4 & 0 & 0 & 0 & 0 \\ -1 & 0 & 4 & 0 & 0 & 0 \end{bmatrix}, \tag{6.111}$$

where the term $4\zeta_1 - 1$ has been rewritten as $4\zeta_1 - (\zeta_1 + \zeta_2 + \zeta_3)$ using Equation (2.142). Similarly, we may work out $\mathbf{G}_2^2$ and $\mathbf{G}_3^2$.

An important observation is that these **G** matrices are independent of the shape of the triangle and so are called universal matrices. Although the matrix $\mathbf{D}_x^n$ does depend on shape, according to Equation (6.109), it is derivable from the universal **G** matrices:

$$\mathbf{D}_x^n = \sum_{i=1}^{3} \frac{b_i}{\Delta} \mathbf{G}_i^n \tag{6.111a}$$

and similarly,

$$\mathbf{D}_y^n = \sum_{i=1}^{3} \frac{c_i}{\Delta} \mathbf{G}_i^n. \tag{6.111b}$$

Consequently, although we are dealing with higher-order elements, if we compute the **G** matrices once and store them, the finite element computation would require only the calculation of the coefficients b and c, as with first–order finite elements.

A further simplification arises from the symmetry of the triangles about the three vertices (Silvester, 1982). That is, referring to Figure 6.7, $\mathbf{G}_2^2$ may be obtained by rotating the triangle of Figure 6.7A so that vertex 2 occupies the location of vertex 1 as in Figure 6.7B, and mapping the values of $\mathbf{G}_1^2$. That is, we are saying that $\mathbf{G}_2^2$ will be the same as $\mathbf{G}_1$ if vertex 2 is replaced by vertex 1. From Figure 6.7A, B, the mapping for the three finite element nodes, corresponding to the three vertices, is:

$$R^1\begin{bmatrix}1 & 2 & 3\end{bmatrix} = \begin{bmatrix}2 & 3 & 1\end{bmatrix}. \tag{6.113}$$

For the second-order example, $\mathbf{G}_2^2$ has six columns corresponding to a second-order triangle and three rows corresponding to a first-order triangle. The nodal mapping R^2 for the second-order triangles is extracted from Figure 6.7C, D:

$$R^2\begin{bmatrix}1 & 2 & 3 & 4 & 5 & 6\end{bmatrix} = \begin{bmatrix}6 & 3 & 5 & 1 & 2 & 4\end{bmatrix}. \tag{6.114}$$

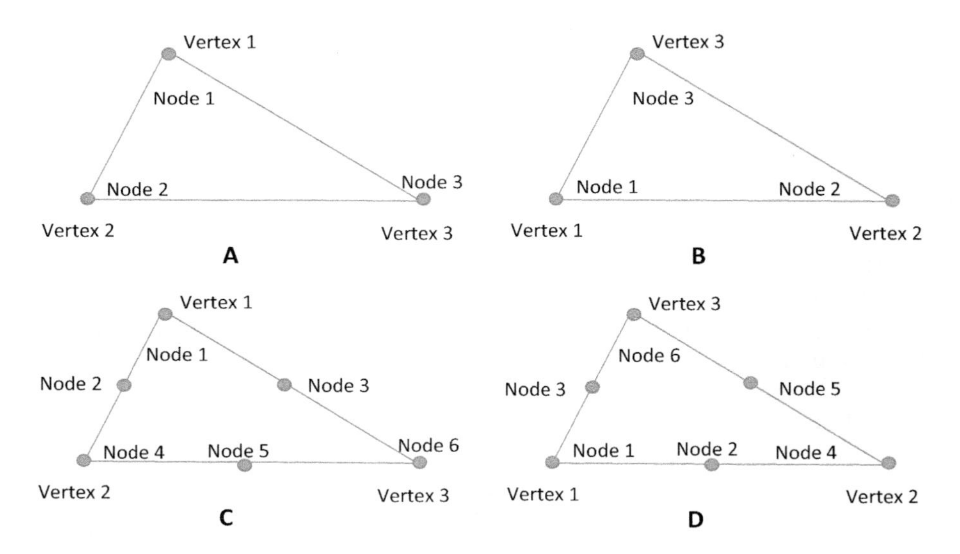

FIGURE 6.7
Rotation of triangles to note symmetry: A and B – A first-order triangle rotated; C and D – A second-order triangle rotated. (Nodes are interpolation points and vertices are simply the corners of a triangle).

The mapping for the **G** matrices, therefore, is in general

$$\mathbf{G}_2^n(i,j)=\mathbf{G}_1^n\left[R^{n-1}(i),R^n(j)\right]. \tag{6.115}$$

And similarly,

$$\mathbf{G}_3^n(i,j)=\mathbf{G}_2^n\left[R^{n-1}(i),R^n(j)\right]=\mathbf{G}_1^n\left[R^{n-1}R^{n-1}(i),R^nR^n(j)\right], \tag{6.116}$$

So that the application of the same rotation defines $\mathbf{G}_3$. For example,

$$\mathbf{G}_2^2(1,2)=\mathbf{G}_1^2\left[R^1(1),R^2(2)\right]=\mathbf{G}_1^2(3,2)=4, \tag{6.117}$$

$$\mathbf{G}_3^2(1,2)=\mathbf{G}_1^2\left[R^1R^1(1),R^2R^2(2)\right]=\mathbf{G}_1^2\left[R^1(3),R^2(2)\right]=\mathbf{G}_1^2(2,3)=0. \tag{6.118}$$

Building thus, we get

$$\mathbf{G}_2^2 = \begin{bmatrix} \mathbf{G}_1^2[R^1(1),R^2(1)] & \mathbf{G}_1^2[R^1(1),R^2(2)] & \mathbf{G}_1^2[R^1(1),R^2(3)] \\ \mathbf{G}_1^2[R^1(2),R^2(1)] & \mathbf{G}_1^2[R^1(2),R^2(2)] & \mathbf{G}_1^2[R^1(2),R^2(3)] \\ \mathbf{G}_1^2[R^1(3),R^2(1)] & \mathbf{G}_1^2[R^1(3),R^2(2)] & \mathbf{G}_1^2[R^1(3),R^2(3)] \end{bmatrix}$$

$$\begin{matrix} \mathbf{G}_1^2[R^1(1),R^2(4)] & \mathbf{G}_1^2[R^1(1),R^2(5)] & \mathbf{G}_1^2[R^1(1),R^2(6)] \\ \mathbf{G}_1^2[R^1(2),R^2(4)] & \mathbf{G}_1^2[R^1(2),R^2(5)] & \mathbf{G}_1^2[R^1(2),R^2(6)] \\ \mathbf{G}_1^2[R^1(3),R^2(4)] & \mathbf{G}_1^2[R^1(3),R^2(5)] & \mathbf{G}_1^2[R^1(3),R^2(6)] \end{matrix} \tag{6.119}$$

$$= \begin{bmatrix} \mathbf{G}_1^2(3,6) & \mathbf{G}_1^2(3,3) & \mathbf{G}_1^2(3,5) & \mathbf{G}_1^2(3,1) & \mathbf{G}_1^2(3,2) & \mathbf{G}_1^2(3,4) \\ \mathbf{G}_1^2(1,6) & \mathbf{G}_1^2(1,3) & \mathbf{G}_1^2(1,5) & \mathbf{G}_1^2(1,1) & \mathbf{G}_1^2(1,2) & \mathbf{G}_1^2(1,4) \\ \mathbf{G}_1^2(2,6) & \mathbf{G}_1^2(2,3) & \mathbf{G}_1^2(2,5) & \mathbf{G}_1^2(2,1) & \mathbf{G}_1^2(2,2) & \mathbf{G}_1^2(2,4) \end{bmatrix}$$

$$= \begin{bmatrix} 0 & 4 & 0 & -1 & 0 & 0 \\ 0 & 0 & 0 & 3 & 0 & 0 \\ 0 & 0 & 0 & -1 & 4 & 0 \end{bmatrix},$$

and similarly:

$$\mathbf{G}_3^2 = \begin{bmatrix} 0 & 0 & 4 & 0 & 0 & -1 \\ 0 & 0 & 0 & 0 & 4 & -1 \\ 0 & 0 & 0 & 0 & 0 & 3 \end{bmatrix}. \tag{6.120}$$

By working out $\mathbf{G}_2^2$ and $\mathbf{G}_3^2$ just the way $\mathbf{G}_1^2$ was in Equation (6.110), the correctness of the above rotations may be verified. What this means is that it is sufficient to store only one of the **G** matrices for each order of element, and the other two may be computed provided the rotations are also stored.

At this point we shall define another universal matrix, which we shall need in the next section. This is called the *metric tensor* and is defined by:

$$\mathbf{T}^{m,n} = A^{-1} \iint \tilde{\boldsymbol{\alpha}}^{nt} \tilde{\boldsymbol{\alpha}}^{n} \, dR, \tag{6.121}$$

where A is the area. Alternatively,

$$\mathbf{T}^{m,n}[i,j] = A^{-1} \iint \alpha_i \alpha_j \, dR, \tag{6.122}$$

Order 1

$$R^1\,[1 \;\; 2 \;\; 3] = [3 \;\; 1 \;\; 2]$$

$$G_1^{\,1} = [1 \;\; 0 \;\; 0]$$

$$T^{1,1} = \left(\frac{1}{12}\right) \begin{bmatrix} 2 & 1 & 1 \\ 1 & 2 & 1 \\ 1 & 1 & 2 \end{bmatrix}$$

Order 2

$$R^2 = [1 \;\; 2 \;\; 3 \;\; 4 \;\; 5 \;\; 6] = [6 \;\; 3 \;\; 5 \;\; 1 \;\; 2 \;\; 4]$$

$$G_1^2 = \begin{bmatrix} 3 & 0 & 0 & 0 & 0 & 0 \\ -1 & 4 & 0 & 0 & 0 & 0 \\ -1 & 0 & 4 & 0 & 0 & 0 \end{bmatrix}$$

$$T^{2,2} = \left(\frac{1}{180}\right) \begin{bmatrix} 6 & 0 & 0 & -1 & -4 & -1 \\ 0 & 32 & 16 & 0 & 16 & -4 \\ 0 & 16 & 32 & -4 & 16 & 0 \\ -1 & 0 & -4 & 6 & 0 & -1 \\ -4 & 16 & 16 & 0 & 32 & 0 \\ -1 & -4 & 0 & -1 & 0 & 6 \end{bmatrix}$$

Order 3

$$R^3[1 \;\; 2 \;\; 3 \;\; 4 \;\; 5 \;\; 6 \;\; 7 \;\; 8 \;\; 9 \;\; 10] = [10 \;\; 6 \;\; 9 \;\; 3 \;\; 5 \;\; 8 \;\; 1 \;\; 2 \;\; 4 \;\; 7]$$

$$G_1^3 = \left(\frac{1}{8}\right) \begin{bmatrix} 44 & 0 & 0 & 0 & 0 & 0 & 0 & 0 & 0 & 0 \\ -1 & 36 & 0 & 9 & 0 & 0 & 0 & 0 & 0 & 0 \\ -1 & 0 & 36 & 0 & 9 & 0 & 0 & 0 & 0 & 0 \\ 8 & -36 & 0 & 72 & 0 & 0 & 0 & 0 & 0 & 0 \\ 8 & -18 & -18 & 9 & 54 & 9 & 0 & 0 & 0 & 0 \\ 8 & 0 & -36 & 0 & 0 & 72 & 0 & 0 & 0 & 0 \end{bmatrix}$$

$$T^{3,3} = \left(\frac{1}{6720}\right) \begin{bmatrix} 76 & 18 & 18 & 0 & 36 & 0 & 11 & 27 & 27 & 11 \\ 18 & 540 & -135 & 162 & -189 & 27 & -54 & -135 & -54 & 27 \\ 18 & -135 & 540 & -135 & 162 & -189 & 27 & -54 & -135 & 0 \\ 0 & 162 & -135 & 540 & 162 & -54 & 18 & 270 & -135 & 27 \\ 36 & -189 & 162 & 162 & 1944 & 162 & 36 & 162 & 162 & 36 \\ 0 & -135 & -189 & -54 & 162 & 540 & 27 & -135 & 270 & 18 \\ 11 & 0 & 27 & 18 & 36 & 27 & 76 & 18 & 0 & 11 \\ 27 & -135 & -54 & 270 & 162 & -135 & 18 & 540 & -189 & 0 \\ 27 & -54 & -135 & -135 & 162 & 270 & 0 & -189 & 540 & 18 \\ 11 & 27 & 0 & 27 & 36 & 18 & 11 & 0 & 18 & 76 \end{bmatrix}$$

FIGURE 6.8
Universal matrices.

In evaluating these terms it is useful to note the two-dimensional equivalent of Equation (6.80) in both of which the terms 1! and 2! are deliberately placed to show that they come from the spatial dimension of the system:

$$\iint\limits_{\text{Triangle}} \zeta_1^i \zeta_2^j \zeta_3^k \, dR = \frac{i!\,j!\,k!\,2!}{(i+j+k+2)!} A \tag{6.123}$$

so that the matrix T is seen to be symmetric. Thus,

$$\mathbf{T}^{1,1} = \frac{1}{A}\iint_{\text{Triangle}} \begin{bmatrix} \zeta_1 \\ \zeta_2 \\ \zeta_3 \end{bmatrix} \begin{bmatrix} \zeta_1 & \zeta_2 & \zeta_3 \end{bmatrix} dR$$

$$= \frac{1}{A}\iint_{\text{Triangle}} \begin{bmatrix} \zeta_1^2 & \zeta_1\zeta_2 & \zeta_1\zeta_3 \\ \zeta_2\zeta_1 & \zeta_2^2 & \zeta_2\zeta_3 \\ \zeta_3\zeta_1 & \zeta_3\zeta_2 & \zeta_3^2 \end{bmatrix} dR \tag{6.124}$$

$$= \frac{1}{12}\begin{bmatrix} 2 & 1 & 1 \\ 1 & 2 & 1 \\ 1 & 1 & 2 \end{bmatrix}$$

using Equation (6.123) for integration. It is seen that dividing the integral by the area again makes the matrix independent of shape, and therefore it is also deservingly given the adjective *universal* since they apply to all triangles regardless of shape. These matrices also may be pre-computed and stored for all orders of interpolation, as may be required using the integral relationship of Equation (6.123). The rotations and **G** and **T** matrices are presented in Figure 6.8 for orders up to three and may easily be verified by the reader. For higher orders, these have already been evaluated and presented by Silvester (1978) and may be obtained from his classic paper. These matrices having been already computed by other workers, we are spared the burden of computing them and may use their results. However, in using their results, it is important to understand the principles that underlie them.

6.6 Functional Minimization

Having defined the universal matrices, we are now ready to return to the functional of Equation (6.108) and employ these matrices:

$$\begin{aligned}
\mathcal{L}[\varphi] &= \sum_{\Delta}\iint\left\{\frac{1}{2}\varepsilon\left[\nabla\tilde{\boldsymbol{\alpha}}^{n}\underset{\sim}{\boldsymbol{\Phi}}\right]\cdot\left[\nabla\tilde{\boldsymbol{\alpha}}^{n}\underset{\sim}{\boldsymbol{\Phi}}\right]-\left[\tilde{\boldsymbol{\alpha}}^{n}\underset{\sim}{\boldsymbol{\Phi}}\right]\left[\tilde{\boldsymbol{\alpha}}^{n}\underset{\sim}{\boldsymbol{\rho}}\right]\right\}dR \\
&= \sum_{\Delta}\iint\left\{\frac{1}{2}\varepsilon\left(\frac{\partial\tilde{\boldsymbol{\alpha}}^{n}\underset{\sim}{\boldsymbol{\Phi}}}{\partial x}\right)^{2}+\left(\frac{\partial\tilde{\boldsymbol{\alpha}}^{n}\underset{\sim}{\boldsymbol{\Phi}}}{\partial y}\right)^{2}-\left[\tilde{\boldsymbol{\alpha}}^{n}\underset{\sim}{\boldsymbol{\Phi}}\right]\left[\tilde{\boldsymbol{\alpha}}^{n}\underset{\sim}{\boldsymbol{\rho}}\right]\right\}dR \\
&= \sum_{\Delta}\iint\left\{\frac{1}{2}\varepsilon\left[\left(\tilde{\boldsymbol{\alpha}}^{n-1}\mathbf{D}_{x}\underset{\sim}{\boldsymbol{\Phi}}\right)^{2}+\left(\tilde{\boldsymbol{\alpha}}^{n-1}\mathbf{D}_{y}\underset{\sim}{\boldsymbol{\Phi}}\right)^{2}\right]-\left[\tilde{\boldsymbol{\alpha}}^{n}\underset{\sim}{\boldsymbol{\Phi}}\right]\left[\tilde{\boldsymbol{\alpha}}^{n}\underset{\sim}{\boldsymbol{\rho}}\right]\right\}dR \\
&= \sum_{\Delta}\left\{\frac{1}{2}\varepsilon\left[\underset{\sim}{\boldsymbol{\Phi}}^{t}\mathbf{D}_{x}^{t}\iint\tilde{\boldsymbol{\alpha}}^{n-1t}\tilde{\boldsymbol{\alpha}}^{n-1}dR\,\mathbf{D}_{x}\underset{\sim}{\boldsymbol{\Phi}}+\underset{\sim}{\boldsymbol{\Phi}}^{t}\mathbf{D}_{y}^{t}\iint\tilde{\boldsymbol{\alpha}}^{n-1t}\tilde{\boldsymbol{\alpha}}^{n-1}dR\,\mathbf{D}_{y}\underset{\sim}{\boldsymbol{\Phi}}\right]\right. \\
&\qquad\left.-\underset{\sim}{\boldsymbol{\Phi}}^{t}\iint\tilde{\boldsymbol{\alpha}}^{nt}\tilde{\boldsymbol{\alpha}}^{n}\,dR\right\}\underset{\sim}{\boldsymbol{\rho}} \\
&= \sum_{\Delta}\left\{\frac{1}{2}\varepsilon A\underset{\sim}{\boldsymbol{\Phi}}^{t}\left[\mathbf{D}_{x}^{t}\mathbf{T}^{n-1,n-1}\mathbf{D}_{x}+\mathbf{D}_{y}^{t}\mathbf{T}^{n-1,n-1}\mathbf{D}_{y}\right]\underset{\sim}{\boldsymbol{\Phi}}-\underset{\sim}{\boldsymbol{\Phi}}^{t}A\mathbf{T}^{n,n}\underset{\sim}{\boldsymbol{\rho}}\right\}.
\end{aligned}\tag{6.125}$$

Minimizing now according to the rules enunciated in Equations (2.36) and (2.40):

$$\frac{\partial\mathcal{L}}{\partial\underset{\sim}{\boldsymbol{\Phi}}}=\sum_{\Delta}\left\{\varepsilon A\left[\mathbf{D}_{x}^{t}\mathbf{T}^{n-1,n-1}\mathbf{D}_{x}+\mathbf{D}_{y}^{t}\mathbf{T}^{n-1,n-1}\mathbf{D}_{y}\right]\underset{\sim}{\boldsymbol{\Phi}}-A\mathbf{T}^{n,n}\underset{\sim}{\boldsymbol{\rho}}\right\}=0. \tag{6.126}$$

Thus, the local right-hand-side element vector is

$$\underset{\sim}{\mathbf{q}}=A\mathbf{T}^{n,n}\underset{\sim}{\boldsymbol{\rho}}, \tag{6.127}$$

and the local matrix is

$$\mathbf{P}=\varepsilon A\left[\mathbf{D}_{x}^{t}\mathbf{T}^{n-1,n-1}\mathbf{D}_{x}+\mathbf{D}_{y}^{t}\mathbf{T}^{n-1,n-1}\mathbf{D}_{y}\right]. \tag{6.128}$$

In Algorithm 2.2 we have a scheme for forming the local matrices for first-order triangles. It uses the subroutine triangle of Algorithm 2.1 to compute the b and c terms and thence forms the local finite element matrix **P** and right-hand-side vector **q**. Here, using Equation (6.124), we extend Algorithm 2.2 to form the local matrix for an n-th-order triangle, as shown in Algorithm 6.2. Algorithm 6.1 first forms the differentiation matrices based on Equation (6.112) and then uses the differentiation matrices to form the local matrices. The rules for adding the local matrix contributions to form a global equation have already been described in Section 3.9. The difference here is that the local matrices of size $N \times N$ may be larger. For second order, for example, the local matrix becomes 6×6, for third order, 10×10, and so on.

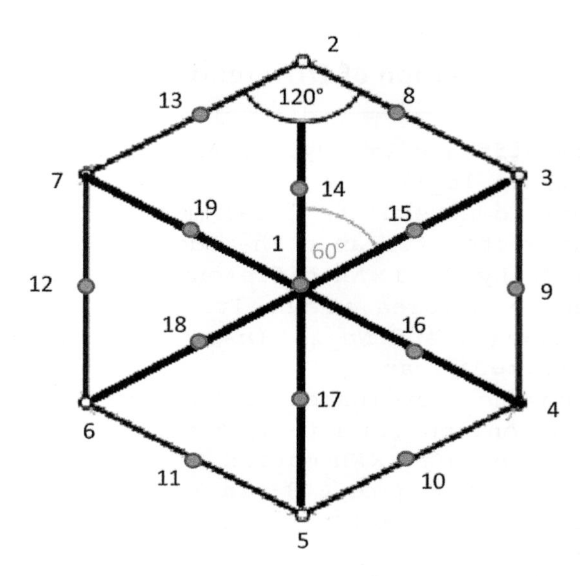

FIGURE 6.9
Six equilateral triangles connecting Node 1 to 18 others.

The only modification required to the algorithm is in redefining the size of the local matrix from 3 to its present size.

For a given set of points connected as, say, second-order triangles, the accuracy will be greater than if they had been connected as first-order triangles. However, the solution of the matrix equation will require more work and storage for the higher-order matrix, because the sparsity of the mesh has been reduced although the matrix size remains the same. For example, as seen in Figure 5.3, if we have six first-order equilateral triangles meeting at a vertex, the vertex node will appear with the six surrounding nodes if it is a first-order mesh. However, if it is a second-order mesh as in Figure 6.9, the vertex node will appear with 18 other nodes so that the row of the matrix equation corresponding to that node will have 18 off-diagonal terms instead of six. Nodes of the second-order mesh such as node 2, which lie at the middle of element edges, will result only in eight off-diagonal terms.

Algorithms 6.1 and 6.2 show how these universal matrices are used to compute the differentiation matrices that will go to form the Dirichlet matrix, [P], the local matrix for a triangle.

Algorithm 6.1: Formation of Differentiation Matrices

```
Procedure DiffMats(Dx, Dy, A, Nn, Nn_1, x, y, Rn,
Rn_1, Gn1, Tn_1n_1)
{Function: Computers the local matrix P and right-
hand-side vector for an n th-order triangle
Outputs: Dx,Dy=Nn_1×Nn differentiation matrices of
the triangle; A=area of the Triangle
Inputs: Nn=N(n) and Nn_1=N(n-1), defined in Equation
(6.95); x, y=3×1 arrays
containing the coordinates of the three vertices; Rn,
Rn_1=Rotations on triangles of order n
and n - 1; Gn1=Nn_1×Nn matrix G1^n - Equations (6.110)
and (6.118) and Figure 6.8; Tn_1n_1=N(n-1)×N(n-1)
metric tensor - Equations (6.120) and (6.121) and
Figure 6.8
Required: Procedure Triangle of Algorithm 2.1}
Begin
   Triangle(b, c, A, x, y)
       {c is already divided by Δ as it comes from
       Triangle}
   For i ← 1 To Nn_1 Do
     For j ← To Nn Do
       i1 ← Rn_1[i]
       i2 ← Rn_1[i1]
       j1 ← Rn[j]
       j2 ← Rn[j1]
       Dx[i, j] ← b[1]*Gn1[i, j]+b[2]*Gn1[i1,
       j1]+b[3]*Gn1[i2, j2]
       Dy[i, j] ← c[1]*Gn1[i, j]+c[2]*Gn1[i1,
       j1]+c[3]*Gn1[i2, j2]    { Equation (6.111a,b)}
End
```

Algorithm 6.2: Formation of Local High-Order Matrices

```
  Procedure LocalMatrix(P, q, Nn, Nn_1, x, y, Eps, Rho,
  Rn, Rn_1, Gn1, Tnn, Tn_1n_1)
  {Function: computes the local matrix P and right-
  hand-side vector for an nth-order triangle
  Outputs: P=Nn×Nn local matrix; q=Nn×1 right-hand-
  side local vector
  Inputs: Nn=N(n) and Nn_1=N(n-1), defined in Equation
  (6.95); x, y=3×1 arrays
  containing the coordinates of the three vertices;
  Eps=material value ε in triangle; Rho=   Nn×1 vector
  containing ρ at the interpolation nodes; Rn,
  Rn_1=rotations on triangles of
  order n and n - 1; Gn1=Nn_1×Nn matrix G1^n - Equation
  (6.109); Tnn, Tn_1n_1=Nn×Nn
  and N(n-1)×N(n-1) metric tensor of Figure 6.8.
  Required: Procedure Diffmats of Algorithm 6.1}
Begin
    DiffMats(Dx, Dy, A, Nn, Nn_1, x, y, Rn, Rn_1, Gn1,
    Tn_1n_1)
    For i ← 1 To Nn Do
      q[i] ← 0.0
      For j ← To Nn Do
        q[i] ← q[i]+Tnn[i, j]*Rho[j] {Equation (6.16)}
        P[i, j] ← 0.0
        For k ← 1 To Nn_1 Do
          For m ← 1 To Nn_1 Do
            P[i, j] ← P[i, j]+Dx[k, i]*Tn_1n_1[k,
            m]*Dx[m, j]
            + Dy[k, i]*Tn_1n_1[k, m]*Dy[m, j]
                                       {Equation (6.128)}
        P[i, j] ← A*Eps*P[i, j]
      q[i] ← A*q[i]
  End
```

6.7 Numerical Integration: Quadrature Formulae

Several physical quantities, such as energy and force, are expressed as integrals. Also, when the function being integrated is complex, it is convenient to perform the integration numerically. If we wish to integrate a function f over a region R, we divide the region into triangular finite elements and perform the integrations over all the triangles and sum the results.

To this end, let us approximate the function f as a polynomial of order n and evaluate f at the interpolation nodes numbering N, as defined in Equation (6.97):

$$f = \tilde{\boldsymbol{\alpha}}^n \underset{\sim}{\mathbf{f}} = \sum_{i=1}^{N} \alpha_i f_1. \tag{6.129}$$

Observe that by the way the interpolation functions α_i are defined (1 at node i and 0 at all other nodes), the approximation will match itself at all the interpolation nodes. The more nodes we have, the more accurate this will be. Now performing the integral,

$$\iint f\, dR = \iint \Sigma \alpha_i f_i = \Sigma f_i \left\{ \iint \alpha_i\, dR \right\} = A^r \Sigma \omega_i f_i. \tag{6.130}$$

where A^r is the area of the triangle and ω_i, called a quadrature weight at node i, is defined by

$$\omega_i = \frac{1}{A^r} \iint \alpha_i\, dR \tag{6.131}$$

and needs to be evaluated only once and stored for use anytime using Equation (6.130), to be recalled at will. For example, for first-order interpolation, as seen from Equation (6.101), α_1, α_2 and α_3 are, respectively, ζ_1, ζ_2 and ζ_3, so that, from Equation (6.123):

$$\omega_1 = \frac{1}{A^r} \iint \zeta_1\, dR = \frac{2!}{3!} = \frac{1}{3} \tag{6.132}$$

Similarly,

$$\omega_2 = \omega_3 = \frac{1}{3}, \tag{6.133}$$

and we obtain

$$\iint f\, dR = A^r \left[\frac{1}{3} f_1 + \frac{1}{3} f_2 + \frac{1}{3} f_3 \right]. \tag{6.134}$$

If we integrate $f=1$ over a triangle, f_1, f_1 and f_3 are all 1 and we get the area of the triangle. Similarly, if we integrate x, the result is the area times the value of x at the centroid, as expected. In both these instances we get exact answers because 1 and x are exactly fitted by a first-order fit Equation (2.42) and no gain in accuracy will be realized by using higher orders. However, as more commonly we integrate general functions not necessarily fitted properly by low-order polynomials, the higher the order we choose in Equation (6.9), the better.

The quadrature weights for various orders have been already worked out and tabulated by Silvester (1970). We may conveniently use his results.

An example of using the integral formulas is in error computation. If we knew the exact solution φ and we wished to find the error in the finite element solution φ^f, let us define the error function as $e = \varphi - \varphi^f$, computable at every node, and the total percentage error

$$\text{Error} = 100 \times \sqrt{\frac{\iint e^2\, dR}{\iint \varphi^2\, dR}}. \tag{6.135}$$

These terms may be easily evaluated. For example,

$$\begin{aligned}\iint \varphi^2\, dR &= \sum_{\Delta} \iint \left[\left(\tilde{\boldsymbol{\alpha}}^n \underset{\sim}{\boldsymbol{\Phi}}\right)\left(\tilde{\boldsymbol{\alpha}}^n \underset{\sim}{\boldsymbol{\Phi}}\right) dR = \sum_{\Delta} \underset{\sim}{\boldsymbol{\Phi}}^t \left[\iint \tilde{\boldsymbol{\alpha}}^{nt} \tilde{\boldsymbol{\alpha}}^n dR \right] \underset{\sim}{\boldsymbol{\Phi}}\, dR \right.\\ &= \sum_{\Delta} A^r \underset{\sim}{\boldsymbol{\Phi}}^t \mathbf{T}^{n,n} \underset{\sim}{\boldsymbol{\Phi}}\end{aligned} \tag{6.136}$$

where the integral over the domain has been split over the elements and Equation (6.122) has been used to express it in terms of the pre-computed universal $\mathbf{T}$ matrix. Thus, the quantity φ must be evaluated at all the interpolation nodes. For first order, for example, $\mathbf{T}^{1,1}$ is defined in Equation (6.121), and if φ is φ_1, φ_2, and φ_3 at the three vertices, over the element of area A^r, Equation (6.136) will work out to $A^r[\varphi_1^2 + \varphi_2^2 + \varphi_3^2 + \varphi_1\varphi_2 + \varphi_2\varphi_3 + \varphi_3\varphi_1]/12$. An application of this follows in the next section.

6.8 Finite Elements and Finite Differences

It may be shown in a rigidly mathematical sense that the finite-difference method is a part of the generalized finite element method in the sense of the Galerkin approach (Zienkiewicz, 1977). The same may be seen, albeit with less generality, from a simple square finite element mesh (Zienkiewicz and

Cheung, 1965). The more interested reader is referred to these classic papers or the more accessible works such as Hoole (1988).

6.9 Sparsity Pattern Computation

By now, particularly after the previous examples of Section 8.4 where we solved for various matrix sizes, it must have been impressed upon the reader that full storage cannot be used indefinitely with impunity. In the solution of practical problems, large matrices are encountered and the sparsity and symmetry of the equations must be exploited, as detailed in Section 6.4. We saw that the coefficients should be ideally stored in a one-dimensional array and that they may be retrieved using the procedure Locate of Algorithm 6.4.2, employing the data structures **Diag** and **Col**. Alternatively, with profile storage, we will replace the vector **Col** by **FC**. In this section we will deal exclusively with sparse symmetric storage, because the principles extend trivially to profile storage.

To effect sparse storage we must be in possession of the arrays **Diag** and **Col**. While these may be manually created by inspection of the mesh, such an approach is neither economic nor feasible for large problems. Much is to be gained by using the computer to create them automatically. In finite element analysis, we may easily devise algorithms to create the relevant arrays by inspection of the mesh, as shown in Algorithms 6.3 and 6.4. In these algorithms we initially assume that the matrix is diagonal. So, for an $n \times n$ matrix we take **Diag**[*i*] as *i* and **Col**[*i*] as *i*, since only the diagonal is occupied. NElems is *n*, since there are *n* coefficients, one on each row. It is time for us now to allocate storage for all off-diagonal elements,

Algorithm 6.3: Procedures Required in Creating Sparse Data Structures

```
Procedure Modify Diag(Diag, RowN, NUnk)
{Function: Modifies Diag, when an element is added to
row number RowN}
Begin
   For i ← RowN To NUnk Do
     Diag[i] ← Diag[i] +1
End

Procedure ModifyCol(Diag, ColN, ij, NElems)
{Function: Modifies Col when an element is added to
location RowN, ColN after the item ij
```

```
in the one-dimensionally stored matrix. Observe that
when Locate returns Found as False, ij
is the location of the previous item before the
missing one}
Begin
   For i ← ij+1 To NElems Do
     j ← ij+1+NElems - i {j goes from NElems down to
     ij+1}
     Col[j+1] ← Col[j]
   Col[ij+1] ← ColN
   NElems ← NElems+1
End
```

Algorithm 6.4: Creating Sparsity Data Structures

```
Procedure Sparsity(Diag, Col, NElems, TriaNum, NUnk, N)
{Function: Creates the sparsity data Diag, Col,
NElems for a finite element mesh.
Input: Number of triangles TriaNum; Number of
Unknowns NUnk, and the number of
interpolation nodes N in the nth-order triangle.
Required: Procedures Locate of Alg. 6.4.2 and
ModifyDiag and ModifyCol above}
Begin
   For i ← 1 To NUnk Do
     Diag[i] ← i
     Col[i] ← i
   NElems ← NUnk
   For i ← 1 To TriaNum Do
     Get v[1], v[2], v[3], . . ., v[N], the N vertices
     of nth-order triangle i
     For j ← 1 To N - 1 Do
      For k ← j+1 To N Do
        If (v[j] ≤ NUnk) And (v[k] ≤ NUnk)
         Then
           If v[j] ≥ v[k]
            Then
              RowN ← v[j]
              ColN ← v[k]
            Else
              RowN ← v[k]
              ColN ← v[j]
```

```
        Locate(ij, Found, Diag, Col, NElems, RowN,
        ColN)
        If Not Found
          Then
           ModifyDiag(Diag, RowN, NUnk)
           ModifyCol(Diag, ColN, ij, NElems)
End
```

using the fact that if two unknown nodes p and q are connected in the mesh (that is, if they are vertices in the same triangle), then location (p, q) will be occupied in the matrix so that storage must be assigned to it.

To this end, we read each triangle in Algorithm 6.4 and, for each unknown vertex, find out using Locate (after ensuring that we are indeed in the lower half of the matrix) if space has been assigned for the column ColN and row RowN given by the vertices (if they are both unknown). If no space has been assigned, as indicated by Found from Locate being returned False, a study of Algorithm 6.4.2 will show that for this case the value ij returned by Locate contains the item after which location (RowN, ColN) should be, but is missing. If we assign space for (RowN, ColN), row RowN, and all rows after, it will terminate one item later, so that **Diag** must be incremented by 1 as in the procedure ModifyDia of Algrithm 6.3. Similarly, **Col** should contain an extra item after item ij returned by Locate. This item reflects column position ColN. Thus, all items after position ij are shifted down an item and the vacancy created at position $ij+1$ is assigned the value ColN in the procedure ModifyCol of Algorithm 6.3. Once the sparsity data are thus automatically created, it will be found that much larger problems may be solved and solved quickly.

6.10 Nonlinear Equations

So far we have been dealing with linear equations, that is equations where the unknown φ appears in order 1; that is raised to the power 1. However, there are many engineering systems where the unknown φ is not of order 1. In such situations we derive a linear equation for the error in φ and solve repeatedly for the error until it vanishes. Consider the solution of the nonlinear equation which is common in solving transistor problems where the charge is a function of the unknown potential itself.

$$-\epsilon\nabla^2\varphi = \rho + \kappa\varphi^\beta \tag{6.137}$$

In linearizing the equation for the small error e in an approximate solution ψ, we have the exact solution $\varphi = \psi + e$. Putting this into the equation, binomially expanding

$$\left(\psi + e\right)^{\beta} = {}^{n}C_0\psi^{\beta} + {}^{n}C_1\psi^{\beta-1}e^1 + + {}^{n}C_2\psi^{\beta-2}e^2 + {}^{n}C_3\psi^{\beta-3}e^3 + \cdots \tag{6.138}$$

and neglecting e^2 and higher order terms, we obtain the linear equation in the error e:

$$-\epsilon\nabla^2\left(\psi + e\right) = \rho + \kappa\left(\psi + e\right)^{\beta} \approx \rho + \kappa\left(\psi^{\beta} + \beta\psi^{\beta-1}e\right) \tag{6.139}$$

Rearranging, we obtain an equation which is like a mixture between the wave and Poisson equations.

$$-\epsilon\nabla^2 e - \kappa\beta\psi^{\beta-1}e = \epsilon\nabla^2\psi + \rho + \kappa\psi^{\beta} \tag{6.140}$$

Happily, this is a Poisson equation for e which we know how to tackle.

6.11 Other Equations and Methods: The Structural Beam and the Bi-Harmonic Equation

We have been dealing mainly with the Poisson and wave equations so far. However, there are also other equations to which the finite element method applies. To demonstrate one such application, keeping the analysis simple, we will look at a structural beam (one-dimensional situation) and the situation where its two ends are embedded in the wall at both ends while it is structurally deformed by a load P(x) varying over the beam (see Figure 6.10). This would apply to a bridge with loads such as traveling vehicles atop it.

$$-\frac{d^2}{dx^2}\left[EI\frac{d^2y}{dx^2}\right] + P\left(x\right) = 0 \tag{6.141}$$

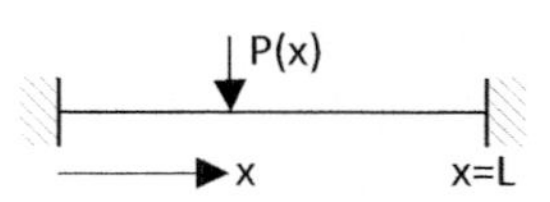

FIGURE 6.10
The beam under load.

P=Load; E=Young's Modulus; I = Area Moment of Inertia.

The cubic Hermite trial function is a good expected fit of the deformation – better with smaller elements:

$$y = a + b\left(\frac{x}{L}\right) + c\left(\frac{x}{L}\right)^2 + d\left(\frac{x}{L}\right)^3 \tag{6.142}$$

The four coefficients a to d may be related to four interpolatory nodes. Instead of four interpolatory points over the element, however, we define $g = dy/dx$ and settle for four end point variables y_1, g_1 at the left-node 1 and y_2, g_2 at the right-node 2. Thus,

$$g = \frac{dy}{dx} = \frac{b}{L} + \frac{2c}{L}\left(\frac{x}{L}\right) + \frac{3d}{L}\left(\frac{x}{L}\right)^2 \tag{6.143}$$

Proceeding as before, we need to replace a, b, c, and d with the interpolatory variables y_1, g_1, y_2, and g_2:

$$y_1 = y(0) = a \tag{6.144}$$

$$y_2 = y(L) = a + b + c = d \tag{6.145}$$

$$g_1 = g(0) = \frac{b}{L} \tag{6.146}$$

$$g_2 = g(L) = \frac{b}{L} + \frac{2c}{L} + \frac{3d}{L} \tag{6.147}$$

This yields the matrix equation

$$\begin{bmatrix} 1 & 0 & 0 & 0 \\ 0 & 1/L & 0 & 0 \\ 1 & 1 & 1 & 1 \\ 0 & 1/L & 2/L & 3/L \end{bmatrix} \begin{bmatrix} a \\ b \\ c \\ d \end{bmatrix} = \begin{bmatrix} y_1 \\ g_1 \\ y_2 \\ g_2 \end{bmatrix} \tag{6.148}$$

Or if

$$A = \begin{bmatrix} 1 & 0 & 0 & 0 \\ 0 & 1/L & 0 & 0 \\ 1 & 1 & 1 & 1 \\ 0 & 1/L & 2/L & 3/L \end{bmatrix} \tag{6.149}$$

Trying to rewrite a, b, c, and d in terms of the new variables just as we rewrote a, b, and c for a triangle in terms of the vertex potentials, the determinant of the matrix of coefficients is

$$\Delta = |A| = 1\left(\frac{1}{L}\right)\left(\frac{3}{L} - \frac{2}{L}\right) = \frac{1}{L^2} \tag{6.150}$$

with cofactors

$$C_{11} = \begin{vmatrix} 1/L & 0 & 0 \\ 1 & 1 & 1 \\ 1/L & 2/L & 3/L \end{vmatrix} = 1/L^2 \quad C_{12} = \begin{vmatrix} 0 & 0 & 0 \\ 1 & 1 & 1 \\ 0 & 2/L & 3/L \end{vmatrix} = 0$$

$$C_{13} = \begin{vmatrix} 0 & 1/L & 0 \\ 1 & 1 & 1 \\ 0 & 1/L & 3/L \end{vmatrix} = -3/L^2 \quad C_{14} = \begin{vmatrix} 0 & 1/L & 0 \\ 1 & 1 & 1 \\ 0 & 1/L & 2/L \end{vmatrix} = -2/L^2$$

$$C_{21} = \begin{vmatrix} 0 & 0 & 0 \\ 1 & 1 & 1 \\ 1/L & 2/L & 3/L \end{vmatrix} = 0 \quad C_{22} = \begin{vmatrix} 1 & 0 & 0 \\ 1 & 1 & 1 \\ 0 & 2/L & 3/L \end{vmatrix} = 1/L$$

$$C_{23} = \begin{vmatrix} 1 & 0 & 0 \\ 1 & 1 & 1 \\ 0 & 1/L & 3/L \end{vmatrix} = 2/L \quad C_{24} = \begin{vmatrix} 1 & 0 & 0 \\ 1 & 1 & 1 \\ 0 & 1/L & 2/L \end{vmatrix} = 1/L$$

$$C_{31} = 0 \quad C_{32} = \begin{vmatrix} 1 & 0 & 0 \\ 0 & 0 & 0 \\ 0 & 2/L & 3/L \end{vmatrix} = 0$$

$$C_{33} = \begin{vmatrix} 1 & 0 & 0 \\ 0 & 1/L & 0 \\ 0 & 1/L & 3/L \end{vmatrix} = 3/L^2 \quad C_{34} = \begin{vmatrix} 1 & 0 & 0 \\ 0 & 1/L & 0 \\ 0 & 1/L & 2/L \end{vmatrix} = 2/L^2$$

$$C_{41} = 0 \quad C_{42} = \begin{vmatrix} 1 & 0 & 0 \\ 0 & 0 & 0 \\ 1 & 1 & 1 \end{vmatrix} = 0 \tag{6.151}$$

$$C_{43} = \begin{vmatrix} 1 & 0 & 0 \\ 0 & 1/L & 0 \\ 1 & 1 & 1 \end{vmatrix} = 1/L \quad C_{44} = \begin{vmatrix} 1 & 0 & 0 \\ 1 & 1/L & 0 \\ 1 & 1 & 1 \end{vmatrix} = 1/L$$

So the inverse of A,

$$B = A^I = \frac{1}{\Delta}\begin{bmatrix} C_{11} & -C_{12} & C_{13} & -C_{14} \\ -C_{21} & C_{22} & -C_{23} & C_{24} \\ C_{31} & -C_{32} & C_{33} & -C_{34} \\ -C_{41} & C_{42} & -C_{43} & C_{44} \end{bmatrix}^t$$

$$= \frac{1}{\frac{1}{L^2}}\begin{bmatrix} 1/L^2 & -0 & -3/L^2 & 2/L^2 \\ -0 & 1/L & -2/L & 1/L \\ 0 & -0 & 3/L^2 & -2/L^2 \\ -0 & 0 & -1/L & 1/L \end{bmatrix}^t \tag{6.152}$$

$$= \begin{bmatrix} 1 & 0 & 0 & 0 \\ 0 & L & 0 & 0 \\ -3 & -2L & 3 & -L \\ 2 & L & -2 & L \end{bmatrix}$$

Alternatively, using MATLAB

```
clc;
clear all;
syms a b c d phi1 phi2 g1 g2 L
phiv= sym([phi1;g1;phi2;g2])
interp = sym([a;b;c;d])
A= sym([1,0,0,0;0, 1/L, 0, 0; 1,1,1,1;0,1/L,2/L,3/L])
Del = sym(det(A))
B=sym(inv(A))
A*B
```

We note that the symbolic algebra capabilities of MATLAB are being invoked as explained next, in Section 6.12, The output of this program verifies our derivation:

```
A =
[ 1,   0,   0,   0]
[ 0, 1/L,   0,   0]
[ 1,   1,   1,   1]
[ 0, 1/L, 2/L, 3/L]

Del =

1/L^2

B =

[ 1,   0, 0, 0]
[ 0,   L, 0, 0]
[ -3, -2*L, 3, -L]
[ 2,   L, -2, L]
```

Therefore, noting that $(x/L) = \zeta_2$

$$
\begin{aligned}
y &= a + b\zeta_2 + c\zeta_2^2 + d\zeta_2^3 = \begin{bmatrix} 1 & \zeta_2 & \zeta_2^2 & \zeta_2^3 \end{bmatrix} \begin{bmatrix} a \\ b \\ c \\ d \end{bmatrix} \\
&= \begin{bmatrix} 1 & \zeta_2 & \zeta_2^2 & \zeta_2^3 \end{bmatrix} \mathrm{B} \begin{bmatrix} y_1 \\ g_1 \\ y_2 \\ g_2 \end{bmatrix} \\
&= \begin{bmatrix} 1 & \zeta_2 & \zeta_2^2 & \zeta_2^3 \end{bmatrix} \begin{bmatrix} 1 & 0 & 0 & 0 \\ 0 & L & 0 & 0 \\ -3 & -2L & 3 & -L \\ 2 & L & -2 & L \end{bmatrix} \begin{bmatrix} y_1 \\ g_1 \\ y_2 \\ g_2 \end{bmatrix} \\
&= \begin{bmatrix} 1 - 3\zeta_2^2 + 2\zeta_2^3 & L\left(\zeta_2 - 2\zeta_2^2 + \zeta_2^3\right) & 3\zeta_2^2 - 2\zeta_2^3 & L\left(-\zeta_2^2 + \zeta_2^3\right) \end{bmatrix} \begin{bmatrix} y_1 \\ g_1 \\ y_2 \\ g_2 \end{bmatrix}
\end{aligned}
\tag{6.153}
$$

Now note that

$$1 - 3\zeta_2^2 + 2\zeta_2^3 = 1 - \zeta_2^2(3 - 2\zeta_2) = 1 - \zeta_2^2(1 + 2\zeta_1) = \beta_1 \text{ say} \tag{6.154a}$$

$$L\left(\zeta_2 - 2\zeta_2^2 + \zeta_2^3\right) = L\zeta_2\left(1 - 2\zeta_2 + \zeta_2^2\right) = L\zeta_2(1 - \zeta_2)^2 = L\zeta_2\zeta_1^2 = \beta_1 \text{ say} \tag{6.154b}$$

$$3\zeta_2^2 - 2\zeta_2^3 = \zeta_2^2(3 - 2\zeta_2) = \zeta_2^2(1 - 2\zeta_1) = \beta_1 \text{ say} \tag{6.154c}$$

$$L\left(-\zeta_2^2 + \zeta_2^3\right) = L\zeta_2^2(-1 + \zeta_2) = -L\zeta_2^2\zeta_1 = \beta_1 \text{ say} \tag{6.155}$$

Therefore,

$$y = \begin{bmatrix} \beta_1 & \beta_2 & \beta_3 & \beta_4 \end{bmatrix} \begin{bmatrix} y_1 \\ g_1 \\ y_2 \\ g_2 \end{bmatrix} = \tilde{\beta}\tilde{Q}^t \tag{6.156}$$

Similarly,

$$
\begin{aligned}
g &= \frac{dy}{dx} = \frac{d\tilde{\beta}}{dx}\tilde{Q}^t \\
&= \frac{1}{L}\left[-6\zeta_2(1-\zeta_2) \quad 1-4\zeta_2+3\zeta_2^2 \quad 6\zeta_2(1-\zeta_2) \quad (-2\zeta_2+\zeta_2^2)L\right]\tilde{Q}^t
\end{aligned}
\tag{6.157}
$$

$$
\begin{aligned}
\frac{dg}{dx} &= \frac{d^2\tilde{\beta}}{dx^2}\tilde{Q}^t \\
&= \frac{2}{L^2}\left[-3+6\zeta_2 \quad (-2+3\zeta_2)L \quad 3-6\zeta_2 \quad (-1+3\zeta_2)L\right]\tilde{Q}^t
\end{aligned}
\tag{6.158}
$$

Now for the solution by Galerkin's method, we weight the residuals. The residual operator

$$A(y) = -\frac{d^2}{dx^2}\left[EI\frac{d^2y}{dx^2}\right] + P(x) = 0 \tag{6.159}$$

But when we substitute the cubic Hermite trial function $y = \tilde{\beta}\tilde{Q}^t$, the residual may not be zero because the trial function need not necessarily be in the form of the actual solution:

$$A\left(\tilde{\beta}\tilde{Q}^t\right) = -\frac{d^2}{dx^2}\left[EI\frac{d^2\tilde{\beta}\tilde{Q}^t}{dx^2}\right] + P(x) \neq 0 \tag{6.160}$$

Galerkin's method minimizes this error in the "direction" of each basis function $\beta_i$of the trial space:

$$\int \beta_i A\left(\tilde{\beta}\tilde{Q}^t\right)dx = 0 \text{ for } i = 1,2,3, \text{ and } 4. \tag{6.161}$$

Expanding by integration by parts twice:

$$
\begin{aligned}
&\int \beta_i\left[-\frac{d^2}{dx^2}\left[EI\frac{d^2y}{dx^2}\right] + P(x)\right]dx \\
&= \int -\beta_i\left[-\frac{d}{dx}\left[\frac{d}{dx}EI\frac{d^2y}{dx^2}\right] + P(x)\right]dx \\
&= -\beta_i\frac{d}{dx}EI\frac{d^2y}{dx^2}\Bigg[{}^2_1 + \int\frac{d\beta_i}{dx}\cdot\left[\frac{d}{dx}EI\frac{d^2y}{dx^2}\right]dx + \int\beta_i P(x)dx \\
&= \int -\frac{d^2\beta_i}{dx^2}\cdot EI\frac{d^2y}{dx^2}dx + \int\beta_i P(x)dx
\end{aligned}
\tag{6.162}
$$

plus two boundary terms that correspond to natural boundary conditions.

Allowing for varying cross section for the beam, linearly from I_1 to I_2:

$$I = I_1\zeta_1 + I_2\zeta_2 = \begin{bmatrix} I_1 & I_2 \end{bmatrix}\begin{bmatrix} \zeta_1 \\ \zeta_2 \end{bmatrix} \tag{6.163}$$

Similarly, a linear load P

$$P = \begin{bmatrix} P_1 & P_2 \end{bmatrix}\begin{bmatrix} \zeta_1 \\ \zeta_2 \end{bmatrix} = \begin{bmatrix} \zeta_1 & \zeta_2 \end{bmatrix}\begin{bmatrix} P_1 \\ P_2 \end{bmatrix} \tag{6.164}$$

Therefore, the stiffness matrix equation follows by weighting the residual by each of the four values of β_i and stacking them on top of each other. Doing so, the left-hand side of the equation is

$$[K]\tilde{Q}^t = \int E\begin{bmatrix} I_1 & I_2 \end{bmatrix}\begin{bmatrix} \zeta_1 \\ \zeta_2 \end{bmatrix}\frac{2}{L^2}\begin{bmatrix} -3+6\zeta_2 \\ (-2+3\zeta_2)L \\ 3-6\zeta_2 \\ (-1+3\zeta_2)L \end{bmatrix}$$

$$\times\frac{2}{L^2}\begin{bmatrix} -3+6\zeta_2 & (-2+3\zeta_2)L & 3-6\zeta_2 & (-1+3\zeta_2)L \end{bmatrix}\tilde{Q}^t dx\tilde{Q}^t \tag{6.165}$$

$$= \frac{E\tilde{I}}{L^2}\begin{bmatrix} \begin{Bmatrix} 6 \\ 6 \end{Bmatrix} & \begin{Bmatrix} 4 \\ 2 \end{Bmatrix}L & \begin{Bmatrix} -6 \\ -6 \end{Bmatrix} & \begin{Bmatrix} 2 \\ 4 \end{Bmatrix}L \\ & \begin{Bmatrix} 3 \\ 1 \end{Bmatrix}L^2 & \begin{Bmatrix} -4 \\ -2 \end{Bmatrix}L & \begin{Bmatrix} 1 \\ 1 \end{Bmatrix}L^2 \\ & & \begin{Bmatrix} 6 \\ 6 \end{Bmatrix} & \begin{Bmatrix} -2 \\ -4 \end{Bmatrix}L \\ SYMM & & & \begin{Bmatrix} 1 \\ 3 \end{Bmatrix}L^2 \end{bmatrix}\tilde{Q}^t$$

and the right-hand side:

$$\int\begin{bmatrix} 1-\zeta_2^2(1+2\zeta_1) \\ L\zeta_2\zeta_1^2 \\ \zeta_2^2(1-2\zeta_1) \\ -L\zeta_2^2\zeta_1 \end{bmatrix}\begin{bmatrix} \zeta_1 & \zeta_2 \end{bmatrix}\begin{bmatrix} P_1 \\ P_2 \end{bmatrix}dx$$

$$= \begin{bmatrix} 18L/24 & 12L/24 \\ 3L^2/24 & 2L^2/24 \\ -2/24 & 0 \\ -2L^2/24 & 3L^2/24 \end{bmatrix}\begin{bmatrix} P_1 \\ P_2 \end{bmatrix} \tag{6.166}$$

6.12 Symbolic Algebra

Many do not realize that MATLAB is as capable of manipulating symbols as much as it can numbers. A simple example has been presented in the previous section 6.11.1 where for the cubic Hermite trial function is having its variables a, b, c, and d replaced by $\varphi_1, g_1, \varphi_2$ and g_2 are the values of φ and its gradient g the ends of the elements:

$$y = a + b\left(\frac{x}{L}\right) + c\left(\frac{x}{L}\right)^2 + d\left(\frac{x}{L}\right)^3 \tag{6.142}$$

Applying this to the two ends at =0 and x=L

$$\begin{bmatrix} 1 & 0 & 0 & 0 \\ 0 & 1/L & 0 & 0 \\ 1 & 1 & 1 & 1 \\ 0 & 1/L & 2/L & 3/L \end{bmatrix} \begin{bmatrix} a \\ b \\ c \\ d \end{bmatrix} = \begin{bmatrix} y_1 \\ g_1 \\ y_2 \\ g_2 \end{bmatrix} \tag{6.148}$$

So inverting the matrix of coefficients by hand, we define $a, b, c, d, \varphi_1, g_1, \varphi_2$ and g_2 and L as elements, and the right-hand side of Equation (6.148) as a symbolic vector B and the coefficient matrix of Equation (6.148) as a symbolic matrix A.

```
syms a b c d phi1 phi2 g1 g2 L

phiv= sym([phi1;g1;phi2;g2])
interp = sym([a;b;c;d])
A= sym([1,0,0,0;0, 1/L, 0, 0; 1,1,1,1;0,1/L,2/L,3/L])
```

Now it is a simple matter to invoke MATLAB as a symbolic exercise to invert A and find

$$\begin{bmatrix} a \\ b \\ c \\ d \end{bmatrix} = \begin{bmatrix} 1 & 0 & 0 & 0 \\ 0 & 1/L & 0 & 0 \\ 1 & 1 & 1 & 1 \\ 0 & 1/L & 2/L & 3/L \end{bmatrix}^{-1} \begin{bmatrix} y_1 \\ g_1 \\ y_2 \\ g_2 \end{bmatrix} \tag{6.167}$$

This is accomplished by the lines

```
Del = sym(det(A))
B=sym(inv(A))
A*B
```

```
A =
[ 1,   0,   0,   0]
[ 0, 1/L,   0,   0]
[ 1,   1,   1,   1]
[ 0, 1/L, 2/L, 3/L]

Del =

1/L^2

B =

[ 1,   0, 0, 0]
[ 0,   L, 0, 0]
[ -3, -2*L, 3, -L]
[ 2,   L, -2, L]
```

There are manuals listing the relevant MATLAB lines for handling symbols. These are very similar to operations with numbers and differ mainly in the declaratory statements of variable as to whether they are numbers or symbols.

Here is our summary of the key declarations and operations in Table 6.2.

6.13 Edge Elements

We have so far been using interpolations on scalars. However, this results in tangential continuity of gradients (the gradient of potential, for example, being the electric field) across elements whereas the normal component of gradient will not be continuous – because the normal component of gradient involves the third vertices of two neighboring triangles.

For example, as shown in Figure 6.11, at the point A on the edge 2-3 that is common to two neighboring triangles, the tangential component of the electric potential φ (which is related to the electric field) will depend only on the potential at nodes 2 and 3 since the homogeneous coordinate for the third vertex is zero along the edge. So the value will be the same whether computed from triangle 1-2-3 or triangle 2-4-3. The normal component at A on the other hand will depend on the potential at 1 if computed from triangle 1-2-3 and on the potential at 4 if computed from triangle 2-4-3. Therefore, the values will be different. However, according to the physics of electric fields, it is the tangential electric field that has to be discontinuous if the materials in the two triangles are different, and it is the normal flux density that needs to be the same (when there is no charge on edge 2-3). This creates non-physical solutions. In wave equation solutions, what we call spurious (non-existing) modes are shown to exist in a waveguide.

TABLE 6.2

Summary of Key Symbolic Algebra Declarations and Operations

Declaring a Symbolic Variable	`syms x`

Addition-Subtraction, Multiplication-Division	`>> syms w x y z`
	`>> (w + x - 2*y) / z`
	
	`ans =`
	
	`(w + x - 2*y)/z`

Derivative	`>> syms x y`
	`>> z = x^2 + y^3;`
	`>> diff(z, x) % Derivative of z with respect to x`
	
	`ans =`
	
	`2*x`

Integration	`>> syms x y`
	`>> z = x^2 + y^3;`
	`>> % Indefinite Integral of z with respect to x`
	`>> int(z, x)`
	`ans =`
	
	`x^3/3 + x*y^3`
	`>> % Definite integral of z with respect to y from 0 to 10`
	`>> int(z, y, 0, 10)`
	
	`ans =`
	
	`10*x^2 + 2500`

Roots of an Algebraic Equation	`>> solve(x^2 + 4*x + 4) % Solve for x`
	`ans =`
	
	`-2`
	`-2`
	
	`>> solve(x^2 + 3*x*y + y^2, y) % Solve for y with respect to x`
	`ans =`
	
	`- (3*x)/2 - (5^(1/2)*x)/2`
	`(5^(1/2)*x)/2 - (3*x)/2`

Solving a System of Equations	`>> syms x y z`
	`>> [x, y, z] = solve(4 * x - 3 * y + z == - 10, 2 * x + y + 3 * z == 0, - x + 2 * y - 5 * z == 17)`
	`x =`
	
	`1`
	
	`y =`
	
	`4`
	
	`z =`
	
	`-2`

(Continued)

TABLE 6.2 (CONTINUED)

Summary of Key Symbolic Algebra Declarations and Operations

Function Evaluation

```
>> syms x y
>> f = x^2*y + 5*x*y + y^2;
>> subs(f, x, 3) % Substitute into f, for variable x, the value 3
ans =

y^2 + 24*y
```

Plot a Function

```
>> syms x y z
>> y = 2*x + 3
>> figure;
>> fplot(y)
>> xlabel('x');
>> ylabel('y');
>> title('y = 2x + 3');
```

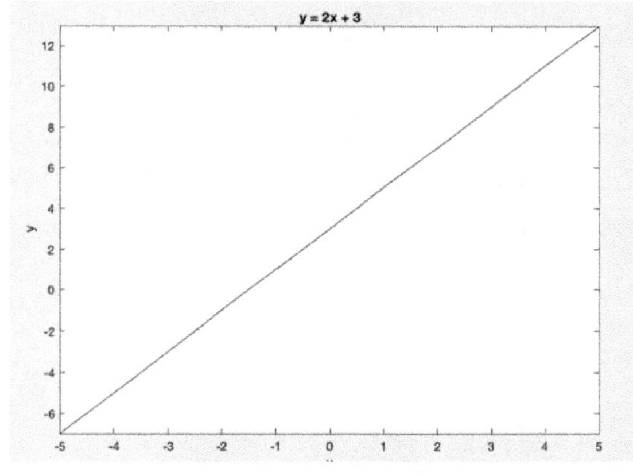


```
>> z = x + y
>> figure;
>> fsurf(z);
>> xlabel('x');
>> ylabel('y');
>> zlabel('z');
>> title('z = x + y');
```

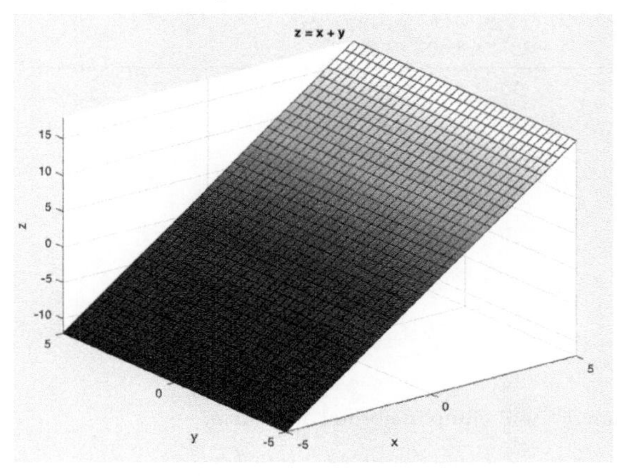

(Continued)

TABLE 6.2 (CONTINUED)

Summary of Key Symbolic Algebra Declarations and Operations

Declaring a Symbolic Matrix

syms x y z
A = [z y z; x y z; z 2 x]

% Matrix operations are the same as any matrix

inv(A)

ans =

[-1/(x - z), 1/(x - z), 0]
[-(x + z)/(2*z - x*y), z/(2*z - x*y), z/(2*z - x*y)]
[(2*x - y*z)/(2*x*z - x^2*y - 2*z^2 + x*y*z), -(2*z - y*z)/(2*x*z - x^2*y - 2*z^2 + x*y*z), -y/(2*z - x*y)]

B = [z y z; x y z; z 2 x];

% Matrix Multiplication

A * B

ans =

[-1/(x - z), 1/(x - z), 0]
[-(x + z)/(2*z - x*y), z/(2*z - x*y), z/(2*z - x*y)]
[(2*x - y*z)/(2*x*z - x^2*y - 2*z^2 + x*y*z), -(2*z - y*z)/(2*x*z - x^2*y - 2*z^2 + x*y*z), -y/(2*z - x*y)]

% Elementwise Multiplication

A .* B

ans =

[z^2, y^2, z^2]
[x^2, y^2, z^2]
[z^2, 4, x^2]

syms a
f = a^2 + 2*a

% Substitution of a matrix into a symbolic expression
% Note that operations are all elementwise operations, not matrix
% operations.
subs(f, a, A)

ans =

[z^2 + 2*z, y^2 + 2*y, z^2 + 2*z]
[x^2 + 2*x, y^2 + 2*y, z^2 + 2*z]
[z^2 + 2*z, 8, x^2 + 2*x]

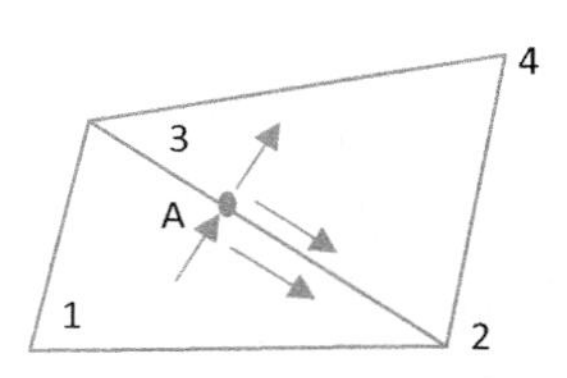

FIGURE 6.11
Neighboring triangles with jump in normal derivative.

To overcome this, vectors are directly used as the variables in what we call vector elements or edge elements.

The equations we solve is the wave equation derived according to

$$\nabla \times \vec{E} = -j\omega\vec{B} \tag{6.168}$$

$$\nabla \times \nabla \times \vec{E} = -j\omega\nabla \times \vec{B} = -j\omega\mu\nabla \times \vec{H} = -j\omega\mu\sigma\vec{E} \tag{6.169}$$

leading to the functional

$$\mathcal{L} = \iint_R \left\{ \frac{1}{2}\left(\nabla \times \vec{E}\right)^2 + \frac{1}{2}\vec{E}^2 \right\} dR \tag{6.170}$$

The interpolations then are (Figure 6.12)

$$A_x = \frac{1}{h}\left[y_c + \frac{h}{2} - y \right] A_1 + \frac{1}{h}\left[y - y_c + \frac{h}{2} \right] A_2 \tag{6.171}$$

$$A_y = \frac{1}{w}\left[x_c + \frac{w}{2} - x \right] A_3 + \frac{1}{w}\left[x - x_c + \frac{w}{2} \right] A_4 \tag{6.172}$$

so that

$$\vec{A} = A_x\vec{u}_x + A_y\vec{u}_y \tag{6.173}$$

Consider the vector function and some of its properties that make it suitable as an interpolatory function for the triangle of Figure 6.13:

$$\vec{w}_{12} = \zeta_1\nabla\zeta_2 - \zeta_2\nabla\zeta_1 \tag{6.174}$$

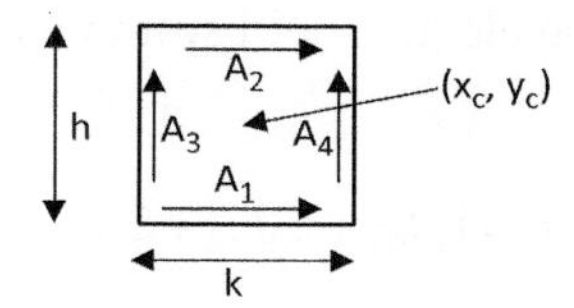

FIGURE 6.12
Edge element interpolation.

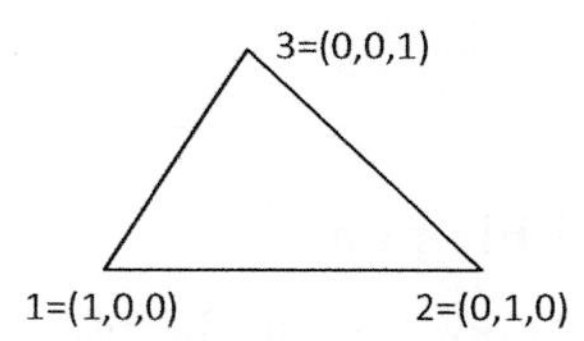

FIGURE 6.13
A first-order triangle.

Property 1

One edge 1-2, $\vec{w}_{12} = \zeta_1 \dfrac{1}{L_{12}} - \zeta_2 \dfrac{(-1)}{L_{12}} = \dfrac{\zeta_1 + \zeta_2}{L_{12}} = \dfrac{1}{L_{12}}$

Property 2

Along edge 1-3, $\zeta_2 = 0$ so $\vec{w}_{12} = 0$ and along edge 2-3, $\zeta_1 = 0$ so $\vec{w}_{12} = 0$

Property 3

$$\begin{aligned} \nabla \cdot \vec{w}_{12} &= \nabla \cdot \zeta_1 \nabla \zeta_2 - \nabla \cdot \zeta_2 \nabla \zeta_1 \\ &= \nabla \zeta_1 \cdot \nabla \zeta_2 + \zeta_1 \nabla^2 \zeta_2 - \nabla \zeta_2 \cdot \nabla \zeta_1 - \zeta_2 \nabla^2 \zeta_1 \\ &= 0 \end{aligned} \tag{6.175}$$

Property 4:

$$\begin{aligned} \nabla \times \vec{w}_{12} &= \nabla \times \zeta_1 \nabla \zeta_2 - \nabla \times \zeta_2 \nabla \zeta_1 \quad (\text{Note: } \nabla \times s\vec{V} = s\nabla \times \vec{V} - \vec{V} \times \nabla s) \\ &= \zeta_1 \nabla \times \nabla \zeta_2 - \nabla \zeta_2 \times \nabla \zeta_1 - (\zeta_2 \nabla \times \nabla \zeta_1 - \nabla \zeta_1 \times \nabla \zeta_2) \\ &= 2\nabla \zeta_1 \times \nabla \zeta_2 \end{aligned} \tag{6.176}$$

The three shape functions are therefore

$$\xi_1 = (\zeta_1 \nabla \zeta_2 - \zeta_2 \nabla \zeta_1) L_{12} \tag{6.177}$$

$$\xi_2 = (\zeta_2 \nabla \zeta_3 - \zeta_2 \nabla \zeta_3) L_{23} \tag{6.178}$$

$$\xi_3 = (\zeta_3 \nabla \zeta_1 - \zeta_1 \nabla \zeta_3) L_{23} \tag{6.179}$$

Therefore, the model for the electric field suitable for eddy current problems is

$$\vec{E} = \xi_1 \vec{E}_{t1} + \xi_3 \vec{E}_{t2} + \xi_3 \vec{E}_{t3} \tag{6.180}$$

It is left to the reader as an exercise to do a simple eddy current problem in a rectangle with edge elements.

6.14 The Quadrilateral Element

The quadrilateral has four nodes while the triangle has three. However, a first-order polynomial has three coefficients making the triangle the natural element of the simple first-order element. However, those started using

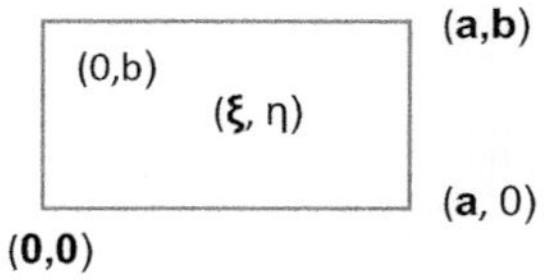

(a,b)
(0,b)
(ξ, η)
(a, 0)
(0,0)

FIGURE 6.14
The quadrilateral element.

finite elements used a quadrilateral element which can fit any shape as well as a triangle can. But fitting a second-order polynomial on a quadrilateral is tough since six nodes are required. Nor is it natural to fit a first-order trial function. What they did was to use a sort of 1.5-order trial function, not of order one and not of order two (see Figure 6.14).

For the four nodes we need four basis functions for the trail functions. Choosing

$$\varphi = c_1 + c_2\xi + c_2\eta + c_4\xi\eta \tag{6.181}$$

Applying this at the four nodes of coordinates (0,0), (a,0), (a,b), and (0,b):

$$\varphi_1 = c_1 \tag{6.182}$$

$$\varphi_2 = c_1 + c_2 a \tag{6.183}$$

$$\varphi_2 = c_1 + c_2 a + c_3 b + c_4 ab \tag{6.184}$$

$$\varphi_4 = c_1 + c_3 b \tag{6.185}$$

Putting these in matrix form and inverting

$$\begin{bmatrix} c_1 \\ c_2 \\ c_3 \\ c_4 \end{bmatrix} = \begin{bmatrix} 1 & 0 & 0 & 0 \\ 1 & a & 0 & 0 \\ 1 & a & b & ab \\ 1 & 0 & b & 0 \end{bmatrix}^{-1} \begin{bmatrix} \varphi_1 \\ \varphi_2 \\ \varphi_3 \\ \varphi_4 \end{bmatrix} = \frac{1}{ab} \begin{bmatrix} ab & 0 & 0 & 0 \\ -b & b & 0 & 0 \\ -a & 0 & 0 & a \\ 1 & -1 & 1 & -1 \end{bmatrix}^{-1} \begin{bmatrix} \varphi_1 \\ \varphi_2 \\ \varphi_3 \\ \varphi_4 \end{bmatrix} \tag{6.186}$$

$$\varphi(\xi,\eta)=\begin{bmatrix}1 & \xi & \eta & \xi\eta\end{bmatrix}\begin{bmatrix}c_1\\c_2\\c_3\\c_4\end{bmatrix}$$

$$=\begin{bmatrix}1 & \xi & \eta & \xi\eta\end{bmatrix}\frac{1}{ab}\begin{bmatrix}ab & 0 & 0 & 0\\-b & b & 0 & 0\\-a & 0 & 0 & a\\1 & -1 & 1 & -1\end{bmatrix}^{-1}\begin{bmatrix}\varphi_1\\\varphi_2\\\varphi_3\\\varphi_4\end{bmatrix} \tag{6.187}$$

$$=\frac{1}{ab}\begin{bmatrix}(ab-b\xi-a\eta+\xi\eta) & (b\xi-\xi\eta) & (\xi\eta) & (a\eta-\xi\eta)\end{bmatrix}\begin{bmatrix}\varphi_1\\\varphi_2\\\varphi_3\\\varphi_4\end{bmatrix}$$

$$=\psi_1\varphi_1+\psi_2\varphi_2+\psi_3\varphi_3+\psi_4\varphi_4$$

The four interpolation functions are

$$\psi_1=\frac{ab-b\xi-a\eta+\xi\eta}{ab}=\left(1-\frac{\xi}{a}\right)\left(1-\frac{\eta}{b}\right) \tag{6.188}$$

$$\psi_2=\frac{b\xi-\xi\eta}{ab}=\left(1-\frac{\xi}{a}\right)\left(1-\frac{\eta}{b}\right)=\frac{\xi}{a}\left(1-\frac{\eta}{b}\right) \tag{6.189}$$

$$\psi_3=\frac{\xi}{a}-\frac{\eta}{b} \tag{6.190}$$

$$\psi_4=\frac{a\eta-\xi\eta}{ab}=\left(1-\frac{\xi}{a}\right)\frac{\eta}{b} \tag{6.191}$$

Note that as to be expected, at node 1(0,0) $\psi_1=1$; at the other three nodes it is zero, as are the other three trial functions. Similarly, at node 2, $\psi_2=1$ while the other three are zero.

It is left as a useful exercise for the reader to work out the local matrices and verify this derivation using the symbolic algebra capabilities of MATLAB which should be along the lines of

```
clc;
clear all;
syms a b phi1 phi2 phi3 phi4 xi eta
phiv= sym([phi1;phi2;phi3;phi4])
interp = sym([1,xi,eta,xi*eta])
A= sym([1,0,0,0;1,a,0,0;1,a,b,a*b;1,0,b,0])
```

```
B=inv(A)
phi = sym(interp*B*phiv)
Integrand1=(diff(phi,xi))
I1Squared=Integrand1*Integrand1
Integrand2=(diff(phi,eta))
I2Squared=Integrand2*Integrand2
```

7

Parametric Mesh Generation for Optimization

7.1 Background and Literature

Today, we can attempt few serious finite element analyses without a mesh generator. It is a complex task to write our own code and we do not advise it when there are so many free open-source mesh generators available, in both two-dimensional and three-dimensional. In this chapter therefore, we will focus on the principles of mesh generation and how best to exploit free open-source codes that are widely available. While the concepts apply to mesh generation even for a single analysis problem, in optimization, a strong focus of this book, mesh generation has to be resorted to in cycles as a parametrically described object changes shape as optimization proceeds.

Figure 7.1 shows the design cycle for a geometric optimization problem. In the beginning, the initial geometric positions are either selected by the subject expert or in the absence of the expert, randomly selected. In the next step, we generate the mesh for the current geometry, measure the object value by a finite element solution and check whether it is at its minimum or not. If this is a minimum, we terminate the loop; otherwise we change the geometric parameters as dictated by our optimization algorithm and repeat the same procedure again.

Mesh generation is therefore a very important part of finite element analysis. However, mesh generators do not support parameter-based mesh generation for optimization. In optimization, the device to be synthesized is described in terms of parameters – the various lengths and materials and currents – and the mesh has to be generated for these parameters describing the present state of design. As optimization proceeds, for nonstop optimization iterations, the mesh has to be regenerated for the new state of design as given by the new parameters. For real-world inverse problems, we need a mesh generator as a library in the sense of an algorithmic function where the design is described by parameters $\bar{h}$ and it takes $\bar{h}$ as input and returns the mesh from iteration to iteration.

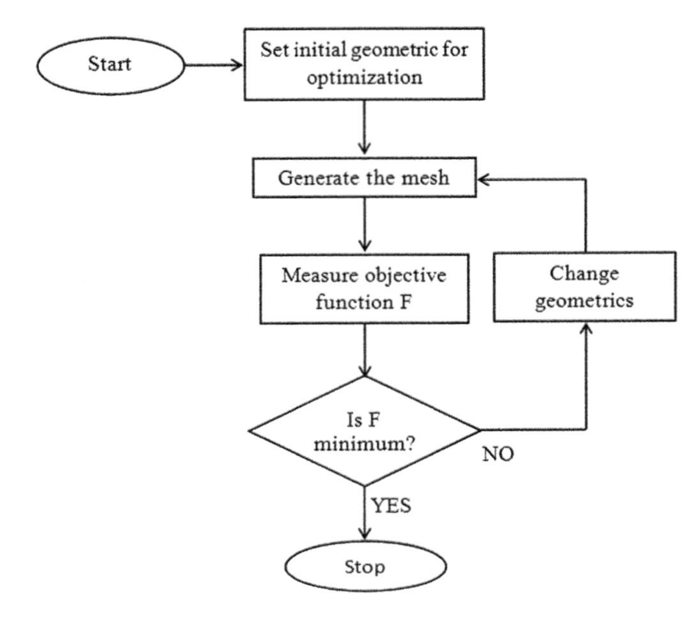

FIGURE 7.1
Design cycle for a geometric optimization.

Figure 7.2 shows the design cycle for an inverse problem. In the first step the design parameter set $\bar{h}$ is randomly selected (or estimated by a subject expert) and thereupon we generate the corresponding mesh, get the finite element solution, and measure the object value (often conveniently defined as a least square difference between design objects desired and computed) and check whether it is minimum or not. If this is a minimum, we terminate the loop; otherwise we change the design parameters and do the same procedure again. This procedure repeats until the object value is acceptably small. For optimization to go on nonstop, the mesh needs to be generated for the new parameters without user intervention. In Non-Destructive Evaluation (NDE) the only difference is that the object function compares measured values with those computed from presumed values of $\bar{h}$ being sought.

In this chapter, Chapter 7, we describe the necessity of a parametric mesh generator that runs nonstop and seamlessly through optimization iterations to convergence. Such mesh generators that do exist are rare, commercial, and not easily available to researchers except at great cost and never with the codes to modify them to suit individual needs. Besides, the typical mesh generator requires some man-machine interaction to define the points and boundary conditions and does not work for nonstop optimization iterations. We will take a regular open-source mesh generator and write a script-based interface as open source to run nonstop for optimization.

There are many mesh generators available on the web (Shewchuk, 1996a,b, 2002; Schoberl, 1997; The CGAL Project, 2013; Geuzaine and Remacle, 2009;

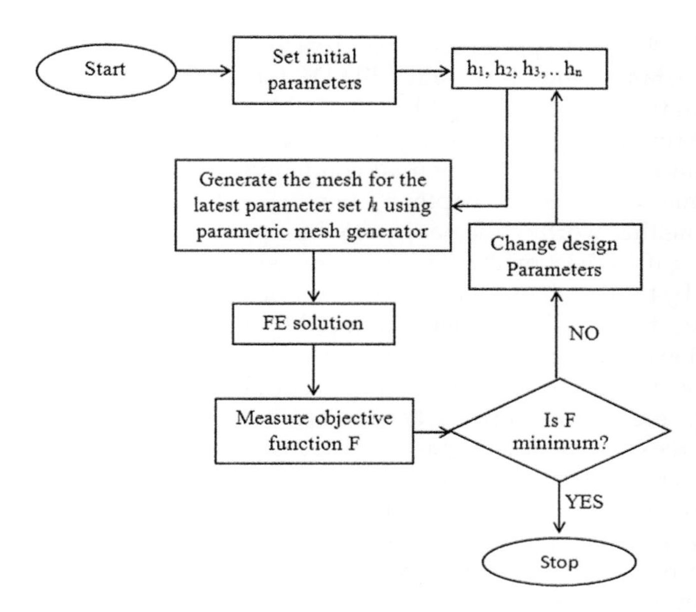

FIGURE 7.2
Design cycle for an inverse problem.

Conroy et al., 2012; Chen et al., 1995; Ma et al., 2011) and in the literature; some packages are open source software and others commercial. But they usually do not support parametric mesh generation. However, they do support features we would like in a parametric mesh generator. To summarize some notable mesh generators, Triangle, which we use, generates exact Delaunay triangulations, constrained Delaunay triangulations, conforming Delaunay triangulations, and Voronoi diagrams to yield high quality triangular meshes without large angles that is suited to finite element analysis (Shewchuk, 1996). AUTOMESH2D generates high quality meshes quickly (Ma et al., 2011). Cardinal's Advanced Mesh INnovation with Octree (Chen et al., 1995), CGAL (The CGAL Project, 2013), ADMESH (Conroy et al., 2012), and Delaundo (Muller, 1996) are all notable for special features. Indeed, there are parametric mesh generators, for example in Niu et al. (2011). However, it is not publicly available. Another such mesh generator is CEDRAT's suite Flux where, as parameters are changed, the mesh is generated and the device analyzed to study the effect of parameters on performance (CEDRAT, n.d.). The same approach has been taken in NDE studies (Xin et al., 2011). However, the latter two works are not intended for nonstop optimization. For that CEDRAT uses a script-based scheme called GOT-It (CEDRAT, n.d. b) which passes parameters to the program Flux and gets the results back for the optimization. Their software and information are mainly in the commercial domain. A 'lightened' version of GOT-It, named FGot, is offered free to students, but there again, the code is not accessible.

Figure 7.3 shows the performance comparison of triangulation using CPU-Delaunay triangulation (DT) and GPU-DT for different numbers of points (Vishnukanthan and Markus, 2012). Here CPU is on a normal computer and GPU is on the Graphics Processing Unit which is discussed in the next chapter. We can see the gain is nominal compared to that from finite element solvers which are given in Chapter 8. Therefore, we conclude that the effort of coding is neither worthwhile nor justified for parallelizing mesh generation.

Moreover, if gradient methods of optimization are to be used, the mesh topology given by the nodal connections need to be held to preserve the C^1 continuity of the object function lest the mesh induced fictitious minima are seen by the optimization algorithm as from the physics of the problem (Hoole et al., 1991).

For these reasons very problem-specific mesh generators are constructed by researchers. For example, when an armored vehicle is targeted by an improvised explosive device, the armor is inspected by an eddy current test probe to characterize the interior damage to determine if the vehicle should be withdrawn from deployment. Figure 7.4 shows a problem-specific parametrically described crack in steel excited by an eddy current probe, where P1–P6 are the lengths that represent the position and shape of the crack, J is current density, and μ_r is relative permeability. In this NDE exercise the parameters need to be optimized to make the computed fields match the measurements. The mesh has been constructed for the specific problem. As the parameters change, the mesh topology is fixed, pulling and crunching triangles as shown in Figure 7.5. Such problem-specific meshes are a headache because they restrict the geometry, lack flexibility, and take time for modifications; hence the need for general-purpose parametric mesh generators.

We can use zeroth-order optimization methods for which C^1 continuity is irrelevant, such as the genetic algorithm, bee colony algorithms, etc., without pulling and crunching meshes for inverse problems; for example,

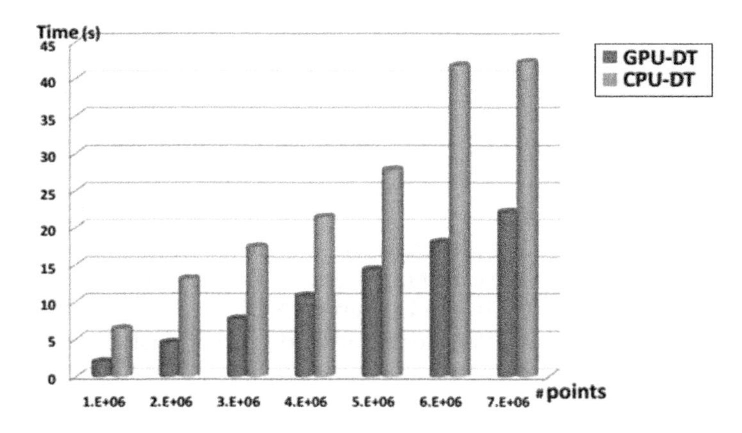

FIGURE 7.3
Performance comparison of triangulation using CPU-DT and GPU-DT for different numbers of points.

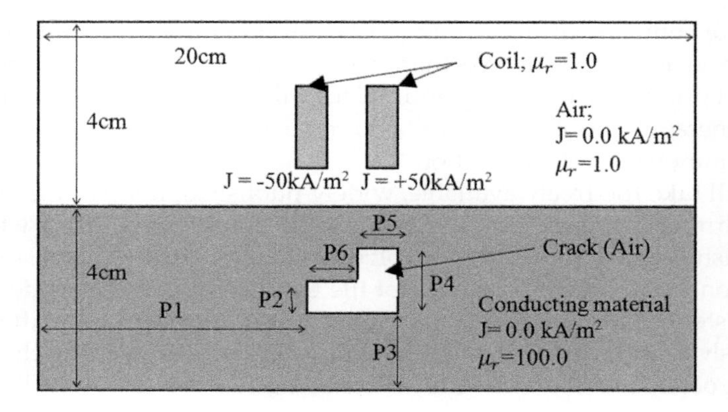

FIGURE 7.4
Problem-specific parametric mesh generators.

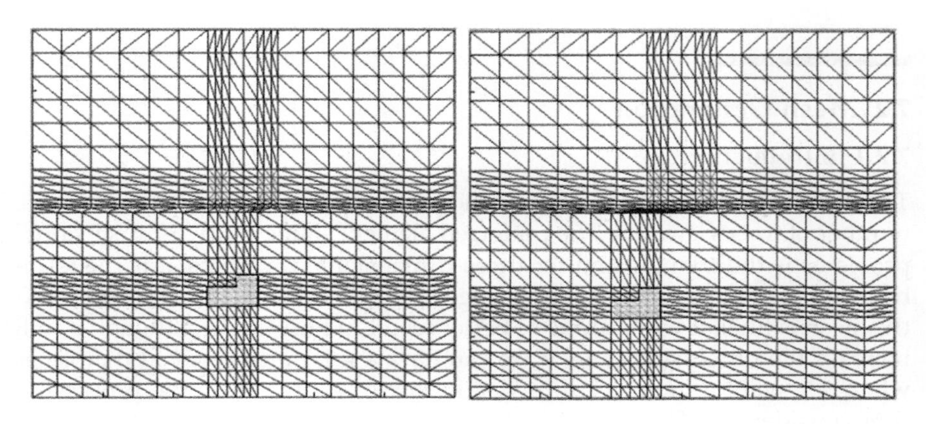

FIGURE 7.5
Elastically deformed problem-specific NDE mesh: As defect moves.

in reconstructing cracks to characterize interior defects, or designing power devices. For nonstop optimization, the commercial code ANSYS offers a gradients-based optimization suite (Vaidya et al., 2006), but gives little information on the techniques employed. That is, although these methods are known within the companies, they are rarely published. There are other companies, particularly from structural engineering, that also offer gradients-based optimization. A huge information lacuna is how they address the problem of mesh-induced minima. These artificial minima are seen as physics-based object function minima and the code tends to get stuck at these. Other approaches like a mathematical distance function to model the geometry lie in the domain of specialized efforts (Talischi et al., 2012).

Like in two-dimensional there are also many three-dimensional mesh generators available on the web and in the literature (Geuzaine and Remacle, 2009; CEDRAT, n.d. b; Si, 2006, 2015). Some are open-source software and others are commercial. TetGen (Xi, 2014, 2015) is for tetrahedral mesh generation and is more effective than previous methods at removing slivers from and producing Delaunay optimal tessellations (Joe, 1995). Each of these mesh generators has its own merits but none of these mesh generators supports parametric mesh generation.

We will take the freely available, widely published, nonparametric, open-source three-dimensional mesh generator TetGen (Si, 2014, 2015), which like all published mesh generators involves user input in the process of mesh generation. Here also we use a script file that uses a parametric description of the system to start the mesh from initial parameters and thereafter runs it seamlessly without stopping as the parameters are updated by the optimization process. Sample three-dimensional meshes are shown in Figures 7.6 and 7.7

7.2 Mesh Generation

7.2.1 Introduction

The finite element method requires the problem space to be split into a finite number of finer elements. The preferred element shape for two-dimensional problems is the triangle and for three-dimensional problems, it is the tetrahedron (Hoole, 1988). This set of triangles/tetrahedrons is called the mesh. If the mesh is finer, it will produce a better result in FEA (ibid.), although it will increase the processing time. If we can have a finer mesh only at the places where we want a more accurate result, then we can reduce the processing time considerably. This is called adaptive mesh generation (ibid.). Apart from

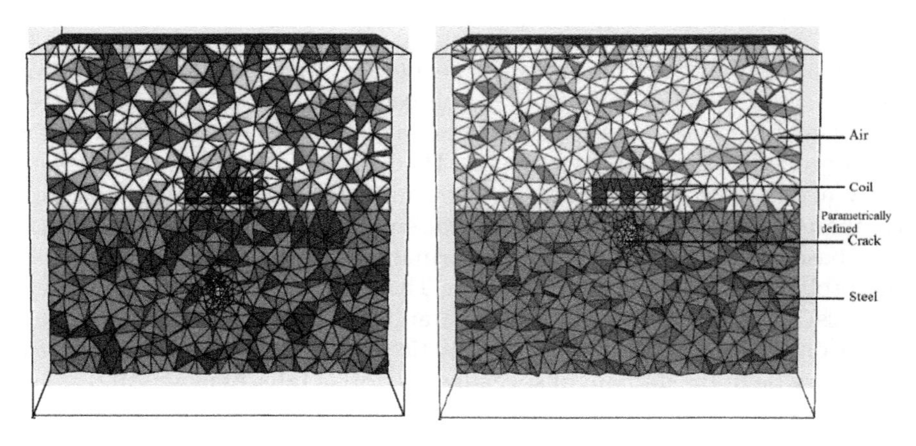

FIGURE 7.6
Three-dimensional mesh for NDE problem: As parameters change.

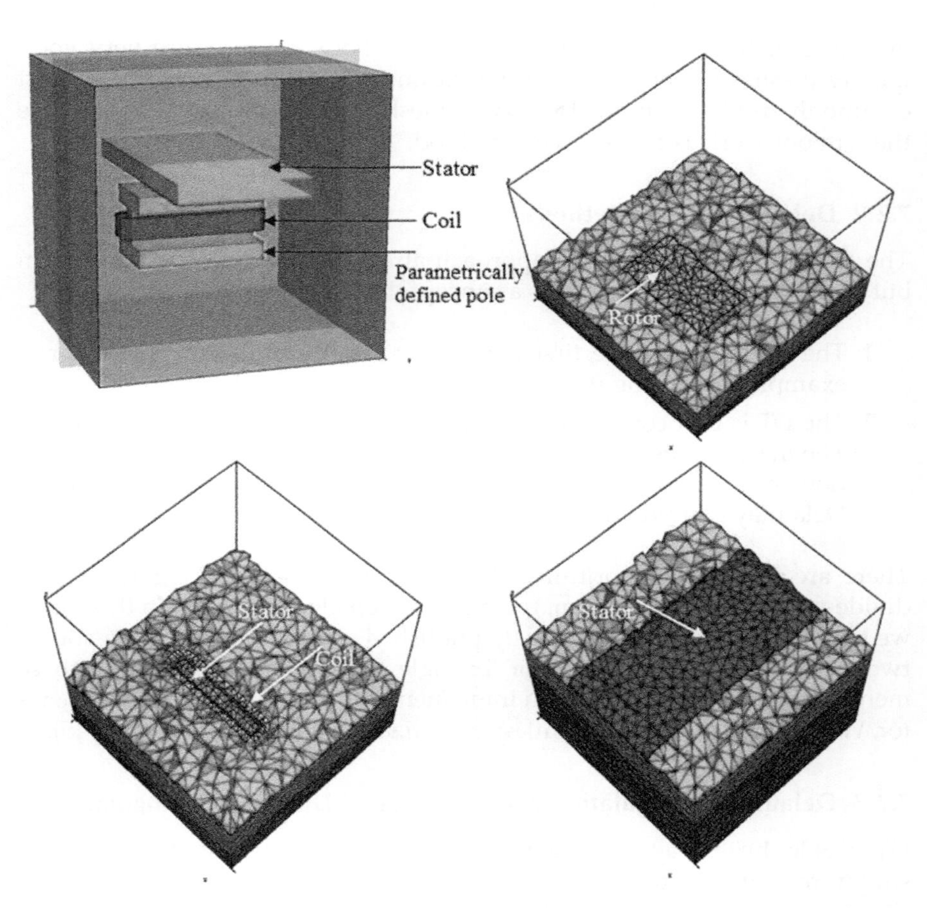

FIGURE 7.7
Three-dimensional mesh for motor problem.

this, the shape of the triangles/tetrahedrons in the mesh has a great effect on the final solution of finite element analysis. If we have very obtuse angles in the triangles of the mesh, they will introduce considerable errors into the final solution. Therefore, all these facts have to be considered when generating a mesh for FEA problems.

There are many mesh generation algorithms available. Basically, we can divide them into two categories. They are,

1. Algorithms that generate a crude mesh to define the basic geometry and then refine it to get a good quality mesh
2. Algorithms that generate a fine mesh from the beginning

The advancing front algorithm (Mavriplis, 1995) and quad-tree algorithm (Schneiders, 2000), Delaunay-based Algorithms (Shewchuk, 1996) are the examples of the second category. These methods can produce very good quality meshes. Delaunay-based algorithms are well-known and the most commonly used algorithms for quality mesh generations. Therefore, we use them in our work as the preferred method.

7.2.2 Delaunay-Based Methods

The Delaunay-based meshing approach already encountered in brief form but expanded more fully here, is a concept that consists of two tasks:

1. The mesh points are first created by a variety of techniques; for example, advancing front, octree, or structured methods.
2. The DT is first computed for the boundary without internal points. The mesh points are then inserted incrementally into the triangulation/tetrahedralization and the topology is updated according to the Delaunay definition.

There are many DT algorithms; the incremental insertion algorithm, the divide-and-conquer algorithm, the plane sweep algorithm, etc. In this work, we take the freely available, widely published, nonparametric, open-source two-dimensional mesh generator Triangle (Shewchuk, 1996). These three mentioned algorithms have been implemented in the Triangle mesh generator. We then adapted it for seamless optimization (Sivasuthan et al., 2015).

7.2.3 Delaunay Triangulation and Constrained Delaunay Triangulation

DT (Hoole, 1988) is a technique used to improve the quality of the mesh by simply rearranging the nodal connections that make triangles. This algorithm ensures that there will be no obtuse angles in the mesh other than in the triangles at boundaries. This is done by rearranging the triangles, if the uncommon point of the neighboring triangle lies inside the inscribing circle of one of the triangles, as shown in Figure 7.8. This can be identified by calculating the two angles corresponding to the uncommon points. By the properties of cyclic quadrilaterals, when the sum of these angles is greater than 180°, the triangles must be rearranged.

In Figure 7.8, the triangle QRS lies inside the inscribing circle of the triangle PQS. This can be recognized by summing the two opposite angles, A and B (These are the angles corresponding to the uncommon points for the two triangles, P and R). Since the sum of A and B is greater than 180° the two triangles are rearranged as PQR and PRS as shown in the figure. Now the vertex of the opposite triangle is not inside the inscribing circles. Constrained DT is a generalization of the DT that forces certain required segments into the triangulation. An example is shown in Figure 7.9. Both triangles are in different regions where each may have different properties. So, they cannot be flipped like in the previous case.

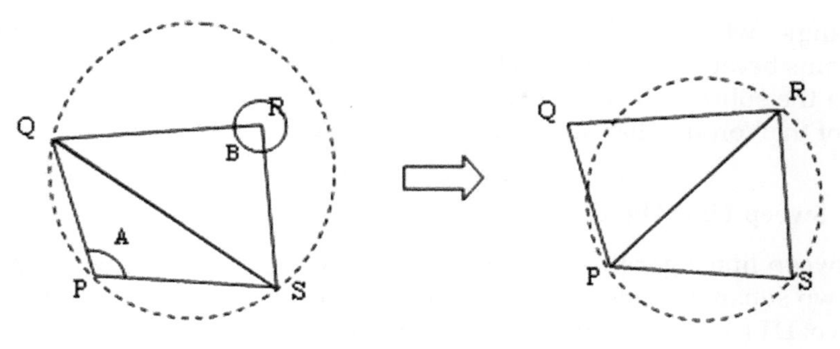

FIGURE 7.8
Applying DT.

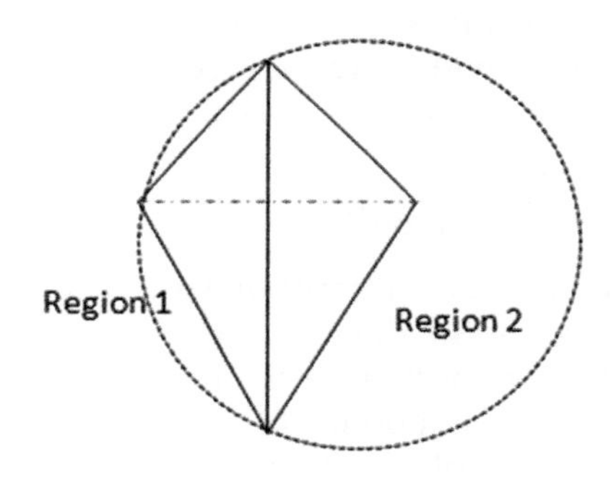

FIGURE 7.9
Example for constrained DT.

7.3 Algorithms for Constructing a Delaunay Triangulation

7.3.1 Speed

There are many DT algorithms; for example, divide-and-conquer (Lee and Schachter, 1980), sweep line (Fortune, 1987), incremental insertion [55], etc. As Su and Drysdale (1997) found, the divide-and-conquer algorithm is fastest; the second fastest is the sweep line algorithm. The incremental insertion algorithm performs poorly, spending most of its time in point location. Su and Drysdale (1997) introduced a better incremental insertion implementation by using bucketing to perform point location, but it still ranks third. A very important development in the divide-and-conquer algorithm is partitioning the vertices with vertical and horizontal cuttings (Lee and Schachter, 1980).

7.3.2 Divide-and-Conquer Algorithm

The point set v is divided into halves until we are left with two or three points in each subset. Then these smaller subsets can be linked with edges or triangles which is called a Voronoi diagram. Now we have a set of Voronoi diagrams because we have a set of smaller subsets. In the conquer step, we merge the subsets to get the whole Voronoi diagram (see Figure 7.10). The dual of the Voronoi diagram is the mesh (Schewchuk, 1996a,b).

7.3.3 Sweep Line Algorithm

The sweep line algorithm uses a sweep line which divides a working area into two subareas. This process constructs the Voronoi diagram – the dual graph of DT (shown in Figure 7.11). This algorithm was introduced by Fortune (1987). Shewchuk (1996) presented a successful algorithm for constructing a higher-dimensional DT. Figure 7.11 explains how the sweep line algorithm works. There is a vertical line which is called a sweep line in Figure 7.11. When this line passes a point, this algorithm creates a Voronoi diagram with other points which have already been passed.

7.3.4 Incremental Insertion Algorithm

Under the incremental insertion algorithm, we generate a fictitious triangle containing all points of V in its interior. The points are then added one by one. Figure 7.12b shows the mesh after the first point is added. Figure 7.12c shows how to handle the insertion of the second or subsequent point. The idea is to draw circumcircles of a particular triangle where the new point is

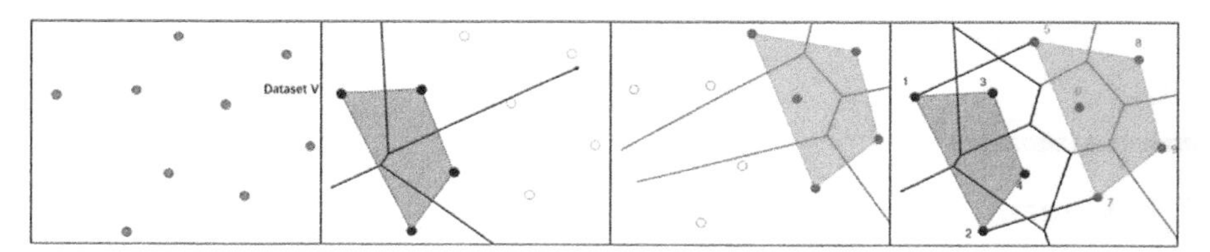

FIGURE 7.10
Divide-and-conquer algorithm.

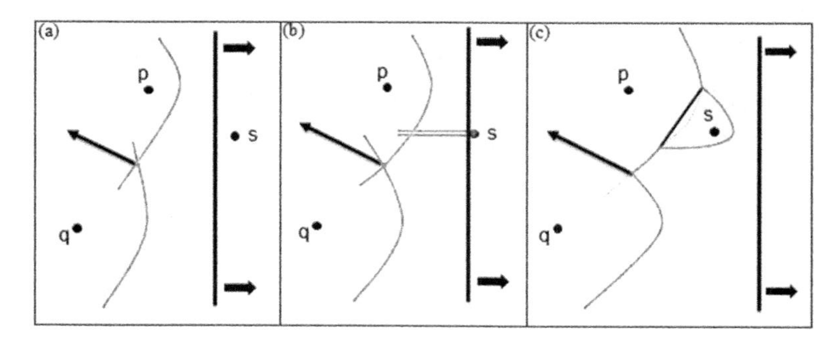

FIGURE 7.11
Sweep line algorithm.

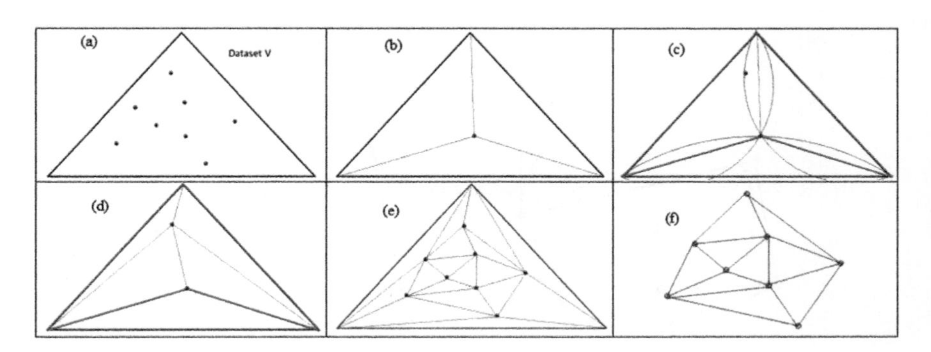

FIGURE 7.12
Incremental insertion algorithm.

located and neighboring triangles, select the triangles whose circumcircles cover the new points, remove the interior edges of selected triangles; and finally, a new point is connected with every point of a created polygon. This algorithm always maintains the DT of the points.

7.4 Mesh Refinement

There are many mesh refinement algorithms available. Most of them are based on Rupert's algorithm (Rupert, 1995). They produce good quality meshes with more nodes at regions where there are finer geometrical shapes and fewer nodes at other regions. The basic idea of the algorithm is to maintain a triangulation, making local improvements in order to remove the skinny triangles.

Figure 7.13 shows the basic idea of avoiding skinny angles.

Both triangles are in different regions (shown in Figure 7.14); each may have different properties. So they cannot be refined like in the previous case. Here the algorithm follows the same idea without violating the boundary between separate regions

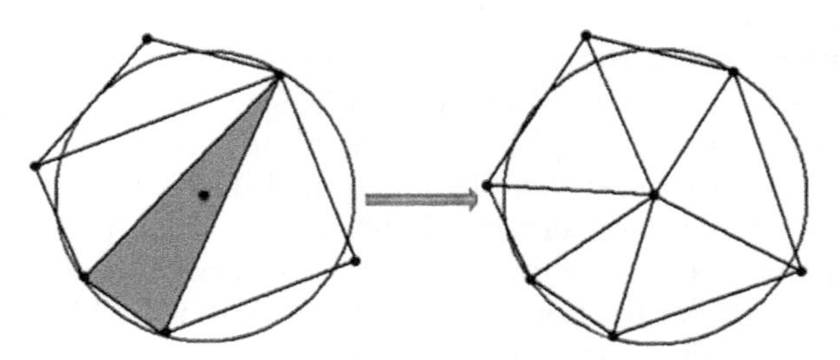

FIGURE 7.13
Delaunay mesh refinement.

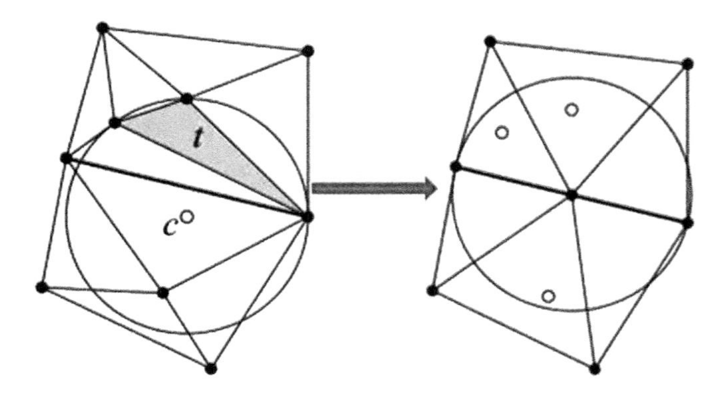

FIGURE 7.14
Delaunay mesh refinement between two regions.

7.5 Three-Dimensional Mesh Generation

Since almost every real-world problem is three-dimensional, we extend our two-dimensional work to three-dimensional geometric parameterized mesh generation for optimization problems in design and NDE. Mesh generators in electrical engineering commonly use Delaunay tetrahedralization and constrained Delaunay tetrahedralization for quality meshes. The incremental insertion algorithm is a well-known algorithm for tetrahedralization (Si, 2015). The worst case runtime of this algorithm is of $O(n^2)$, but the expected runtime for this algorithm is of $O(n \log n)$ (Shewchuk, 1996). Constrained Delaunay tetrahedralization was first considered by Shewchuk (1996). Gmsh (Geuzaine and Remacle, 2009) and TetGen (Si, 2015) are the better-known free, open-source three-dimensional mesh generators. TetGen uses a constrained Delaunay refinement algorithm which guarantees termination and good mesh quality.

A three-dimensional DT is called a Delaunay tetrahedralization. Ideas for Delaunay operation, constrained DTs, and mesh refinements are the same but only the dimension is different. Three-dimensional objects are usually represented by Piecewise Linear Complexes (PLCs) (Si, 2015). The design goal of TetGen is to provide a fast, light, and user-friendly meshing tool with parametric input and advanced visualization capabilities. Even though TetGen and Gmsh (Geuzaine and Remacle, 2009) are great open-source mesh generators, from an inspection of the code, it is very hard to use for nonstop optimization problems. For the nonstop optimization that ANSYS offers (Vaidya et al., 2006), it gives little information on the techniques employed. CEDRAT uses a script-based scheme called GOTIt (CEDRAT, n.d. c). FGot is offered for free to students although the code is not accessible and therefore will not permit modification nor work for industry-scale problems. Here TetGen is used as a backend for parameterized meshes for optimization. Therefore, we will develop it on our own and make it available as open source.

It is always possible to tetrahedralize a polyhedron if points are not vertices of the polyhedron. Two types of points are used in TetGen:

1. The first type of points is used in creating an initial tetrahedralization of PLCs.
2. The second type of points is used in creating quality tetrahedral meshes of PLCs.

The first type of points is mandatory in order to create a valid tetrahedralization. While the second type of points is optional, they may be necessary in order to improve the mesh quality.

7.6 Parameterized Mesh Generation – A New Approach

As noted, parametric mesh generation is a very important part of finite element optimization problems. In optimization problems, parameters describe the device in terms of materials, currents, and dimension. During optimization, as these parameters are changed to minimize an object function, a new mesh has to be generated and a new finite element solution obtained to re-evaluate the object function. At each iteration of an optimization algorithm, given the variables as input, the mesh is generated without user intervention. Finite element mesh generators exist in the public domain, a few even based on a parametric device description. The typical mesh generator requires some man-machine interaction to define the points and boundary conditions, and does not work for nonstop optimization iterations for which we need a mesh dynamically evolving through the iterations with optimization variables as changing parameters. Such mesh generators as do exist are rare, commercial, and not easily available to researchers except at great cost and never with the code to modify them to suit individual needs.

We take the freely available, widely published, nonparametric, open-source two-dimensional mesh generator Triangle (Schewchuk, 1996) and three-dimensional mesh generator TetGen Si (2006, 2015) which like all published mesh generators (with the exception of commercially restricted ones whose methodology is not published) involves user input in the process of mesh generation. However, for use in optimization we cannot stop the iterations to make input (Sivasuthan et al., 2015). To address these problems we use a script file which uses a parametric description of the system to start the mesh from initial parameters and thereafter runs it seamlessly without stopping as the parameters are updated by the optimization process (ibid.). The

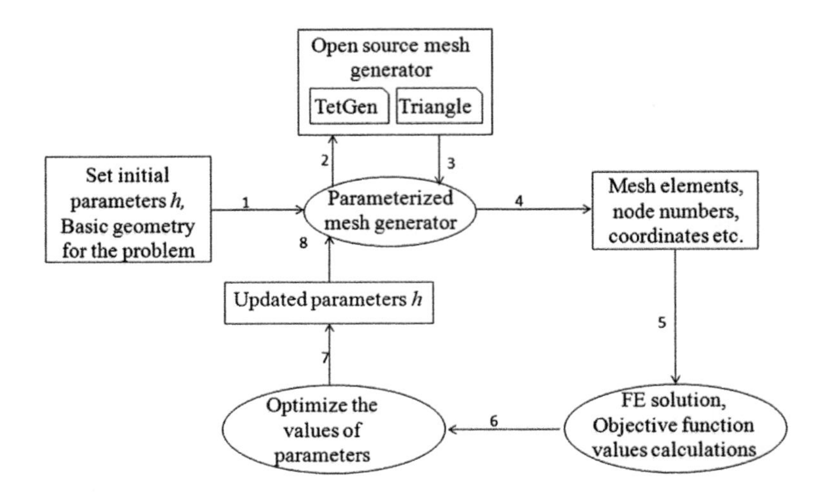

FIGURE 7.15
The new approach to parameterized mesh generation.

script file provides the user input while the code is iteratively running, with input that is normally made in the mesh generator being used, but for which the optimization iterations cannot stop. Figure 7.15 explains our approach to parametrized mesh generation. In the first step, the initial input which is described in the following sections is given to our mesh generator. Next, the mesh generator calls our chosen open-source mesh generator to generate the mesh. After that, the FEM solver uses the mesh to solve the problem. Next, the optimization algorithm updates the parameters which are accepted by our new mesh generator to generate the new mesh.

7.7 Data Structure and User Interface

7.7.1 Data Structure

The data structure used in these mesh generator software suites contains the following collections of objects:

1. Points list
2. Regions list
3. Properties list
4. Variable points list
5. Measuring points list
6. Segments list

7. Mesh details list (triangles/tetrahedrons)
8. Holes list
9. Boundary conditions

To explain what these are:

- Points List: The point collection contains all the points used in the problem definition and solving process. Each point contains the coordinates of the relevant finite element node. For the three-dimensional mesh generator, the list has x, y, and z coordinates. For the two-dimensional mesh generator, it should have x and y coordinates only.
- Regions List: The region list contains all the regions used in the problem definition. Region here means a material-source combination which has different physical characteristics.
- Properties List: The properties list contains all the properties of each region. Each region has a set of properties.
- Variable points list: The variable points list contains the information about the points to be moved according to the changes of the parameters, during optimization.
- Measuring points list: The measuring points list contains the points in the solution space where we want to find the potentials, flux density, etc.
- Segments list: The segments list contains the edges of the problem model. Each problem may have many segments.
- Mesh details list: The mesh details list contains the triangles/tetrahedrons of the mesh. This collection is empty until the mesh is generated. For a two-dimensional mesh generator each triangle contains references to its three vertex points. For a three-dimensional mesh generator each tetrahedron contains references to its four vertex points.
- Holes list: Holes are a special kind of region where we do not need to generate the mesh. A hole list contains all the holes used in the problem definition.
- Boundary conditions: There are two types of boundary conditions that are usually used in FEA problems. They are:
    1. Neumann boundary conditions
    2. Dirichlet boundary conditions

A Dirichlet boundary condition means the potential along the given boundary is fixed and a Neumann boundary condition means the derivative of the potential along the given boundary is fixed, and usually zero. Dirichlet boundary conditions can be implemented by keeping the potential of all the points on the given boundary to be fixed at their given value. The user can select any segment and define the potential of that segment. If the potential of the segment is set, then all the points which will be added onto that line will get this potential automatically. The boundaries that do not implement Dirichlet conditions will automatically act as Neumann boundaries during the FEA process. This is because it is natural to the finite element formulation (Hoole, 1988). Therefore, no special provisions are needed to define Neumann boundaries.

7.7.2 User Interface and Defining Geometry

A proper user interface is very important for good software. If the user interface is not friendly to use, even if it is very powerful, most users will not be able to use it effectively. Therefore, the user interface is carefully designed and used in this software as described in the next subsection.

Since we are providing the code as open source, it is necessary to describe it for other users to reengineer the code when necessary. This software is made to import drawings from the text file format. Figure 7.16 shows the sample input file of our mesh generator adapting Triangle using a script file to run nonstop from iteration to iteration without manual intervention.

1. The mesh generator code does not care about lines of code that start with #. We can write comments using the # sign.
2. First interpreted row: <a number, a number> – The first number represents the number of nodes in the domain; the second number represents the number of variable points. These variable points are also nodes but their coordinates may vary with the optimization iterations.

```
# A set of pointsin 2D(* WITHOUT VARIABLE POINTS).
# Number of nodes is 9 number of variables is 5
9 5
# And here are the nine points.
1 0.0 0.0
2 10.0 0.0
3 20.0 0.0
4 10.0 10.0
5 0.0 10.0
6 2.0 2.0
7 4.0 2.0
8 4.0 4.0
9 2.0 4.0
# variable points
# number of points in first draw. Then coordinates
10 20.0 3.0
```

FIGURE 7.16
Sample input file for mesh generator.

3. From the second row to row number 10 (that is, 9+1) in the domain, there are, associated with that row,

   <an integer number, a floating-point number, a floating-point number>

   The integer represents the node number. These must be numbered consecutively, starting from one. The two floating-point numbers represent the coordinates of this node. For the three-dimensional mesh generator, the three floating-point numbers represent the coordinates of this node.
4. The next segment of this input file represents variable points. These are also in the same format as the previous segment.
5. In the third part of the input file we have segment details. The first row of this file has the number of segments. From the second row onwards, it has four columns. The first column includes the segment numbers which must be numbered consecutively, starting from one. The next two columns are node numbers. Each row represents a segment. The fourth column is a marker. A marker has different integer values; it can be used to define the boundary condition. Here -1 means it is not on a boundary. If we have to define the boundary condition, we have to give any positive integer to the marker. Then we can assign the boundary value using these markers. For the three-dimensional mesh generator, the first row of this segment of this file has the number of faces and a boundary marker. Each face has a list. The first part of the list has the number of polygons in the face, number of holes on the surface, and the boundary marker. From the second row onwards, polygons and holes of the surface are defined.
6. The next segment of this file is the definition of boundaries. The first row contains the number of boundary conditions. From the second row onwards the first column represents the numbering; the second column is a marker number which has been already defined in the previous part (the segment part). The third column represents the boundary value for a particular marker.
7. The subsequent segment is a definition of the regions of the problem. Here a different region means different materials, so it has different properties. The first row represents the number of regions in the domain. From the second row, each row has five columns. The first column represents the numbering as usual. The second and the third columns represent the coordinates. These coordinates are used to identify the region. The point may be any point in the particular region. The fourth column is an integer which starts from 1. It can be used to assign properties to these regions. The next column is not used here because it is an area constraint coming from Triangle but, as mentioned, it is not used by us.

8. The next segment of this input file represents the number of holes. The hole is a region in which we do not want to generate the mesh. In the first step we define the number of holes in the problem domain. Fromm the next row, there are three columns. The first column represents the hole number. The second and third columns represent the x and y coordinates of any point within that hole.
9. The segment thereafter is for the measuring point list for object function evaluation. The first part is the number of measuring points where we are going to calculate the solution to get the target solution. Our source code helps users to identify the errors in the input file. It works very efficiently for any shape of problem domain. The sample inputs files are included in this book as appendices at the end of this chapter. We found that users customize their own problem very efficiently as was shown when tried out in our lab. This software is easy to use. This software is well supported in any operating system, that is, Linux/Unix, Windows, etc.

7.7.3 Post-Processing of Meshing

Once we triangulate/tetrahedralize the problem domain, we have to define the boundaries and boundary values. Upon triangulation/tetrahedralization, we have an element list (node numbers), the properties of regions, and point list (for FE solution). We do not need to calculate the solution of known nodes so we have to separate the known and unknown nodes. This step is known as renumbering the nodes which is also used to reduce the profile of the matrix (Hoole, 1988). In this process, we:

1. Define the boundary (generally using segment numbers in two-dimensional form, faces in three-dimensional form)
2. Get the boundary values (different boundaries may have different boundary values)
3. Get all nodes which are on boundaries (We used the segment marker list to determine the boundary nodes)
4. Separate boundary elements from non-boundary elements; and separate unknown nodes from known nodes
5. Give the first set of numbers for the unknown nodes and the last set of numbers for the known nodes
6. Renumber the whole point list based on new numbering.
7. Renumber the node entries in the triangle list based on the new numbering system.
8. Get all properties for the particular regions
9. Assign these properties to all corresponding triangles

Since real-world problem size is typically very large, this renumbering process takes a very long time. Algorithm 7.1 describes the regular renumbering process. This algorithm is very inefficient because each node will be searched for in an index array. The order of this algorithm is $O(n^3)$. This step can be improved. In this work the traditional algorithms have been improved based on the merge sort technique.

Algorithm 7.1: Renumbering

```
1: {t old triangle/tetrahedron list (node numbers), n
   number of elements (triangles/tetrahedrons), index-
   rearranged node numbers(unknowns first; knowns last)
   and (u renumbered triangle/tetrahedron list (node
   numbers}
2: for i = 0 to n - 1 do
3:   for j = 0 to (3or4) - 1 do
4:    temp ← t(i + n*j)
      {t is a one-d array}
5:    for k = 0 to n - 1 do
6:     if temp == index(k) then
7:       u(i + n * j) ← k
8:       break
9:     end if
10:   end for
11:  end for
12: end for
```

7.7.4 Approach to Renumbering

Instead of searching for every element from the whole list, this thesis tracks the node number changes and updates the node numbers of the mesh. Figure 7.17 shows a simple example finite element problem. The boundary elements are circled. Figure 7.18a shows the node numbers which are assigned in the mesh generation process. Figure 7.18b presents the rearranged nodes which are separated based on whether the nodes are of known or unknown values. Figure 7.18c shows the new numbering system. The boundary elements are shown in the gray boxes in Figure 7.18. The corresponding old number list (c) is renumbered in (b). Figure 7.19 shows the new numbering of the nodes. Let us take the array of Figure 7.18a and use a new index which has 1, 2, up to the number of elements. We sort the array of Figure 7.18b using the merge sort algorithm which will be described in the following section. We apply every operation of the sort algorithm to a new index array. The resultant arrays are shown in Figure 7.20. Now if we want to

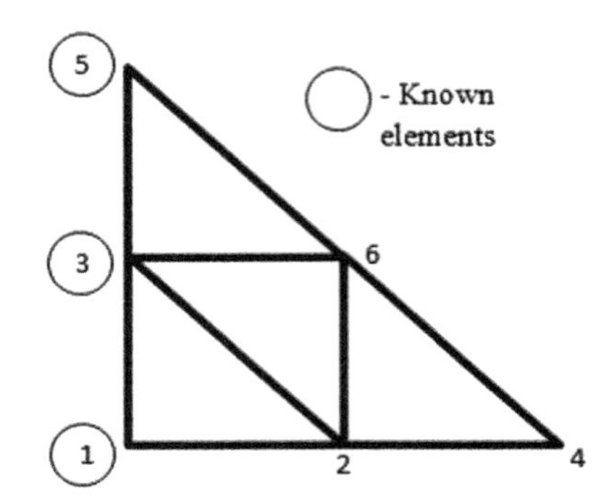

FIGURE 7.17
Simple example problem.

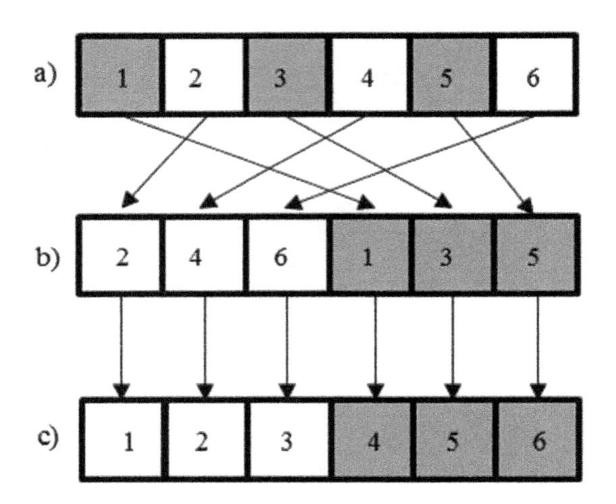

FIGURE 7.18
Numbering and renumbered nodes.

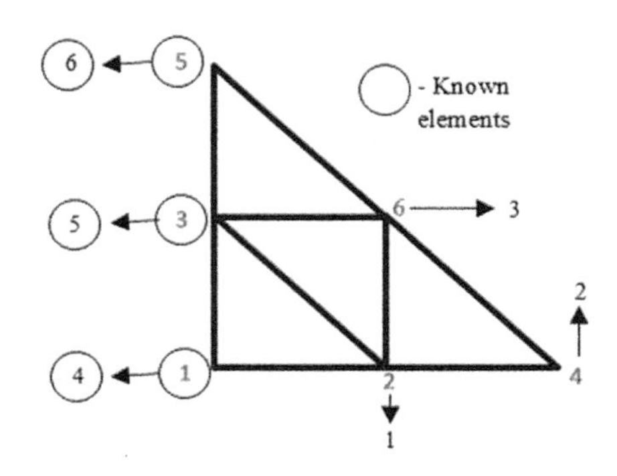

FIGURE 7.19
Renumbered nodes.

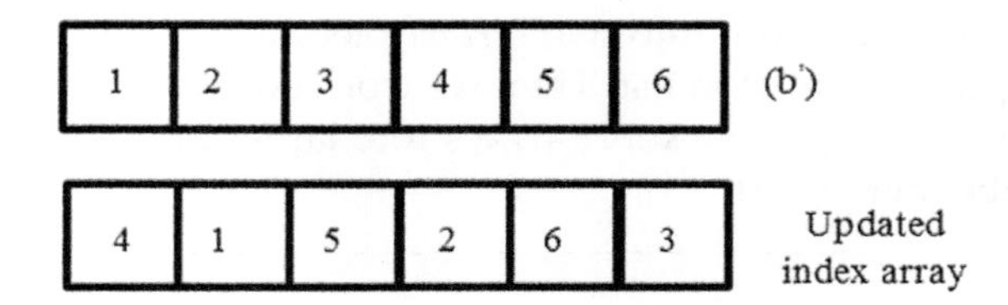

FIGURE 7.20
Sorted version of (b) and corresponding index changes.

get a renumbered value of a particular node number we can directly access it from the updated index array. For example, 1 is replaced by 4, 2 by 1, 3 by 5, and so on.

7.7.5 Merge Sort

Sorting is a technique that arranges the elements in a certain order. There are many sorting algorithms such as counting sort, bucket sort, radix sort, merge sort, heapsort, quicksort, etc. (Cormen et al., 2011). Each algorithm has its own advantages. Merge sort is one of the best sort algorithms which has *n log (n)* time complexity for each average, best, and worst case time complexities (ibid.). Another very important reason for selecting merge sort is that this algorithm is easily parallelizable – parallelization on the GPU is the main theme taken up in a subsequent chapter. Figure 7.21 (Merge Sort, n.d.) describes the merge sort in a graphical way. First, the list is split into two

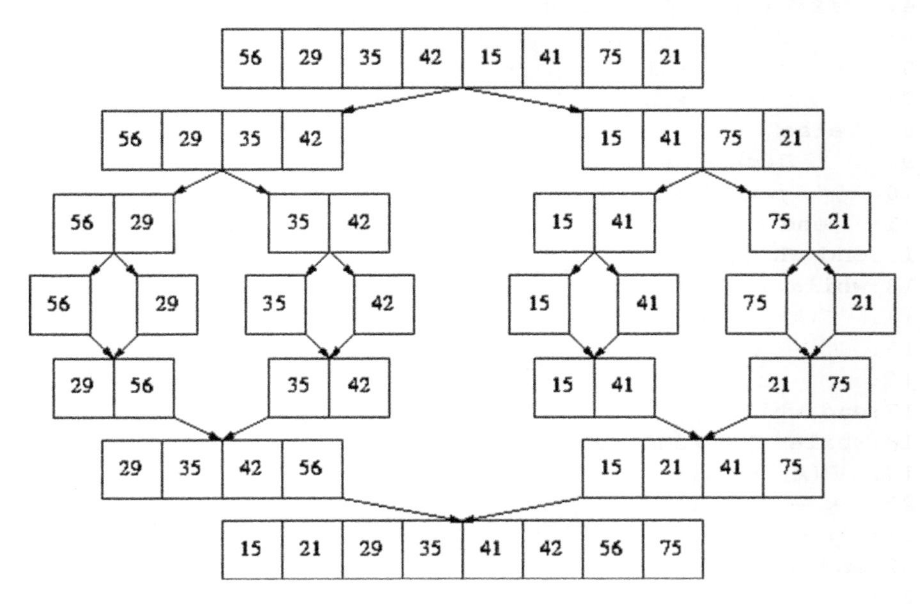

FIGURE 7.21
Merge sort.

pieces and then each is split into two *further* pieces. This process is repeated until arriving at the singleton list. Then we work our way back up the recursion by merging together the short arrays into larger arrays. Algorithms 7.2 and 7.3 describe merge sort.

Algorithm 7.2: Merge Sort

```
1: MergeSort(A, right, left)
2: if left < right then
3:     mid ← floor((left + right)/2)
4:     MergeSort(A, left, mid)
5:     MergeSort(A, mid + 1, right)
6:     Merge(A, left, mid, right)
7: end if
```

Algorithm 7.3: Merge

```
1: Merge(A, left, mid, right)
2: i ← left, j ← mid + 1, k ← 0,
3: while i ≤ mid and j ≤ right do
4:   if A(i) < A(j) then
5:      B(k) ← A(i)
6:      k ← k + 1
7:      i ← i + 1
8:   else
9:      B(k) ← A(j)
10:     j ← j + 1, k ← k + 1
11:   end if
12:end while
13:while i ≤ mid do
14:  B(k) ← A(i)
15:  k ← k + 1
16:  i ← i + 1
17:end while
18:while j ≤ right do
19:  B(k) ← A(j)
20:  k ← k + 1
21:  j ← j + 1
22:end while
23:  k ← k - 1
24:while k ≥ 0 do
```

```
25:    A(left + k) ← B(k)
26:    k ← k - 1
27: end while
```

7.7.6 Modified Form of Merge Sort for Renumbering

As we described in Section 7.7.5, we track the node number changes using a form of merge sort instead of searching for every element from the whole list. This new algorithm has *n log (n)* time complexity but the traditional method has time complexity of O(n3). We define an array of size given by the number of nodes in the mesh. The array has 0, 1, 2, up to (number of nodes – 1). We applied every operation of the sort algorithm to this newly defined array. Algorithms 7.4 and 7.5 describe the idea underlying what we have used.

Algorithm 7.4: Merge Sort – Modified

```
1: MergeSort(A, IdxArray, right, left)
2: if left < right then
3:    mid ← floor((left + right)/2)
4:    MergeSort(A, IdxArray, left, mid)
5:    MergeSort(A, IdxArray, mid + 1, right)
6:    Merge(A, IdxArray, left, mid, right)
7: end if
```

Algorithm 7.5: Merge-Modified

```
1: Merge(A, IdxArray, left, mid, right)
2: i ← left, j ← mid + 1, k ← 0,
3: while i ≤ mid and j ≤ right do
4:   if A(i) < A(j) then
      B(k) ← A(i)
      C(k) ← IdxArray(i)
      k ← k + 1
      i ← i + 1
    else
      B(k) ← A(j)
      C(k) ← IdxArray(j)
      j ← j + 1
      k ← k + 1
    end if
  end while
```

```
while i ≤ mid do
  B(k) ← A(i)
  C(k) ← IdxArray(i)
  k ← k + 1
  i ← i + 1
end while
while j ≤ right do
  B(k) ← A(j)
  C(k) ← IdxArray(j)
  k ← k + 1
  j ← j + 1
end while
k ← k - 1
while k ≥ 0 do
  A(left + k) ← B(k)
  IdxArray(left + k) ← C(k)
  k ← k - 1
end while
```

Appendix 1: Sample Input File: Two-Dimensional

```
#A set of points in 2D(* WITHOUT VARIABLE POINTS).
#Number of nodes is 9 number of variables is 5 9 5
#And here are the nine points. 10.0 0.0
210.0 0.0
320.0  0.0
410.0 10.0
50.0  10.0
62.0  2.0
74.0  2.0
84.0  4.0
92.0  4.0
#variable points
#number of points in first draw. Then coordinates 10
20.0 1.85 11 18 1.90
1215.5 2.10
1313 2.40
1410  3
#segments, 1st line --number of segments following lines
are segments (node numbers, each segment has two node
numbers and a marker to identify the boundary elements.
```

```
#number of segments
15
#segments (two nodes) and a marker
1   3   10  1
2   4   5   2
3   6   7   -1
4   7   8   -1
5   1   2   -1
6   2   3   -1
7   5   1   -1
8   14  4   -1
9   10  11  -1
10  11  12  -1
11  12  13  -1
12  13  14  -1
13  2   14  -1
14  8   9   -1
15  9   6   -1
#segment markers used to identify the boundaries and set
boundary
#conditions. do not give 0 or 1 to segment marker
because already
#fixed as a default, 3rd column is boundary condition
value
2
#asdf
1 2 0
2 3 0
#regions, number of regions x y coordinates of region,
regional
#attribute (for whole mesh), Area constraint that will
not be used #we can leave one region without any
assignments we have to assign for #this case 0 0 0 0 but
we can give properties to this region
2
1 1  0.5 1 0.1
2 12 3   2 0.9
#
#properties of regions, first number of properties then
property
#values
2
1   1     1.32
2  1.90   9.312
# holes, number of holes x y coordinates of the hole
```

```
1
1   3   3
#type
0
#measuring points
10
#point coordinates
1  1.1   1.10
2  2.1   1.19
3  3.1   1.18
4  3.3   1.17
5  3.6   1.16
6  3.9   1.15
7  4.1   1.14
8  4.4   1.13
9  4.9   1.12
10 5.1   1.1
```

Appendix 2: Sample Input File: Three-Dimensional

```
#3DMesh  Input File
#Number of points <--> Number of variable points
36   22
#13  3     0     1
1    0     0     0
2    100   0     0
3    100   0     100
4    0     0     100
5    0     50    0
6    100   50    0
7    100   50    100
8    0     50    100
9    0     100   0
10   100   100   0
11   100   100   100
12   0     100   100

#coil
13 40.0  52.0  48.0
14 44.0  52.0  48.0
15 48.0  52.0  48.0
```

```
16 52.0  52.0  48.0
17 56.0  52.0  48.0
18 60.0  52.0  48.0
19 40.0  52.0  52.0
20 44.0  52.0  52.0
21 48.0  52.0  52.0
22 52.0  52.0  52.0
23 56.0  52.0  52.0
24 60.0  52.0  52.0
25 44.0  56.0  48.0
26 48.0  56.0  48.0
27 52.0  56.0  48.0
28 56.0  56.0  48.0
29 44.0  56.0  52.0
30 48.0  56.0  52.0
31 52.0  56.0  52.0
32 56.0  56.0  52.0
33 40.0  60.0  48.0
34 60.0  60.0  48.0
35 60.0  60.0  52.0
36 40.0  60.0  52.0

#variables
37 52.85316955   44     50.92705098
38 51.76335576   42     52.42705098
39 50            43     53
40 48.23664424   45     52.42705098
41 47.14683045   42     50.92705098
42 47.14683045   43     49.07294902
43 48.23664424   45     47.57294902
44 50            44     47
45 51.76335576   43     47.57294902
46 52.85316955   43     49.07294902
47 51.42658477   46     50.46352549
48 50            46     51.5
49 48.57341523   47     50.46352549
50 49.11832212   46     48.78647451
51 50.88167788   46.5   48.78647451
52 50            48     50
53 51.42658477   40     50.46352549
54 50            40     51.5
55 48.57341523   40     50.46352549
56 49.11832212   40     48.78647451
57 50.88167788   40     48.78647451
58 50            38     49
```

```
#number of faces, boundary markers
53  1
#no of polygons> no of holes, boundary marker
1  0  3
4  1  2   3   4
1  0  3
4  5  6   7   8
1  0  3
4  9  10  11  12
1  0  3
4  1  2   6   5
1  0  3
4  5  6   10  9
1  0  3
4  2  3   7   6
1  0  3
4  6  7   11  10
1  0  3
4  1  4   8   5
1  0  3
4  5  8   12  9
1  0  3
4  8  7   11  12
1  0  3
4  4  3   7   8
1  0  -1
12 13 14  25  26  15  16  27  28  17  18  34  33
1  0  -1
12 19 20  29  30  21  22  31  32  23  24  35  36
1  0  -1
4  13 19  36  33
1  0  -1
4  18 24  35  34
1  0  -1
4  33 34  35  36
1  0  -1
4  13 14  20  19
1  0  -1
4  14 20  29  25
1  0  -1
4  25 26  30  29
1  0  -1
4  15 21  30  26
1  0  -1
4  15 16  22  21
```

```
1  0   -1
4  16 22  31  27
1  0   -1
4  27 28  32  31
1  0   -1
4  17 23  32  28
1  0   -1
4
#new
1  17
0  18
-1 24  23
3  37  38  47
1  0   -1
4  38  39  48  47
1  0   -1
4  39  40  49  48
1  0   -1
3  40  41  49
1  0   -1
4  41  42  50  49
1  0   -1
3  42  43  50
1  0   -1
4  43  44  51  50
1  0   -1
3  44  45  51
1  0   -1
4  45  46  47  51
1  0   -1
3  46  37  47
1  0   -1
4  47  48  52  51
1  0   -1
3  48  49  52
1  0   -1
3  49  50  52
1  0   -1
3  50  51  52
1  0   -1
3  37  38  53
1  0   -1
4  38  39  54  53
1  0   -1
4  39  40  55  54
```

```
1  0    -1
3  40   41   55
1  0    -1
4  41   42   56   55
1  0    -1
3  42   43   56
1  0    -1
4  43   44   57   56
1  0    -1
3  44   45   57
1  0   -1
4  45   46   53   57
1  0    -1
3  46   37   53
1  0    -1
4  53   54   58   57
1  0    -1
3  54   55   58
1  0    -1
3  55   56   58
1  0    -1
3  56   57   58

#boundary conditions
1

#conditions
1 3 0

# 2 regions
3
1 10 10 10 1 0.1
2 46 50 50 2 0.01
3 -1 -1 -1 3 1.2

#number of properties
2
1 1.90 2.20
2 2.213.30
3 3.214.30

#number of holes
0
```

```
#measuring points
5
1 52.85316955 51 50.92705098
2 51.76335576 51 52.42705098
3 50 51 53
4 48.23664424 51 52.42705098
5 47.14683045 51 50.92705098

#mesh area constraint
10
```

8

Parallelization through the Graphics Processing Unit

8.1 Parallelization

Parallelization is to reorder our algorithms so that steps of computation that do not involve precedence may be computed by different processors. So, for example, in adding 100 numbers, two persons add 50 of those numbers each and then their separate results are added together. That step of adding the results together is an extra step compared to adding all 100 numbers one after the other but can be done in almost half the time.

In the early days of field computation we dealt with small problems requiring the solution of equations with no larger than a 1000×1000 matrix. As late as 1983, this writer could not fit a 300×300 matrix on an IBM PC. To solve such problems, special methods were resorted to.

To give an example of one such method, note that in finite element matrix equation $Ax = B$, the coefficient matrix is formed by assembling A from a 3×3 matrix from a first order element. In the successive over-relaxation (SOR) scheme of solution, we repeatedly subtract from every $B(i)$, the off-diagonal terms $A(i,j)^*x(j)$, $j = 1$ to n, $j <> i$, and then divide the result by $A(i,i)$ for the latest update $x(i)$. Here the update of $x(i)$ uses all updates of $x(i)$s preceding i; that is, in updating say $x(10)$, the most up to date vaues of $x(1)$ to $x(9)$ are used. However, what this means is that until $x(i)$ is updated, $x(i+1)$ cannot be updated. It is a sequential algorithm. We have seen how by reverting to the "less efficient" Gauss algorithm where $x(i)$ in an iteration is improved using only the values from the last iteration, it can be made "more efficient" for element-by-element processing which is essentially a parallel agorithm where all $x(i)$ values may be improved at the same time. This is parallelization (Hoole, 1990).

In design optimization (Arora and Hang, 1976), the product is defined by parameters and the design criterion is defined as an object function to be minimized with respect to the parameters (like lengths, currents, and material values). The object function, computed from the finite element solution, now becomes a function of the parameters and needs to be minimized.

The process is therefore as follows: define the design object by parameters and from the parametrically defined object create a mesh, solve the finite element problem, measure the object function, and change the geometry through optimization to reduce the object function. This is an iterative process so stop if the object function is a minimum or go back to the first step.

There are a few bottlenecks to be noted. First, to mesh generate repeatedly as the shape changes, the optimization cannot be stopped to feed the new shape inputs to the mesh generator as required by single problem mesh generators. Therefore, the mesh generation has to be parameter-based (Sivasuthan et al., 2015). Codes exist for this but are in the commercial domain. A means is provided for overcoming this using open source code (ibid.). Second, the matrix computation is the biggest load in field computation and parallelization has been the means of overcoming this, particularly when repeated solutions are required. However, although parallel computers with n processors can cut down solution time by almost $(n-1)$ times (Hoole, 1991a,b; Adeli and Kumar, 1995), they have a memory bottleneck making more than 16 or 32 processors very costly. And then in optimization, in zeroth order methods like the genetic algorithm (GA) or even simple search methods, for several sets of parameters, the object function value needs to be evaluated. This increases the load. Multiple object function evaluations may be launched in parallel if the memory limit with shared memory can be overcome; for when we embark on many finite element solutions in parallel, the memory load will go up immensely.

8.2 Optimization with Finite Elements

It was left to engineers dealing with stress analysis and fluids to couple optimization with the finite element method (Arora and Hang, 1976; Vanderplaats, 1984; Marrocco and Pironneau, 1978), and the second half of the 1970s and 1980s would be the time for true synthesis – solving for geometric shape and material values from design criteria. The earliest persons to automate this cycle in magnetics were Marrocco and Pironneau (1978). They attempted to optimize the shape of the magnetic pole of a recording head so that the fringing effect at the edges of a pole could be countered so as to realize the object of constant flux density B in a recording head. They approached this problem by defining an object function F consisting of the square of the difference between the computed and desired flux densities. Thus, the problem is one of optimizing – that is, minimizing – F which is a function of parameters $p_1\, p_2 \ldots$ defining the geometry and which parameters are computed (i.e., optimized) so as to minimize F.

In magnetics the Ecole Nationale Supérieure d'Ingénieurs Electriciens de Grenoble (ENSIEG) group led by J.C. Sabonnadiere, would bring mathematical optimization of magnetic systems to bear on finite element analysis design in 1989 (Gitosusastro et al., 1989). This early work flowing from the ENSIEG group used gradients-based methods, steepest descent in particular (Vanderplaats, 1984). In 1990, the concept of newness by Russenschuck (1990) was in the methods of optimization, for example on the Rosenbrock and sequential unconstrained minimization technique (SUMT) optimization algorithms for searching for the object function's minimum by gradient methods (Vanderplaats, 1984). Schafer-Jotter and Muller (1990) gave more details and introduced the zeroth order, statistics-based simulated annealing method to magnetics. Later in 1990, an Austrian group (Preis et al., 1990) used 4th, 6th, and 8th order polynomials to model the shape and optimize the polynomial coefficients. A Japanese group under Nakata (Nakata et al., 1991) optimized a magnetic circuit by minimizing a least-square object function using Rosenbrock's search method. An early Italian paper from 1992 (Drago et al., 1992) had very powerful results in 3-D but offered little information in terms of method or the object function except to say they used pattern search with constraints. Other papers introduced parallel computation on shared memory systems for finite element optimization (Hoole, 1991, 1995; Hoole and Agarwal, 1997). Genetic-algorithm-based finite element optimization in magnetics has been carried out by many researchers over the past 20 year period (Uler et al., 1994; Juno et al., 1996; Dong-Joon et al., 1997; Mohammad et al., 2013; Saludjian et al., 1998; Enomoto et al., 1998). Though the GA is practicable and gives a faster solution when parallelized, it is slow, in our experience, as a single process when compared with the gradient optimization methods.

8.3 Finite Element Computation in CUDA C

Wu and Heng (2004) were the first to exploit the compute unified device architecture (CUDA) architecture and graphics processing unit (GPU) for finite element field computation focusing on the conjugate gradients solver. After a lapse of some time it was picked up by others (Kakay et al., 2010; Okimura et al., 2013; Godel et al., 2010; Kiss et al., 2012; Fernandez et al., 2012; Dziekonski et al., 2012; Attardo and Borsic, 2012; Richter et al., 2014).

Godel et al. (2010) implemented the discontinuous Galerkin finite element method discretization using the multi-GPU programming model. It got great attention because of element level decomposition. A speed-up (defined as central processing unit (CPU) computation time/GPU computation time) of 18 is reported by them. Kiss et al. (2012) have recently applied element by element (EbE) processing to solve their finite element equations from a first order tetrahedral mesh using the bi-conjugate gradients algorithm. The bi-conjugate gradients algorithm is reported to give good speed-up in their paper. Fernandez et al. (2012) decoupled the solution of a single element from that of the whole mesh (see Figure 8.1), thus exposing parallelism at the element level. The hp in Figure 8.1 refers to mesh accuracy improvement (Fernandez et al. 2012) by either refining triangles (h-refinement) or increasing their order (p-refinement). The SOR in Figure 8.1 refers to the SOR iterative algorithm. Individual element solutions are then superimposed nodewise using a weighted sum over concurrent nodes (Fernandez et al., 2012). Dziekonski et al. (2012) have used GPU computation for both numerical integration and matrix assembly in the higher order finite element method. It was reported that GPU implementation of matrix generation allows one to achieve speed-ups by a factor of 81 and 19 over the optimized single and multi-threaded CPU-only implementations. Attardo and Borsic (2012) introduced the wavelet-based algebraic multigrid method as a preconditioner for solving Laplace's equation with the domain discretized by the finite element method on a GPU. Richter et al. (2014) presented an algebraic multigrid preconditioner for accelerating large sparse problems. A speed-up of 5.2 is reported by Fernandez et al. (2012). The results indicate that the speed-up of the algebraic multigrid solving phase will be especially benefited in transient quasi-static simulations.

Many of these researchers focus on matrix solvers since they take a huge amount of time and can be easily parallelized for best time gains. Other researchers as may be seen, focus on EbE finite element solvers to overcome GPU memory limitations.

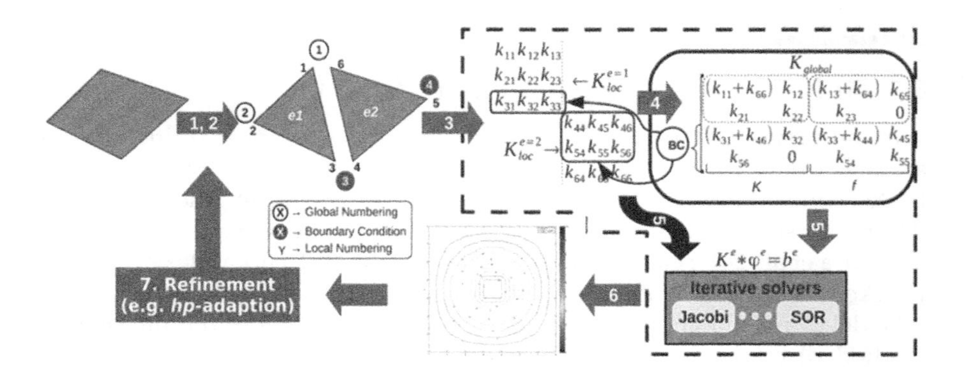

FIGURE 8.1
Steps in the classic finite element method (FEM) and the proposed changes for single element solution (SES) enclosed in dashed line (Fernandez et al., 2012).

8.4 Solution of Sparse, Symmetric Finite Element Equations

As noted, the biggest load in finite element field computation is in solving the matrix solution after assembly. GPU capabilities are now exploited and we have unlimited speed-up (for the large matrix size we have tried, although for even larger matrix sizes communication problems could possibly lead to saturation) by resorting to EbE processing begun in the mid-1980s (Carey et al., 1988) to overcome limitations on the PC which we now exploit on the GPU. Figure 8.2 shows speed-up with Jacobi conjugate gradients and Gauss iterations. Speed-up with the former saturates but the latter method provides unlimited speed-up which is not shown in Figure 8.2 only because we stopped at a matrix size of 100,000 as a result of our being unable to get a comparison on a CPU for larger sizes (Hoole et al., 2014).

The iterative conjugate gradients (CG) algorithm is usually used for solving a symmetric, positive definite system of linear equations (Nazareth, 2009). Algorithm 1 describes the CG method for the system of linear equations $[A]\{x\} = \{b\}$; where $[A]$ is a real, positive definite, symmetric matrix and $\{x\}$ is the initial solution of the system which is improved in each iteration k. In the case of implementation of this algorithm, we use CUDA C on the GPU for parallel implementation and C++ on the CPU for sequential implementation. CG has a time complexity of $O(n^{3/2})$ and $O(n^{4/3})$ for two-dimensional and three-dimensional problems respectively (Painless, n.d.).

When EbE processing was developed for the Gauss algorithm, in solving $[A]\{x\} = \{b\}$; the older displaced Gauss iterations are brought back because they use the old iteration k's value for computing every x_i in iteration $k+1$ and are therefore essentially parallelizable unlike the Gauss Seidel iterations which strictly involve sequential processing. Therefore, the computation of a particular x_i value is independent of the computation of all other x_i values for that iteration and is therefore parallelizable:

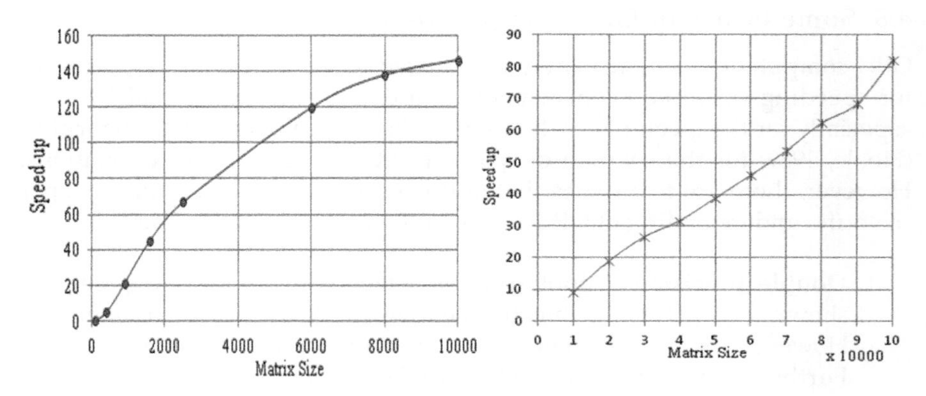

FIGURE 8.2
Speed-up vs matrix size. Left: Jacobi conjugate gradients; right: Gauss.

$$x_i^{k+1} = \frac{1}{A_{ii}} \left\{ b_i - \sum_{k=1}^{i-1} A_{ij} x_i^k - \sum_{k=i+1}^{n} A_{ij} x_i^k \right\} \tag{8.1}$$

This is inefficient in the context of sequential computations for which Gauss–Seidel is better. But in the case of parallelization it is highly efficient as was laid out by Mahinthakumar and Hoole (1990) and Carey et al. (1988) for shared memory systems. Speed-ups were just below $(p-1)$ where p is the number of processors. If $[D]$ is the matrix $[A]$ with all off-diagonal elements eliminated, then the Gauss iterations yield

$$[D]\{x\}^{k+1} = \{b\} - [A - D]\{x\}^k \tag{8.2}$$

Algorithm 8.1

```
1:  r_0 = b - Ax_0
2:  p_0 = r_0   k = 0
3:  while_stopping_condition_do
4:  α_0 = r_k^T r_k / P_k^T AP^k
5:  x_{k+1} = x_k + α_k P_k
6:  r_{k+1} = r_k + α_k AP_k
7:  if -r_{k+1} is - sufficiently - small - then - exit - loop
8:  β_0 = r_{k+1}^T r_{k+1} / r_k^T r_k
9:  P_{k+1} = r_{k+1} + β_k P_k;   k = k + 1
10: end - while
```

8.5 Some Issues in GPU Computation

GPU computation is increasingly recognized as a magnificent opportunity for speeding up finite element field computation through parallelization, especially for the incomplete Cholesky conjugate gradient (ICCG) algorithm (Hoole, 1989) for the solution of sparse positive definite matrix equations. However, there are five issues that still need to be addressed for a proper scientific understanding of GPU computation and its effective use.

1. **Double precision**: Double precision work on the GPU has been shown to take twice as much time than single precision arithmetic. However, there are papers where this is not so (Bogdan et al., 2012). Further, even as we increase the number of processes, the speed-up tends to saturate as in Figure 8.2. Communications is a factor but the exact nature is still not known.

2. **Accuracy**: It is agreed that GPU computations give answers different from CPU computation in the least significant digits. But contrasting claims exist on whether CPU computations or GPU computations are superior (Whitehead and Fit-Florea, 2011). We have already noted a credible explanation on why parallel computation answers are less accurate – because processes in the course of parallelization are broken up and put back together, thereby leading to greater round-off error (Miriam et al., 2012). However, Devon Yablonski, in his master's thesis analyzes numerical accuracy issues that are found in many scientific GPU applications due to floating-point computation. Accumulating values serially (see Figure 8.3a) will sometimes result in a large value that each successive small value is added to, resulting in diminished accuracy of the results. The reduction style of computation (see Figure 8.3b) avoids the issue in accumulating floating-point values in a way that is similar to binning (Devon, 2013). Numbers surrounded by a box in Figure 8.3 represent the actual result's floating-point value with seven digits. In GPU addition, the use of different threads reduces overflow.

As that author puts it, two widely held myths about floating points on GPUs are that the CPU's answer is more precise than the GPU version and that computations on the GPU are unavoidably different from the same computations on a CPU. He appears to have dispelled both myths by studying a specific application: digital breast tomosynthesis (DBT). He described situations where the general-purpose

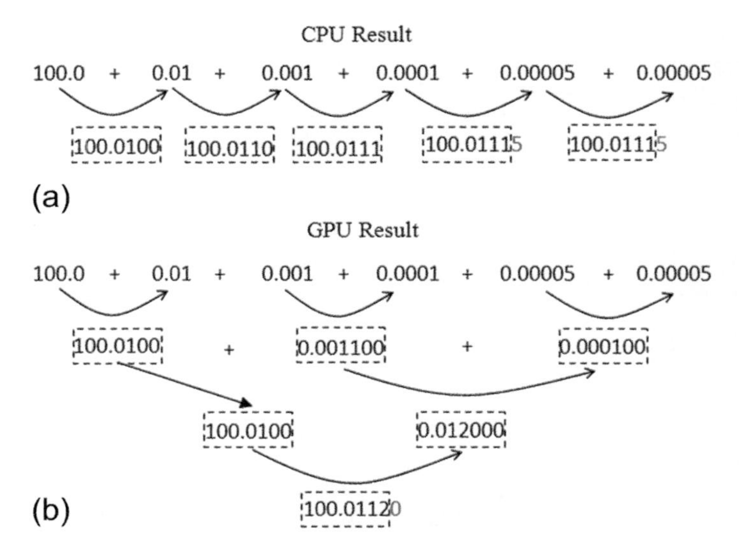

FIGURE 8.3
Serial addition: losing precision.

computing on a GPU provides greater precision in the application. DBT was analyzed and differences between the CUDA C (GPU) and C++ (CPU) implementations were completely resolved (Devon, 2013). However, if GPU computations are truly more accurate, one would expect the vendors of GPU platforms to say so loudly and clearly. Given these countervailing claims, the subject must be kept open until further evidence emerges from wider sources.

3. **Unexpected Ups and Downs in Speed-up**: There is sometimes an up and down speed-up (Krawezik and Poole, 2009). The causes need to be investigated as to whether this occurs because of padding of arrays in memory, cache action, memory coalescing, etc. In a study we did of GPU computation for finite element optimization by the GA (Karthik et al., 2015), the speed-up showed an unexpected erratic up and down. This result is seen in other works too such as that of Krawezik and Poole (2009) as shown in Figure 8.4 from Kiss et al. (2012) and Figure 8.5 which also is from Krawezik and Poole (2009). In the absence of an explanation we carry on but a real understanding of the method to obtain the best speed-up requires some investigation.
4. **Speed-up**: The general impression from reading the GPU literature is that GPU computation offers much better speed-up than multi-computing-element machines. The technical guide by NVIDIA is very explicit that the actual gains with GPU depend on the application. But even this claim is a little rosier than warranted by facts because as widely reported, some engineering applications like

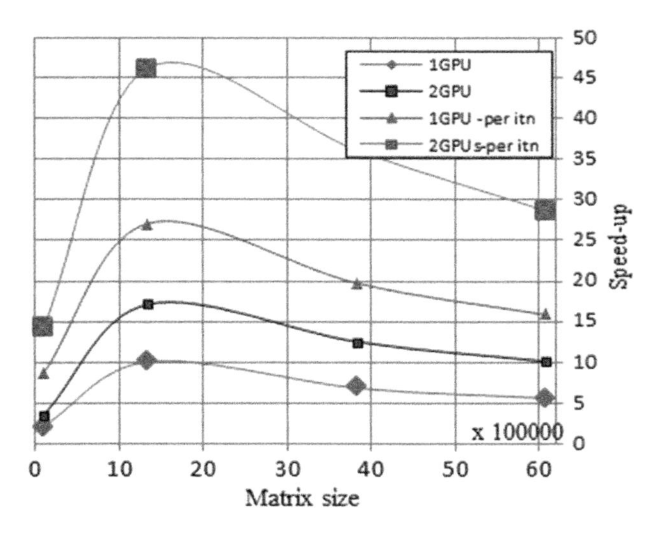

FIGURE 8.4
Speed-up vs matrix size (Kiss et al., 2012).

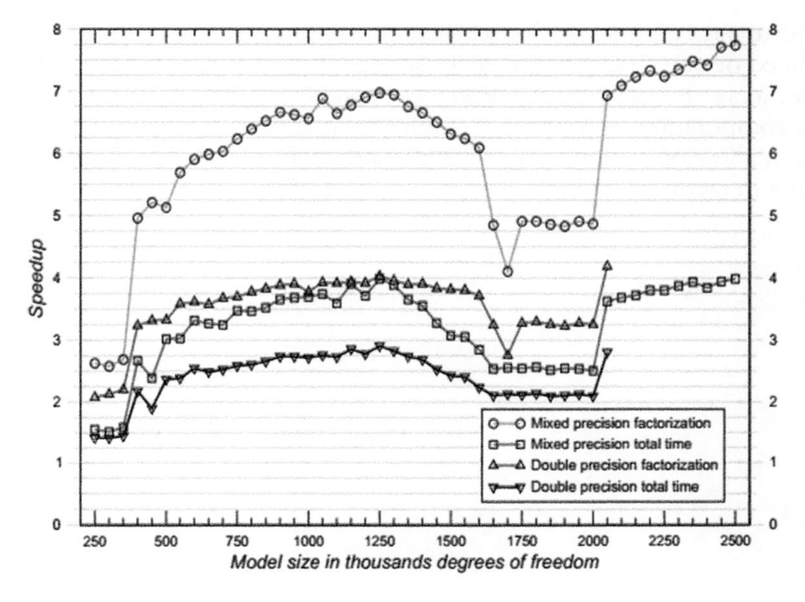

FIGURE 8.5
Unexpected behavior of gain (Krawezik and Poole, 2009).

finite element matrix assembly and (lower upper) LU decomposition give only a gain of 2 or so (Bogdan et al., 2012) while sparse CG in our implementation has given us a gain of 147 (see Figure 8.2).

5. **Compiler**: While expert programmers have programmed their problems in CUDA C to reap the benefits of speed-up, it is to be noted that the compiler is very difficult to work with. Error messages are still difficult to use in debugging. When memory is violated in one function, the program crashes in another without a proper error message. Debugging is therefore more difficult in CUDA C, particularly because we cannot print intermediate outputs directly from the GPU. All programming languages go through these developmental stages and we may expect that once better compilers come onto the scene, matters will be much easier. Future CUDA C development must focus on the ease of using compilers.

8.6 Conclusions

The GPU offers an invaluable platform for efficiency and speed-up in finite element optimization. GPU computations are emerging only now as a science and are yet to be understood fully. As such some applications see large speed-up and others low. Some results from the same application when parallelized on the GPU give excellent speed-up while the results of other programmers of the same application yield poor speed-up. While the low cost of GPU computations is a strong factor in its favor vis-à-vis multi-CE computation, in a few operations like finite element matrix assembly the advantages are not in favor of GPUs.

9

Coupled Problems

9.1 The Electrothermal Problem

In electromagnetic field problems, optimization methods have been successfully developed and applied efficiently. But, most of these methods only deal with the direct, single-field problem. Real problems on the other hand, are more complex and often coupled, with two or more physical systems interacting.

Heavy currents always lead to heating through the joule effect. This heat is often undesirable as in electrical machinery like alternators where the heat not only diminishes the efficiency of the generator but also can damage the insulation. In other cases this heat can be beneficial as in a) the metallurgical industry where the heat is used to melt the ore and mix it through electromechanical forces or b) hyperthermia treatment in oncology where cancerous tissue is burned off albeit with lower currents, achieving the heating by stronger eddy currents through a higher 1 kHz frequency (Vanderplaats, 1984; Brauer et al., 2001; Hoole et al., 1990). Whatever the situation, it is often desirable to accomplish a particular thermal distribution – whether to save an alternator from overheating or to accomplish the necessary melting of the ore or to burn cancerous tissue without hurting healthy tissue.

As shown in Figure 9.1, the design process involves setting the parameters {p} that describe the electro-heat system (consisting of geometric dimensions, currents in magnitude and phase, and material values), and solving the eddy current problem for the magnetic vector potential $\boldsymbol{A}$ (Marinova and Wincenciak, 1998):

$$-\frac{1}{\mu}\nabla^2 \boldsymbol{A} = \boldsymbol{J} = \sigma_e \boldsymbol{E} = \sigma_e\left[-j\omega \boldsymbol{A} - \nabla\varphi\right] \tag{9.1}$$

where μ is the magnetic permeability, σ_e is the electrical conductivity, $\boldsymbol{E}$ the electrical field strength and $-\nabla\varphi$ is the externally imposed electric field driving the current (Pham and Hoole, 1995). The frequency ω is relatively low (50 Hz to 1 kHz) so that the current density $\boldsymbol{J}$ has only the conduction term $\sigma_e \boldsymbol{E}$

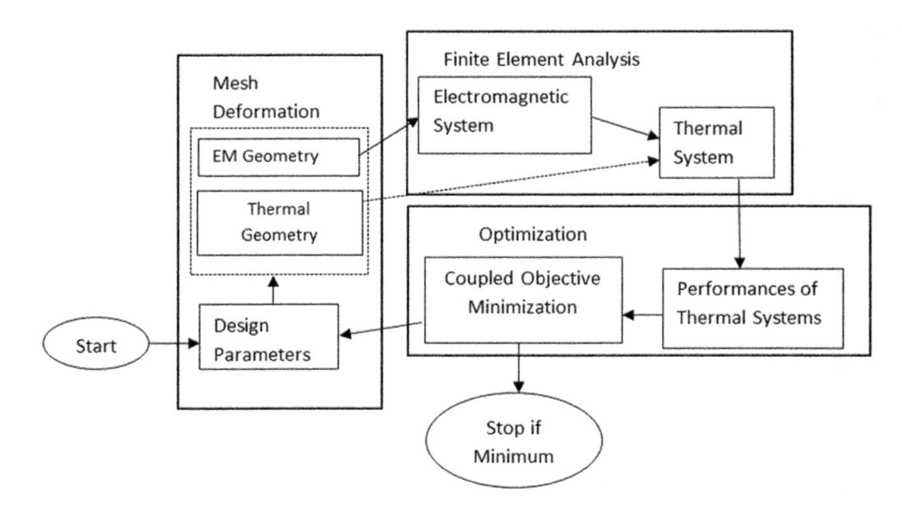

FIGURE 9.1
Finite element analysis and optimization of coupled magneto-thermal problems.

and no displacement term $j\omega\epsilon \boldsymbol{E}$. After finding $\boldsymbol{A}$, we compute the joule heating density q from

$$\boldsymbol{E} = -j\omega\boldsymbol{A} - \nabla\varphi \tag{9.2}$$

and

$$q = \frac{\sigma_e}{2}\boldsymbol{E}^2 \tag{9.3}$$

Once we have the heat source distribution q, the second problem of finding the resulting temperature is addressed by solving (Marinova and Mateev, 2012),

$$-\sigma_t \nabla^2 T = q \tag{9.4}$$

where σ_t is the thermal conductivity. Here we assume that σ_t is isotropic (meaning that heat flow in all directions bears the same constant of proportionality to $-\nabla T$). We also assume q is only from joule heating. We avoid more exact details because our purpose here is to establish the feasibility of the numerical methods we use and to demonstrate it.

For solving Equation (9.4) for which the same methodology as used for the Poisson equation for the electric scalar potential is applicable with the added complexity of having to solve the Poisson equation for A and then for T, the electrothermal finite element computational work is depicted in Figure 9.1.

Since the problem began with defining the parameters of system description {p}, we note that T = T({p}) since the computed T will depend on the values of {p}. When a particular temperature distribution $T_0(x, y)$ is desired, the problem is one of finding that {p} which will yield

$$\mathrm{T}\left(\{\mathrm{p}\}\right) = T_0. \tag{9.5}$$

This is recognized as inverting Equation (9.5) to find {p}, and therefore it is referred to as the inverse problem, which is now well understood in the literature, particularly when we are dealing with one branch of physics like electromagnetics.

In multi-branch, coupled physics problems like the electro-heat problem under discussion, {p} is often defined in the electromagnetic system and the object function F in the thermal system (Pham and Hoole, 1995). Further, when dealing with numerical methods such as the finite element method, T is given at the nodes (although technically $T(x,y)$ may be derived from the finite element trial functions using the nodal values) and T_0, rather than being a function of x and y, is more conveniently defined at measuring points i, numbering say m. The design desideratum then may be cast as an object function F to be minimized with respect to the parameters $\{p\}$

$$F = F\left(\{p\}\right) = \frac{1}{2}\sum_{i=1}^{m}\left[T^i - T_0^i\right]^2 \tag{9.6}$$

where T^i is T as computed at the same m points i where T_0 is defined. The optimization process, by whatever method (Vanderplaats, 1984), keeps adjusting $\{p\}$ until F is minimized, making T^i come as close to T_0^i as it can – as close as it can rather than exactly to T_0^i because our design goal as expressed in Equation (9.6) may not always be realistic and achievable. At that point $\{p\}$ would represent our best design. The computational process in inverse problem solution as it relates to Figure 9.1 is shown in Figure 9.2. It requires solving for the vector of design parameters $\{p\}$. We first, generate the mesh from the latest parameter set $\{p\}$ and get the corresponding finite element solution for $\boldsymbol{A}$. Then from $\boldsymbol{A}$, we compute the rate of heating Q. Thereafter, we find the finite element solution for the temperature T. From T we evaluate the object function F. The method of optimization used will dictate how the parameter set of device description $\{p\}$ is to be changed depending on the computed F.

To set this work in context, we note that this two-part optimization problem has been worked on before and we need to state what is novel here. In our work the design vector $\{p\}$ includes geometric dimensions. This is what is new (Karthik et al., 2015). That is, we do shape optimization by an accurate finite element eddy current problem followed by another accurate finite element

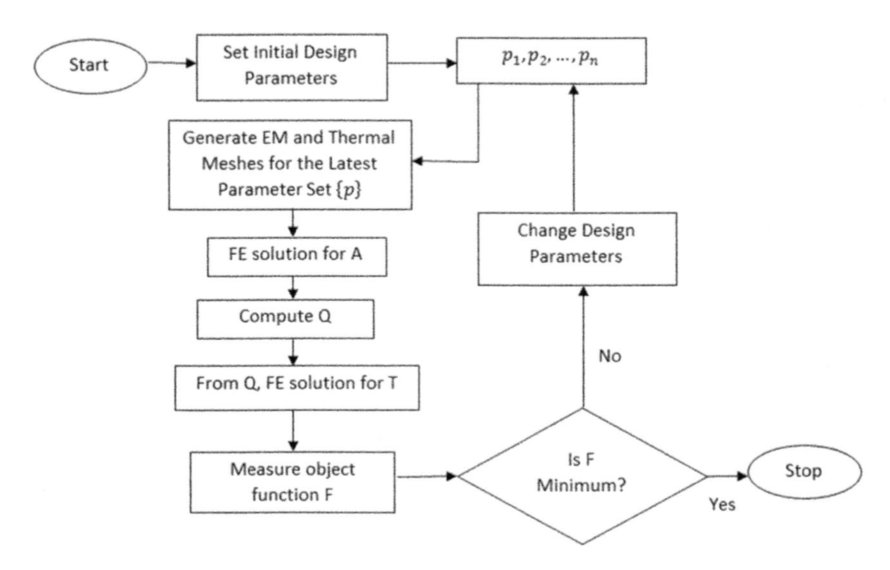

FIGURE 9.2
The design cycle for the coupled magneto-thermal problem.

temperature problem and this two-part finite element problem is cyclically iterated on by the inherently parallel genetic algorithm (GA). The inherent parallelism of the GA was recognized as far back as in 1994 by Henderson (1994) and done on parallel computers. In 2009, Wong and Wong (2009) and Robilliard et al. (2009) took that parallelization work on single field problems to the Graphics Processing Unit (GPU) (described in Chapter 8), working still with single field systems. We apply their work to the coupled field problem where solutions can take very long and reducing solution times is critical in creating practicably quick engineering design systems without sacrificing accuracy.

The coupled electro-heat optimization problem has indeed been tackled before as we found in a Web of Science Core Collection literature search, but with key differences. Pham and Hoole (1995) have used two finite element solutions and done shape optimization. But as pointed out below, that process is difficult to build general-purpose software with because of difficulties with the gradient optimization; that deficiency is being rectified here. Siauve et al. (2004) solve eddy current and thermal problems sequentially but, what they optimize is not shape but the antenna currents to obtain a specific absorption rate (SAR) for hyperthermia treatment. It is not shape optimization that they do. Thus, the same mesh suffices for every optimization iteration and they do not need to tackle the problem of changing meshes in every iteration and along every line search within an iteration. Likewise, in the electro-heat optimization work of Nabaei et al. (2010) too, the current distribution in furnace transformers is being optimized using an impedance model on MATLAB and there is no shape optimization. Similarly, Battistetti et al. (2001) avoid a two-part finite element solution by using an analytical solution for the electromagnetic part and a finite difference solution for the thermal part.

9.2 Finite Element Computation for the Electrothermal Problem

Finite element computation for the electrothermal problem is a little more complex than when the heat source q is given. The difference from Section 2.9 is that q is given there. But here we have to find q in terms of $\boldsymbol{J}$. We demonstrate the required procedure in the hand calculation below.

By combining $\boldsymbol{J} = \sigma_e \boldsymbol{E}$ of Equation (9.1) and $q = (\sigma_e/2)\boldsymbol{E}^2$,

$$q = \frac{\boldsymbol{J}^2}{2\sigma_e} \tag{9.7}$$

For the thermal problem governed by $-\sigma_t \nabla^2 T = q$ of Equation, (9.4), the functional will be

$$\mathcal{L}(T) = \iint \left(\left[\frac{1}{2}\sigma_t (\nabla T)^2 - qT \right] \right) dR = \sum_{\text{Elements}} \left\{ \frac{1}{2}\sigma_t \left(b^2 + c^2\right) A^r - \iint qTdR \right\} \tag{9.8}$$

where the terms b, c, and A^r are exactly as in Section 2.9.

Let us consider $\iint qTdR$, substituting Equation (9.7) using the first-order interpolation for $\boldsymbol{J}$

$$\begin{aligned}
\iint qTdR &= \iint \underline{T}^t \frac{\boldsymbol{J}^2}{2\sigma} dR = \iint \underline{T}^t \frac{1}{2\sigma} \underline{\boldsymbol{J}}^t \underline{\boldsymbol{J}} dR \\
&= \iint \frac{1}{2\sigma} \begin{bmatrix} T_1 & T_2 & T_3 \end{bmatrix} \begin{bmatrix} \zeta_1 \\ \zeta_2 \\ \zeta_3 \end{bmatrix} \begin{bmatrix} \boldsymbol{J}_1 & \boldsymbol{J}_2 & \boldsymbol{J}_3 \end{bmatrix} \begin{bmatrix} \zeta_1 \\ \zeta_2 \\ \zeta_3 \end{bmatrix} \begin{bmatrix} \zeta_1 & \zeta_2 & \zeta_3 \end{bmatrix} \begin{bmatrix} \boldsymbol{J}_1 \\ \boldsymbol{J}_2 \\ \boldsymbol{J}_3 \end{bmatrix} dR \\
&\iint \tfrac{1}{2\sigma} \underline{T}^t \begin{bmatrix} \zeta_1 J_1 & \zeta_1 J_2 & \zeta_1 J_3 \\ \zeta_2 J_1 & \zeta_2 J_2 & \zeta_2 J_3 \\ \zeta_3 J_1 & \zeta_3 J_2 & \zeta_3 J_3 \end{bmatrix} \begin{bmatrix} \zeta_1^2 & \zeta_1\zeta_2 & \zeta_1\zeta_3 \\ \zeta_2\zeta_1 & \zeta_2^2 & \zeta_2\zeta_3 \\ \zeta_3\zeta_1 & \zeta_3\zeta_2 & \zeta_3^2 \end{bmatrix} \begin{bmatrix} J_1 \\ J_2 \\ J_3 \end{bmatrix} dR \\
&= \frac{1}{2\sigma} T^r \frac{A^r}{60} \begin{bmatrix} 6J_1^2 + 2J_2^2 + 2J_3^2 + 4J_1J_2 + 4J_1J_3 + 2J_2J_3 \\ 2J_1^2 + 6J_2^2 + 2J_3^2 + 4J_2J_1 + 4J_2J_3 + 2J_1J_3 \\ 2J_1^2 + 2J_2^2 + 6J_3^2 + 4J_3J_1 + 4J_3J_2 + 2J_1J_2 \end{bmatrix}
\end{aligned} \tag{9.9}$$

Now substituting Equation (9.9) in Equation (9.8),

$$\sum_{\text{Elements}} \left\{ \frac{1}{2}\underline{T}^t \sigma_t \frac{A^r}{60} \begin{bmatrix} b_1^2 + c_1^2 & b_1 b_2 + c_1 c_2 & b_1 b_3 + c_1 c_3 \\ b_2 b_1 + c_2 c_1 & b_2^2 + c_2^2 & b_2 b_3 + c_2 c_3 \\ b_3 b_1 + c_3 c_1 & b_3 b_2 + c_3 c_2 & b_3^2 + c_3^2 \end{bmatrix} \underline{T} \right.$$

$$\left. -\frac{1}{2\sigma}\underline{T}^t \frac{A^r}{60} \begin{bmatrix} 6J_1^2 + 2J_2^2 + 2J_3^2 + 4J_1J_2 + 4J_1J_3 + 2J_2J_3 \\ 2J_1^2 + 6J_2^2 + 2J_3^2 + 4J_2J_1 + 4J_2J_3 + 2J_1J_3 \\ 2J_1^2 + 2J_2^2 + 6J_3^2 + 4J_3J_1 + 4J_3J_2 + 2J_1J_2 \end{bmatrix} \right\} \tag{9.10}$$

$$= \sum_{\text{Elements}} \left\{ \frac{1}{2}\underline{T}^t [P] \underline{T} - \underline{T}^t \underline{Q} \right\}$$

where, in corresponding notation, the local matrices

$$[P] = \sigma_t A^r \begin{bmatrix} b_1^2 + c_1^2 & b_1 b_2 + c_1 c_2 & b_1 b_3 + c_1 c_3 \\ b_2 b_1 + c_2 c_1 & b_2^2 + c_2^2 & b_2 b_3 + c_2 c_3 \\ b_3 b_1 + c_3 c_1 & b_3 b_2 + c_3 c_2 & b_3^2 + c_3^2 \end{bmatrix} \tag{9.11}$$

and

$$\underline{Q} = \frac{1}{2\sigma}\frac{A^r}{60} \begin{bmatrix} 6J_1^2 + 2J_2^2 + 2J_3^2 + 4J_1J_2 + 4J_1J_3 + 2J_2J_3 \\ 2J_1^2 + 6J_2^2 + 2J_3^2 + 4J_2J_1 + 4J_2J_3 + 2J_1J_3 \\ 2J_1^2 + 2J_2^2 + 6J_3^2 + 4J_3J_1 + 4J_3J_2 + 2J_1J_2 \end{bmatrix} \tag{9.12}$$

Finite element analysis provides the solution to Equation (9.4) by applying certain boundary conditions. The local matrices of elements will be added to the corresponding position of the global matrix to be solved for $\underline{T}$. This leads to the finite element matrix equation

$$\left[P^g\right]\underline{T} = \underline{Q}^g \tag{9.13}$$

Equation (9.13) is solved in the same way as explained for single-field problems, calculating local matrices and adding them to the global matrices of Equation (9.13) to be solved for $\underline{T}$.

9.3 GPU Computation for Genetic Algorithms for Electro-Heat Problems

Naturally, the computational work in two-physics electro-heat problems in finite element GA optimization is far beyond that for a single finite element solution. We have a two-part coupled problem, and we have to solve the magnetic field for **A** and then the thermal problem where we solve for the temperature T. For realistic problems this has to be done several times – indeed tens of thousands of times – in searching the solution space for the minimum object function. Wait times can be excessive, making optimization practicably infeasible.

To cut down solution time, parallel processing needs to be resorted to (Hoole, 1990, 1991a,b). From the 1990s, multiprocessor computers have been tried out. Typically with n processors (or computing elements), solution time could be cut down by almost a factor of $(n-1)$ – that is $(n-1)$ rather than n because one processor is reserved for controlling the other $(n-1)$, and *almost* a factor of $(n-1)$ rather than *exactly* $(n-1)$ because of the additional operations of waiting while one processor accesses the data being changed by another. Although much of this parallelization work was moved to cheaper engineering workstations by the late 1990s (Hoole and Agarwal, 1997), the restrictions on the number of processors remained.

Recently, the GPU, endowed with much computing prowess to handle graphics operations, has been exploited to launch a computational kernel as several parallel threads as explained in the previous chapter (Hoole et al., 2014). This is ideally suited for object function evaluation as the kernel so that multiple threads can perform the finite element analyses and evaluation of F for each $\{p\}$ in parallel. NVIDIA Corp's GPUs invented in 1999 and the Compute Unified Device Architecture (CUDA) (NVIDIA, n.d.) are today available on practically every PC as a standard and are increasingly exploited with more and more applications being ported thereto. Significantly, the number of parallel threads is not limited as on a shared memory supercomputer. Wong and Wong (2009), Robilliard et al. (2009), and Fukuda and Nakatani (2012) have shown that the GA with its inherently parallel structure may be efficiently implemented on the GPU to optimize magnetic systems. We extend that here to coupled problems.

Cecka et al. (2012) have also created and analyzed multiple approaches in assembling and solving sparse linear systems on unstructured meshes. The GPU coprocessor using single-precision arithmetic achieves speedups of 30 or so in comparison with a well-optimized double-precision single core implementation (Hoole et al., 1991). We see that this is far better than the factor of just below seven that is possible on a very expensive eight processor supercomputer. So this is the way we will go, using the GPU to process the GA algorithm in parallel as described in the previous chapter.

We fork the fitness value computations as in Figure 9.3. This fitness value calculation is the time-consuming part as it involves both mesh generation, and finite element calculation. This forms a kernel that will be launched in parallel threads. Therefore, we divide the GPU threads and blocks of the same number as the population size and compute fitness values simultaneously (see Figure 9.3). Since we have 65,536 blocks (2^{16}) and 512 threads in a general GPU, we can go up to a population size of $65{,}536 \times 512 = 33{,}554{,}432$

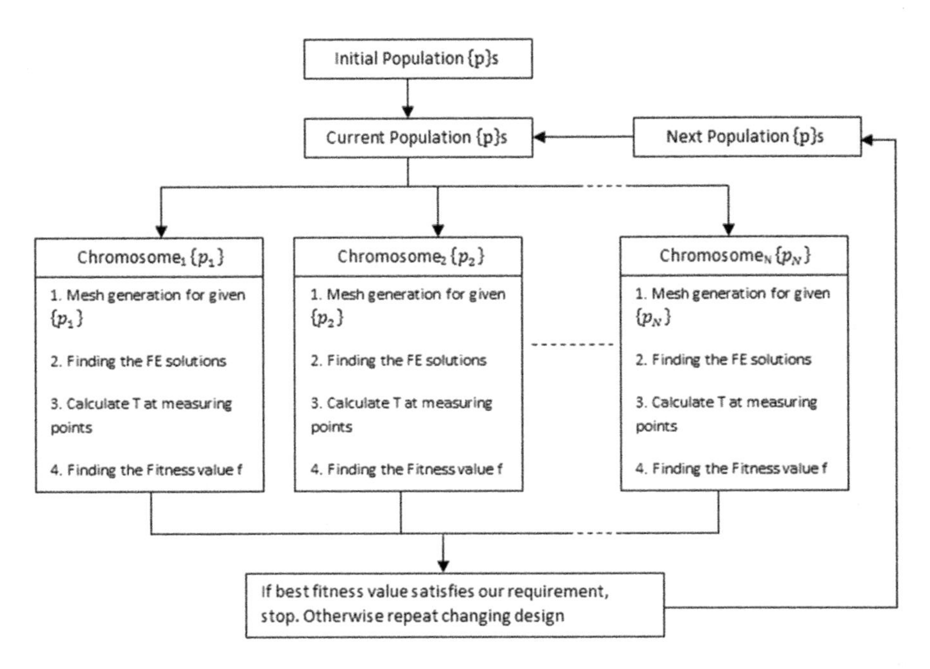

FIGURE 9.3
The parallelized process of the GPU.

using the one GPU card on a PC. Since we do not need such a large population size for effective optimization, this is not restrictive. All the finite element calculation parts were programmed on the GPU in the CUDA C language. So when, given each {*p*}, we launch the fitness computation kernel as several threads – one for each {*p*}. The fitness score for all chromosomes is thereupon calculated at the same time in parallel. But for each chromosome, the finite element calculation will be done sequentially (see Figure 9.3) and not parallelized along the lines of Cecka et al. (2011) because that would be attempting to fork within a fork. (For the ability to fork an already forked thread, see the discussion by Hoole et al., 2014).

9.4 Shaping an Electro-Heated Conductor

The test problem chosen (Pham and Hoole, 1995) is a simple one on which the method can be demonstrated and its feasibility established. Shown in Figure 9.4 a is a rectangular conductor which is heated by a current through it. The equi-temperature profiles would be circle-like around the

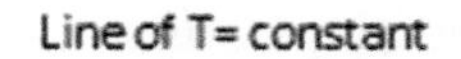

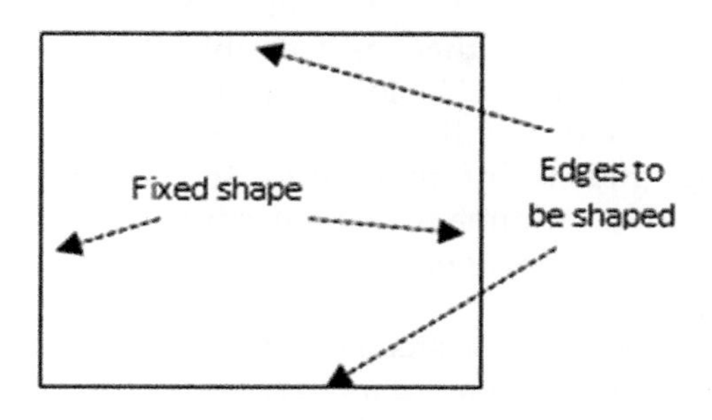

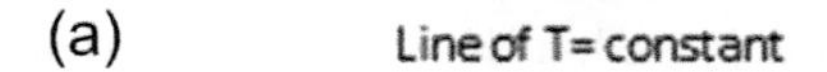

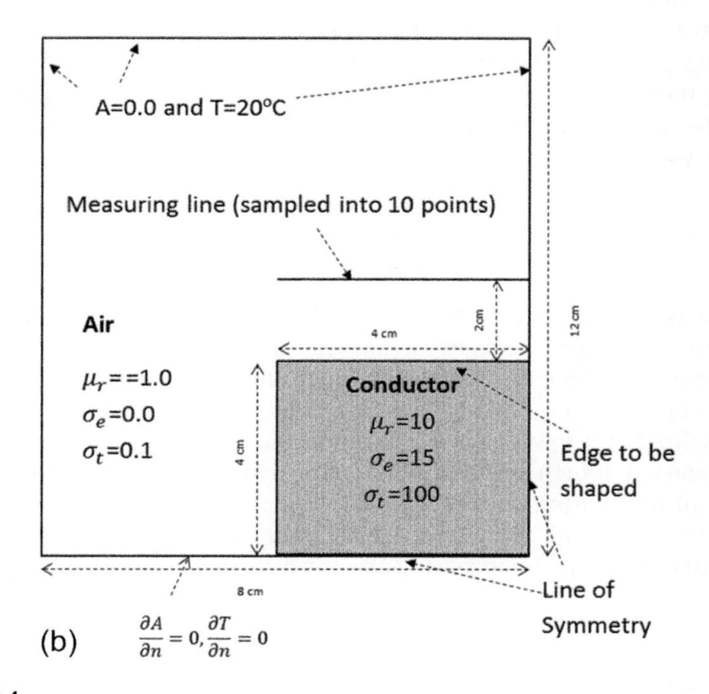

FIGURE 9.4
Numerical model for coupled electro-heat problem. (a) Electrically heated conductor: the actual geometry, (b) Symmetric quarter: boundary value problem (nominal values shown for magnetic permeability μ, electrical conductivity σ_e, and thermal conductivity σ_t).

conductor. But we want a constant temperature along two lines parallel to the pre-shaping rectangular conductor's two opposite edges which are to be shaped. The question is this: how should that edge be reshaped to accomplish a constant temperature along the lines on either side of the conductor? This is the same problem that has been solved by the gradient method which, as noted by Pham and Hoole (1995), needs an alternative computational process because of the difficulties in constructing general-purpose software yielding gradient information for the coupled problem and the mesh-induced fictitious minima which cause problems (Weeber and Hoole, 1991) and need special mesh generators to address them (Krishnakumar and Hoole, 2005).

Figure 9.4b presents the associated boundary value problem formed from a quarter of the minimal system for analysis consisting of a square conductor (with $\mu_r = 10, \sigma_e = 15\,\text{S/m}$, and $\sigma_t = 0.1$ W/m/°C). A current density $\boldsymbol{J}_0 = 10 + j_0\ \text{A/m}^2$ has a relatively low frequency of $\omega = 10/\text{s}$, which has been kept deliberately low to avoid a very fine mesh, our purpose here being to investigate and establish methodology rather than to solve large problems in their full complexity. The top edge of the conductor has to be shaped to get a constant temperature profile of 60°C, at $y = 6$ cm, and at ten equally spaced measuring points in the interval $4\ \text{cm} \leq x \leq 8\ \text{cm}$ as shown in Figure 9.4 along the measuring line. We define the object function from Equation (9.12), and for testing we set the desired temperature as 60°C. Therefore, $T_0^i = 60$ and the F will be

$$F = \frac{1}{2}\sum_{i=1}^{10}\left[T^i - 60\right]^2 \tag{9.14}$$

An erratic undulating shape with sharp edges arose when Pironneau (1984) optimized a pole face to achieve a constant magnetic flux density (and this was overcome by the others through constraints (Subramaniam et al., 1994). Haslinger and Neittaanmaki (1996) suggest Bezier curves to keep the shapes smooth with just a few variables to be optimized, while Preis et al. (1990) have suggested fourth order polynomials which when we tried gave us smooth but undulating shapes. As such we follow a variant of the method by Subramaniam et al. (1994) and extend their principle, so as to maintain a non-undulating shape by imposing the constraints

$$h_1 > h_2 > h_3 > h_4 > h_5 > h_6 > h_7 \tag{9.15}$$

to ensure a smooth shape. Penalties were imposed by adding a penalty term to the object function F in Equation (9.14) whenever it fails to satisfy the conditions for constraints (Vanderplaats, 1984). Tolerance boundaries of each h_i were set to

$$1.5\ \text{cm} \leq h \leq 5.5\ \text{cm} \tag{9.16}$$

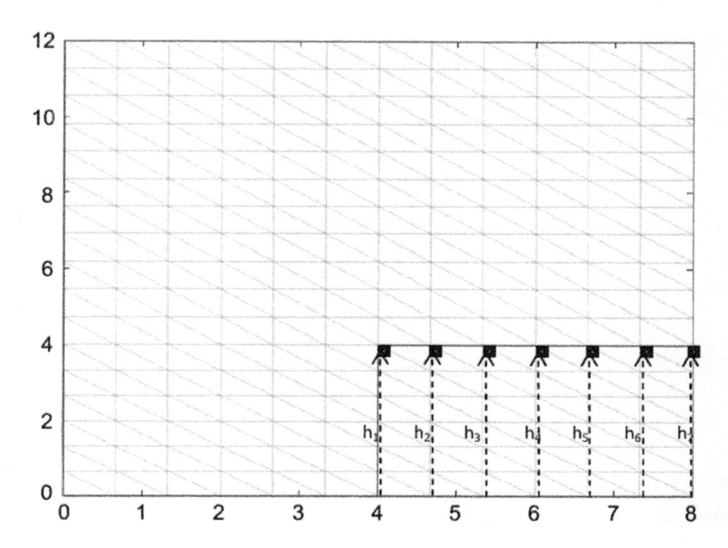

FIGURE 9.5
The parameterized geometry.

The parameterized problem-specific mesh is shown in Figure 9.5 where the device descriptive parameter set $\{p\}$ consists of the seven heights h_i. The numerical model was uniformly meshed with 234 nodes and 408 elements. This was deliberately kept crude to control debugging after succeeding with the method and establishing that it works as a method and as programmed on the GPU. Thereafter, the mesh was refined for greater accuracy.

In the process of optimization, as these heights h_i change, the mesh connections remain the same but the element sizes and shapes change. For the specific example shown in Figure 9.5, the heights h_i are divided into six pieces so the locations of the seven equally spaced seven nodes along the height h_i would change smoothly (Krishnakumar and Hoole, 2005). Accordingly, the length from above the edge of the conductor being shaped to the vertical boundary will be adjusted and divided into 11 equal lengths as shown in Figure 9.5.

Figure 9.6 shows temperature profile at different iterations. By the images we can see that when the iteration number increases, we get the desired constant temperature along the measuring line and move toward the optimum shape. Figure 9.7 shows the optimum shape of the conductor and temperature profile after 40 iterations for a population size of 512. The corner of the conductor rising toward the line where the constant 60°C temperature is desired is as to be expected. For as seen in Figure 9.8 (Karthik et al., 2015) (which shows the design goals being accomplished), the constant 60°C temperature is perfectly matched. The lower graph giving the initial temperature shows that the temperature drops above the corner of the conductor. Therefore, to address this, the corner has to rise close to the line of measurement to heat

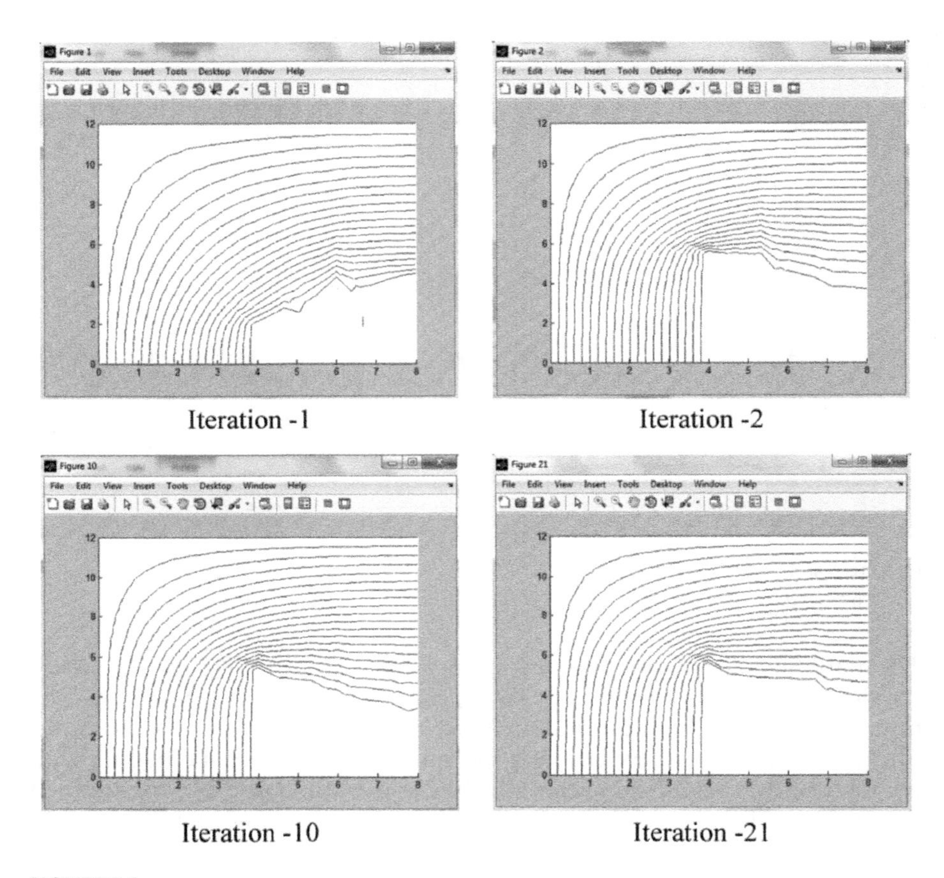

FIGURE 9.6
Equi-temperature lines at different iterations progressing to optimum.

the line above the corner and Figure 9.8 shows that this is what the optimization process has accomplished.

Significant speedup was accomplished with GPU computation as seen in Figure 9.9. No gains in speedup beyond a factor of 28 were seen after a population size of 500. The meandering nature of the gain after that may be attributed to the happenstance inherent to a statistical method like the GA.

9.5 Shape Optimization of Two-Physics Systems: Gradient and Zeroth-Order Methods

In the optimization of a single physics problem as in magnetostatics or eddy current analysis (Starzynski and Wincenciak, 1998) where the design

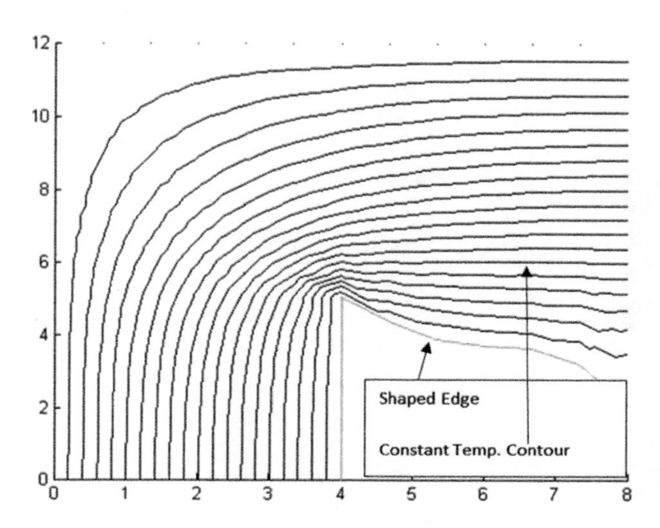

FIGURE 9.7
Optimal shape by the genetic algorithm.

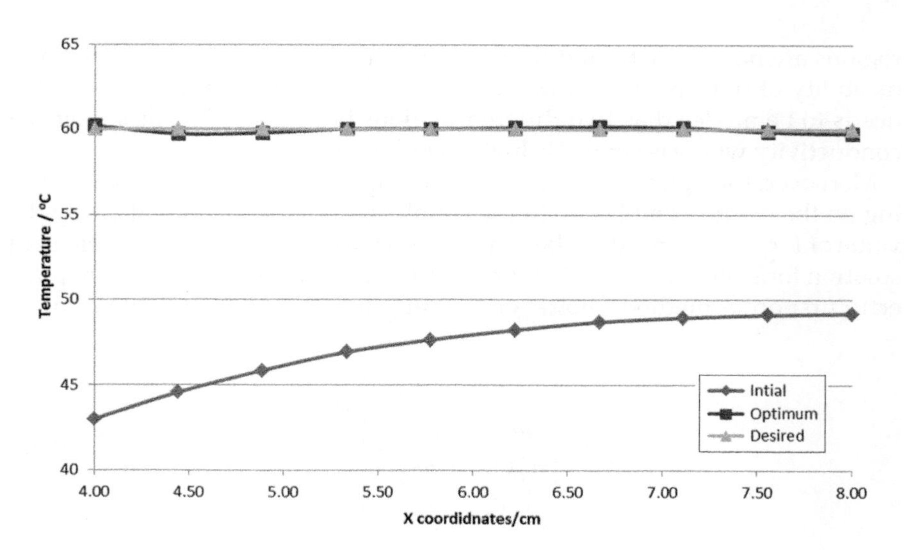

FIGURE 9.8
Temperature distribution: desired, initial, and optimized.

variables are defined in the vector $\{p\}$, we construct one finite element mesh, solve for the magnetic vector potential A, and then change $\{p\}$. The method by which we change $\{p\}$ depends on the method of optimization we employ (Vanderplaats, 1984). In coupled problems like the electro-heat problem under discussion, two different meshes are often required (Pham and Hoole, 1995). For example, at a copper-air boundary in magnetics, both

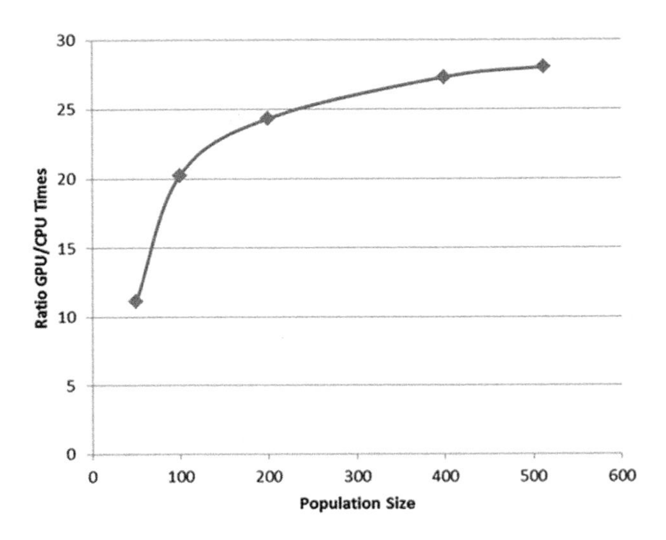

FIGURE 9.9
Speedup: GA optimization GPU time: CPU time.

regions are nonmagnetic and therefore have the same permeability, the permeability of free space μ_0. However, for the thermal problem, the geometry needs to be modeled as two different regions because air has little thermal conductivity whereas copper is highly conductive.

Moreover, the optimization process too imposes huge difficulties depending on the method employed. In the simpler zeroth-order methods, only the value of *F*, given {*p*}, needs to be computed. This takes simply a finite element solution for a mesh constructed for the present value of {*p*}. In the more powerful first-order methods, however, defining symbols, the gradient vector

$$\nabla F = \frac{\partial F}{\partial \{p\}} = \left\{ \frac{\partial F}{\partial p_i} \right\} = \begin{bmatrix} \frac{\partial F}{\partial p_1} \\ \vdots \\ \frac{\partial F}{\partial p_n} \end{bmatrix} \tag{9.17}$$

needs to be computed (Vanderplaats, 1984; Venkataraman, 2009). This may be by finite differences – that is, to get the derivative of F with respect to p_i, we need to evaluate F(p_i), then adjust p_i by a very small amount δp_i, redo the finite element solution (which means a new mesh has to be generated for the changed geometry), and then evaluate F, which would give us

$$\delta F(p_i + \delta p_i) = F(p_i + \delta p_i) - F(p_i) \tag{9.18}$$

thereby leading to the derivate by finite difference, $\delta F/\delta p_i$. This is computationally expensive because of having to be done for all m parameters p_i, which means the first finite element solution at {p} and evaluation of $F(\{p\})$ has to be followed by m finite element solutions with only a particular p_i adjusted by δp_i and then the evaluation of m values of δF. That is, m + 1 finite element meshes and solutions are required at great cost. Furthermore, even after all that work, it is known that the derivative by this finite difference process has poor accuracy (Hoole and Hoole, 1985). The less approximate way to obtain accurate derivatives is from the derivative information inherent to the finite element solution through the finite element trial function; that is, although the solution for the magnetic vector potential $\boldsymbol{A}$ is explicitly in terms of the values of $\boldsymbol{A}$ at the finite element nodes, we are really solving for $A(x,y)$ as given by the trial function which is expressed through the nodal values of $\boldsymbol{A}$ (Hoole, 1989a). That is, we solve the finite element Dirichlet matrix equation

$$[P]\{A\}=\{Q\} \tag{9.19}$$

As defined in Equations (2.116) and (2.117), although we explicitly solve for the nodal values {A}, it is really for the trial function $\mathbf{A}$(x,y) that we are solving. Both the Dirichlet matrix [P] and right-hand side {Q} in Equation (9.19) are expressed as known functions of the vertex coordinates of the finite elements of the mesh, permeability, and current density. After solving for {A} in Equation (9.19), differentiating the equation with respect to p_i we obtain (Pironneau, 1984; Coulomb et al., 1982; Gitosusastro et al., 1982; Hoole, 1995):

$$[P]\frac{d\{A\}}{dp_i}=\frac{d\{Q\}}{dp_i}-\frac{d[P]}{dp_i}\{A\} \tag{9.20}$$

where {A} has already been solved for, and $d\{Q\}/dp_i$ and $d[P]/dp_i$ are computable (Hoole, 1995).

The computational efficiency lies in the fact that the main work in Cholesky's scheme for matrix equation solution is in finding [L] by solving Equation (9.19). Thereafter, the forward elimination and back-substitution in solving Equation (9.20) for every i are trivial (Hoole et al., 1991).

Be that as it may, while solving Equation (9.3) is trivial, forming it is not because computing $\partial\{Q\}/\partial p_i$ and $\partial[P]/\partial p_i$ is, in terms of programming, an arduous task that is not easily amenable to building up as general-purpose software. As a particular p_i changes in value, some vertices of a few triangles will move (ibid.) and the analyst needs to keep track of whether one, two, or all three of the vertices of a triangle move by δp_i. Very complex coding is required that is problem-specific rather than general purpose. In a coupled problem, computing the derivatives of the finite element equations for temperature with respect to parameters in the magnetic problem is prohibitively complex, although, there are ways around it (Weeber and Hoole, 1992). Programming the problem-specific computations of $\partial\{Q\}/\partial p_i$ and $\partial[P]/\partial p_i$ is ill-advised because of the complexities.

Therefore, zeroth-order optimization methods, which are more slowly convergent than gradient methods, are the best route to go for optimizing coupled electro-heat problems. Simkin and Trowbridge (1991) aver that simulated annealing and the evolution strategy (a variant of the GA (Haupt, 1995)) take many more function evolutions. Although they are computationally intensive, they are far easier to implement, especially as general-purpose software. Indeed commercial codes that need to be general purpose use zeroth-order methods not necessarily because they are superior to gradient methods, but because once a finite element analysis program is developed by a company, giving it optimization capabilities only takes coupling it with an optimization package to which object function evaluations can be fed – whereas feeding both the object function and its gradient (as would be required when gradient methods are in use) would take extensive code development. In the context of single physics problems, Haupt (1995) advises that the GA is best for many discrete parameters and the gradient methods for where there are but a few continuous parameters. We rationalize this position on the grounds that gradient computation though difficult is more manageable when there are fewer parameters to optimize to minimize an object function. Indeed, we have gone up to 30 continuous parameters using gradient methods without problems.

But we are now dealing with multi-physics electro-heat problems to which these considerations based on single physics systems do not apply.

9.6 Electroheating Computation for Hyperthermia

The prefix hyper- (from the Greek over-above as in hyperactive) lies in contrast to hypo- (under as in hypodermic, under the skin). Hyperthermia refers to cancer treatment by heating tissue above the level where the tissue can survive.

In Section 9.4, the electrothermal problem was explained including how it would be solved and the shape of the current source optimized to get the desired temperature at the locations of interest. In this chapter, this electrothermal problem is extended onto its application side, particularly to the treatment of tumors using hyperthermia.

Historically, the treatment of cancerous tumors with hyperthermia can be traced back to 3000 B.C. when soldering sticks were inserted in tumors. Coley's toxin (C.R. UK, 2012) was introduced in the 19th century. This produced whole body hyperthermia which resulted in tumor regression. In recent years, using hyperthermia or related forms of therapy has increased tremendously.

Currently hyperthermia is an experimental treatment and is usually applied to late-stage patients. There are many heating methods such as whole-body heating using wax, hot air, hot water suits, infrared or partial body heating using radio-frequency (RF), microwave, ultrasound, and hot blood or fluid perfusion. Clinical and experimental results show a promising future for hyperthermia. However, the main problem is the generation and control of heat in tumors. An extremely important issue is to control the temperature distribution in the treated area to avoid excessive temperatures in the normal tissues surrounding the tumor (Chou, 1988). There are many studies on the treatment of cancer using hyperthermia which demonstrate that this aspect is still important and more research is needed in this matter (Kurgan, 2011).

Successful applications of electrothermal stimulation with the aid of a low frequency field to treat the tumors have been reported (Chou, 1988). There is evidence that hyperthermia can significantly reduce/treat the tumor, but it is not clear how it works and how it should be applied, (Stea et al., 1994). Some clinical studies have demonstrated the efficiency of thermal therapy in suspending tumor growth (Chou, 1988). Therefore, numerical modeling to distribute the optimal temperature in hyperthermia can be helpful in identifying the better treatments.

Hyperthermia treatment involves heating the tumor tissue to a temperature greater than 42°C without exceeding the normal physiologic temperature which is lower than 44°C–45°C. Therefore, the working temperature margin is very small. If the temperature at the tumor is lower than 42°C, there is no therapeutic effect. On the other hand, if the temperature is greater than 44°C–45°C, both healthy and tumorous cells are damaged (Das et al., 1999). The blood vessels in tumor cells have a greater diameter than in healthy cells, and therefore occupy a greater volume. The temperature at a tumor is greater than at the surrounding tissue during hyperthermia treatment. This is caused by the fact that the healthy cells have usually greater conductivity than the cancerous cells. In other words, we may say that tumor tissues are more sensitive to heat. The temperature rise in tumors and tissues is determined by the energy deposited and the physiological responses of the patient. When electromagnetic methods are used, the energy deposition is a complex function of frequency, intensity, the polarization of the applied fields, geometry and size of the applicator, and the geometry and size of the tumor. The final temperature elevations are not only dependent on the energy deposition but also on blood perfusion which carries away heat and affects thermal conditions in tissues. Generally, it is not easy to obtain an accurate temperature distribution over the entire treatment region during clinical hyperthermia treatment. Furthermore, to ensure that the temperature is within the desired range, the clinician usually monitors the temperature every few seconds. Thus, it is desirable to develop a mathematical model that can determine design parameters to optimize the temperature distribution in the target region before treatment. In this way the treatment efficiency can be assessed more precisely (Tsuda et al., 1996).

9.7 The Hyperthermia Model

Let us consider a cross section of the human thigh, shown in Figure 9.10 (Kurgan, 2011). It is assumed that the human thigh and bone inside have an ellipsoidal shape and that the tumor has a circular form. In their model, the tumor inside the human thigh was heated by external RF hyperthermia excitation at the frequency of 100 MHz. A similar model was used by Tsuda et al. (1996). Here they optimize the electrode configuration, particularly driving voltages for RF of 10 MHz using the gradient method.

In our work, this model is extended to a low frequency field such as at 1 kHz to 5 kHz. Here we optimize the shape of the current-carrying conductor to get the desired temperature distribution at the tumor. The numerical model of the problem to be solved is shown in Figure 9.11 (Sullivan, 1990). The geometrical dimensions of the numerical model are described in Table 9.1. The dimensions are taken from Gabriel et al. (1996).

Near the human thigh a coil with the exciting current is placed. The exciting current in the coil generates a sinusoidal electromagnetic field which induces eddy currents in the human body. These currents act as sources of joule heat and after some transient time a temperature distribution in the body is established. This final, steady state temperature distribution is what we shall seek to solve for.

As discussed in Section 1, for a given current $\boldsymbol{J}$, magnetic vector potential $\boldsymbol{A}$ is found by solving the eddy current problem

$$-\nabla\times\frac{1}{\mu}\nabla\times\boldsymbol{A}=\boldsymbol{J}_0-j\omega\sigma_e\boldsymbol{A} \tag{9.21}$$

The frequency ω is relatively low (1 kHz to 5 kHz) so that the current density J has only a conduction term $\sigma_e\boldsymbol{E}$ and no displacement term $j\omega\epsilon\boldsymbol{E}$. Solving

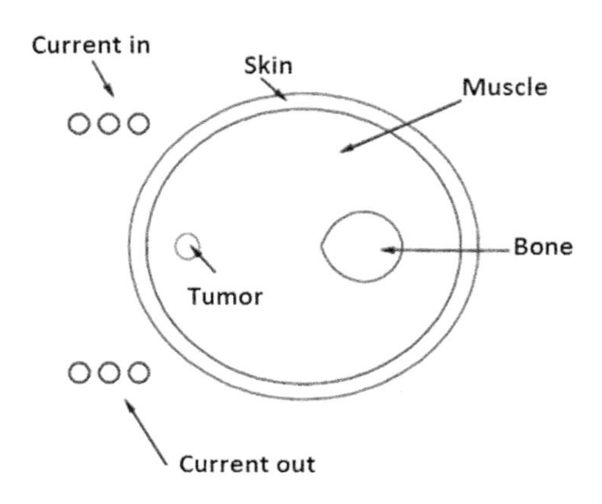

FIGURE 9.10
Cross section of the human thigh with a tumor.

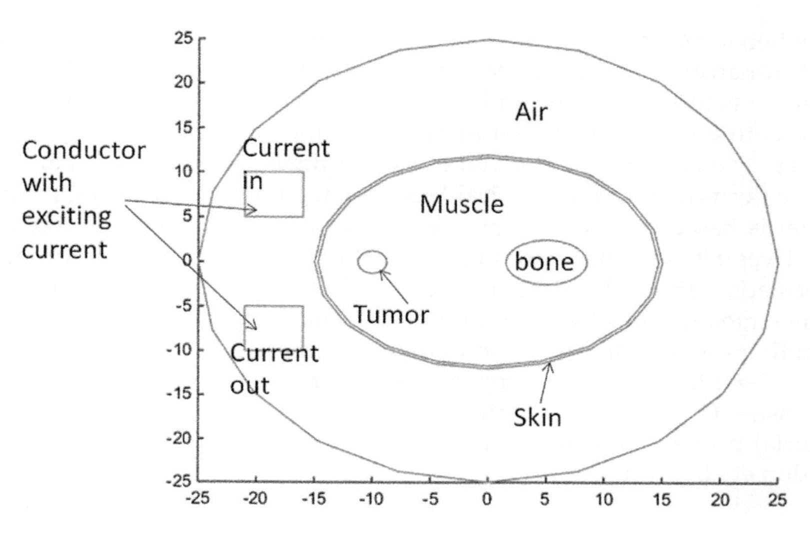

FIGURE 9.11
Numerical model of the problem.

TABLE 9.1

Geometrical Dimensions of the Model

Human body-length of semi-axes	$A = 15$ cm, $B = 12$ cm
Tumor radius	$R = 1.25$ cm
Skin thickness	$d = 1.5$ cm
Bone-length of semi-axes	$a = 3.5$ cm, $b = 2.5$ cm
Current-carrying conductor	Square shaped $l = 5$ cm

for $\boldsymbol{A}$ using the finite element method is explained in Sections 9.1 and 9.2. After finding $\boldsymbol{A}$ we compute the joule heating density q from

$$\boldsymbol{E} = -j\omega\boldsymbol{A} - \nabla\varphi \tag{9.2}$$

and

$$q = \frac{\sigma_e}{2}\boldsymbol{E}^2 \tag{9.3}$$

Once we have the heat source distribution q, the second problem is finding the resulting temperature. In Section 9.1 for the electrothermal problem, we solved the heat equation (Marinova and Mateev, 2012).

$$-\sigma_t\nabla^2 T = q \tag{9.4}$$

But when it comes to electric currents in a human body, no clear consensus exists for an appropriate mathematical model for the evaluation of the temperature field distribution in biological tissues. An extremely important work is that by Pennes (1948) in modeling the heat transfer in biological tissues. The equation he derived is named the bio-heat equation and this can be derived from the classical Fourier law of heat conduction. Pennes's model is based on the assumption of the energy exchange between the blood vessels and the surrounding tumor tissues. It may provide better information about the temperature distributions in the whole body. The Pennes model states that the total heat exchange between the tissue surrounding a vessel and the blood flowing in it is proportional to the volumetric heat flow and the temperature difference between the blood and the tissue. The expression of the bio-heat equation in a body with uniform material properties in transient analysis is given by (Kroeze et al., 2003; Gordon et al., 1976):

$$\rho C\frac{\partial T}{\partial t} - \sigma_t \nabla^2 T = \rho_b C_b \omega_b \left(T_b - T\right) + q + Q_{met} \tag{9.22}$$

where T is the body temperature, σ_t is the thermal conductivity, ρ is the density, C is the specific heat, T_b is the blood vessel temperature, ρ_b is the blood density, C_b is the blood-specific heat, ω_b is the blood perfusion rate, q is the heat generation by the external heat source which is responsible for the changing of temperature inside the body according to the Equations (9.21), (9.2), and (9.3), and Q_{met} is the metabolic heat generation rate. Metabolic heat production is caused by the chemical factors in the body. The specific dynamic action of food is often mentioned, especially protein, that results in a rise of metabolism; and a high environmental temperature that, by raising temperatures of the tissues, increases the velocity of reactions and thus increases heat production.

In our problem, we only consider steady state analysis. Therefore, the transient part of the bio-heat equation will be neglected. For the rest of the bio-heat equation, the functional will be

$$\begin{aligned}\mathcal{L}(T) &= \iint\left(\left[\frac{1}{2}\sigma_t\left(\nabla T\right)^2 - qT - Q_{met}T - \rho_b C_b \omega_b T\left(T_b - \frac{T}{2}\right)\right]\right)dR \\ &= \sum_{\text{Elements}}\left\{\frac{1}{2}\sigma_t\left(\mathrm{b}^2+\mathrm{c}^2\right)A^r - \iint qTdR - \iint Q_{met}TdR\right. \\ &\qquad \left. - \iint \rho_b C_b \omega_b T\left(T_b - \frac{T}{2}\right)dR\right\}\end{aligned} \tag{9.23}$$

In Equation (9.23), the first two parts we have already calculated in the finite element computation for the electrothermal problem in Section 9.1. Therefore, in this section, only heat generation due to the metabolic effects and heat removal because of the blood circulation will be addressed.

Let us consider heat generation due to the metabolic effects,

$$-\iint Q_{met} T dR = -\iint Q_{met} \begin{bmatrix} T_1 & T_2 & T_3 \end{bmatrix} \begin{bmatrix} \zeta_1 \\ \zeta_2 \\ \zeta_3 \end{bmatrix} dR$$
$$= -\underline{T}^t Q_{met} \frac{A^r}{3} \begin{bmatrix} 1 \\ 1 \\ 1 \end{bmatrix} \tag{9.24}$$

Let us now look further at heat removal because of blood circulation,

$$-\iint \rho_b C_b \omega_b T \left(T_b - \frac{T}{2} \right) dR = -\iint \rho_b C_b \omega_b T\, T_b dR + \iint \frac{1}{2} \rho_b C_b \omega_b T^2 dR$$

$$-\iint \rho_b C_b \omega_b T \left(T_b - \frac{T}{2} \right) dR$$
$$= -\iint \rho_b C_b \omega_b T_b \begin{bmatrix} T_1 & T_2 & T_3 \end{bmatrix} \begin{bmatrix} \zeta_1 \\ \zeta_2 \\ \zeta_3 \end{bmatrix} dR \tag{9.25}$$
$$+ \iint \frac{1}{2} \rho_b C_b \omega_b \begin{bmatrix} T_1 & T_2 & T_3 \end{bmatrix} \begin{bmatrix} \zeta_1 \\ \zeta_2 \\ \zeta_3 \end{bmatrix} \begin{bmatrix} \zeta_1 & \zeta_2 & \zeta_3 \end{bmatrix} \begin{bmatrix} T_1 \\ T_2 \\ T_3 \end{bmatrix} dR$$

By applying

$$\iint \left(\zeta_1^{\,i} \zeta_1^{\,j} \zeta_1^{\,k} \right) dR = \frac{i!j!k!2!}{(i+j+k+2)} A^r,$$

$$-\iint \rho_b C_b \omega_b T \left(T_b - \frac{T}{2} \right) dR = -\underline{T}^t \rho_b C_b \omega_b T_b \frac{A^r}{3} \begin{bmatrix} 1 \\ 1 \\ 1 \end{bmatrix}$$
$$+ \frac{1}{2} \underline{T}^t \rho_b C_b \omega_b \frac{A^r}{12} \begin{bmatrix} 2 & 1 & 1 \\ 1 & 2 & 1 \\ 1 & 1 & 2 \end{bmatrix} \underline{T} \tag{9.26}$$

By substituting Equations (9.24) and (9.26) in Equation (9.23),

$$L(T)=\sum_{\text{Elements}}\left\{\frac{1}{2}\sigma_t\left(b^2+c^2\right)A^r-\iint qTdR-\underline{T}^tQ_{\text{met}}\frac{A^r}{3}\begin{bmatrix}1\\1\\1\end{bmatrix}\right.$$
$$\left.-\underline{T}^t\rho_bC_b\omega_bT_bT_3^r\frac{A^r}{3}\begin{bmatrix}1\\1\\1\end{bmatrix}+\frac{1}{2}\underline{T}^t\rho_bC_b\omega_b\frac{A^r}{12}\begin{bmatrix}2&1&1\\1&2&1\\1&1&2\end{bmatrix}\underline{T}\right\} \tag{9.27}$$

The full equation using Equation (9.25) then is,

$$\mathcal{L}(T)=\sum_{\text{Elements}}\left\{\frac{1}{2}\underline{T}^t\sigma_tA^r\begin{bmatrix}b_1^2+c_1^2&b_1b_2+c_1c_2&b_1b_3+c_1c_3\\b_2b_1+c_2c_1&b_2^2+c_2^2&b_2b_3+c_2c_3\\b_3b_1+c_3c_1&b_3b_2+c_3c_2&b_3^2+c_3^2\end{bmatrix}\underline{T}\right.$$
$$-\frac{1}{2\sigma}\underline{T}^t\frac{A^r}{60}\begin{bmatrix}6J_1^2+2J_2^2+2J_3^2+4J_1J_2+4J_1J_3+2J_2J_3\\2J_1^2+6J_2^2+2J_3^2+4J_2J_1+4J_2J_3+2J_1J_3\\2J_1^2+2J_2^2+6J_3^2+4J_3J_1+4J_3J_2+2J_1J_2\end{bmatrix}$$
$$\left.-\underline{T}^tQ_{met}\frac{A^r}{3}\begin{bmatrix}1\\1\\1\end{bmatrix}-\underline{T}^t\rho_bC_b\omega_bT_b\frac{A^r}{3}\begin{bmatrix}1\\1\\1\end{bmatrix}+\frac{1}{2}\underline{T}^t\rho_bC_b\omega_b\frac{A^r}{12}\begin{bmatrix}2&1&1\\1&2&1\\1&1&2\end{bmatrix}\underline{T}\right\} \tag{9.28}$$

By arranging the same coefficients at one place,

$$\mathcal{L}(T)=\sum_{\text{Elements}}\left\{\frac{1}{2}\underline{T}^t\left(\sigma_tA^r\begin{bmatrix}b_1^2+c_1^2&b_1b_2+c_1c_2&b_1b_3+c_1c_3\\b_2b_1+c_2c_1&b_2^2+c_2^2&b_2b_3+c_2c_3\\b_3b_1+c_3c_1&b_3b_2+c_3c_2&b_3^2+c_3^2\end{bmatrix}\underline{T}\right.\right.$$
$$\left.+\rho_bC_b\omega_b\frac{A^r}{12}\begin{bmatrix}2&1&1\\1&2&1\\1&1&2\end{bmatrix}\right)$$
$$-\underline{T}^t\left(\frac{1}{\sigma}\frac{A^r}{120}\begin{bmatrix}6J_1^2+2J_2^2+2J_3^2+4J_1J_2+4J_1J_3+2J_2J_3\\2J_1^2+6J_2^2+2J_3^2+4J_2J_1+4J_2J_3+2J_1J_3\\2J_1^2+2J_2^2+6J_3^2+4J_3J_1+4J_3J_2+2J_1J_2\end{bmatrix}\right. \tag{9.29}$$
$$\left.\left.-Q_{met}\frac{A^r}{3}\begin{bmatrix}1\\1\\1\end{bmatrix}-\rho_bC_b\omega_bT_b\frac{A^r}{3}\begin{bmatrix}1\\1\\1\end{bmatrix}\right)\right\}$$
$$=\sum_{\text{Elements}}\left\{\frac{1}{2}\underline{T}^t[P]\underline{T}-\underline{T}^t\underline{Q}\right\}$$

where, in corresponding notation, the local matrices

$$[P] = \sigma_t A^r \begin{bmatrix} b_1^2 + c_1^2 & b_1 b_2 + c_1 c_2 & b_1 b_3 + c_1 c_3 \\ b_2 b_1 + c_2 c_1 & b_2^2 + c_2^2 & b_2 b_3 + c_2 c_3 \\ b_3 b_1 + c_3 c_1 & b_3 b_2 + c_3 c_2 & b_3^2 + c_3^2 \end{bmatrix} + \rho_b C_b \omega_b \frac{A^r}{12} \begin{bmatrix} 2 & 1 & 1 \\ 1 & 2 & 1 \\ 1 & 1 & 2 \end{bmatrix} \quad (9.30)$$

and

$$\underline{Q} = \frac{1}{\sigma} \frac{A^r}{120} \begin{bmatrix} 6J_1^2 + 2J_2^2 + 2J_3^2 + 4J_1 J_2 + 4J_1 J_3 + 2J_2 J_3 \\ 2J_1^2 + 6J_2^2 + 2J_2^2 + 4J_2 J_1 + 4J_2 J_3 + 2J_1 J_3 \\ 2J_1^2 + 2J_2^2 + 6J_3^2 + 4J_3 J_1 + 4J_3 J_2 + 2J_1 J_2 \end{bmatrix} + Q_{met} \frac{A^r}{3} \begin{bmatrix} 1 \\ 1 \\ 1 \end{bmatrix} + \rho_b C_b \omega_b T_b \frac{A^r}{3} \begin{bmatrix} 1 \\ 1 \\ 1 \end{bmatrix} \quad (9.31)$$

Finite element analysis provides the solution minimizing (9.29) by applying certain boundary conditions such as the Dirichlet and Neumann conditions (Hoole, 1989). The local matrices of elements will be added to the corresponding position of the global matrix to be solved for $\underline{T}$. This leads to the finite element matrix equation

$$\left[P^g\right]\underline{T} = \underline{Q}^g \quad (9.32)$$

Equation (9.32) is solved in the same way as explained in Section 2.9, calculating local matrices and adding them to the global matrix equation to be solved for $\underline{T}$.

9.8 A Note on Electrical and Thermal Conductivity Changes

9.8.1 Electrical Conductivity

In general, when temperature increases, electrical conductivity reduces. The electrical conductivity (σ_e) of most materials changes with temperature. If the temperature T does not vary too much, we typically use the approximation (Ward, 1971):

$$\sigma_e(T) = \frac{\sigma_{e0}}{\left(1 + \alpha(T - T_0)\right)} \tag{9.33}$$

where α is the temperature coefficient of resistivity, T_0 is the fixed reference temperature (20°C), and σ_{e0} is the electrical conductivity at T_0. Isaac Chang (2003) has done a study with temperature-dependent electrical conductivity $\sigma_e(T)$ and constant electrical conductivity σ_{e0} for the electrothermal coupled problem using finite element analysis. His results show that in temperatures below 45°C, the change in electrical conductivity is less than 10% (Chang, 2003). In hyperthermia treatment, the temperature does not exceed 43–45°C (Allen et al., 1988). Therefore, in our computation, we approximat electrical conductivity as a constant.

9.8.2 Thermal Conductivity

The effect of temperature on thermal conductivity is different for metals and nonmetals. In metals, the conductivity is primarily due to free electrons. The thermal conductivity in metals is proportional to a multiple of the absolute temperature and electrical conductivity (Jones and March, 1985):

$$\sigma_t(\mathrm{T}) \propto |\mathrm{T}|\sigma_e(\mathrm{T}) \tag{9.34}$$

In metals σ_e (T) decreases when the temperature increases. Thus, the product of Equation (9.34).$|\mathrm{T}|\sigma_t(\mathrm{T})$, stays approximately constant. σ_t (T) in nonmetals is mainly due to lattice vibrations (phonons). Except for high quality crystals at low temperatures, the phonon mean free path is not reduced significantly at higher temperatures. Thus, the thermal conductivity of nonmetals is approximately constant at low temperatures. At low temperatures well below the Debye temperature, the thermal conductivity decreases, as does the heat capacity. Therefore, the thermal conductivity in both metals and nonmetals is approximately constant at low temperatures. So, in our hyperthermia problem too we consider thermal conductivity as a constant.

9.9 The Algorithm for the Inverse Method for Electroheating

In this problem we define the shape of the conductor as a design parameter $\{p\}$. In our two-dimensional model, we have two conductors, both carrying current in opposite directions to each other. For practical implementation purposes, we have the same symmetrical design parameters for both conductors as shown in Figure 9.12. The design desideratum then may be cast as an object function F to be minimized with respect to the parameters $\{p\}$

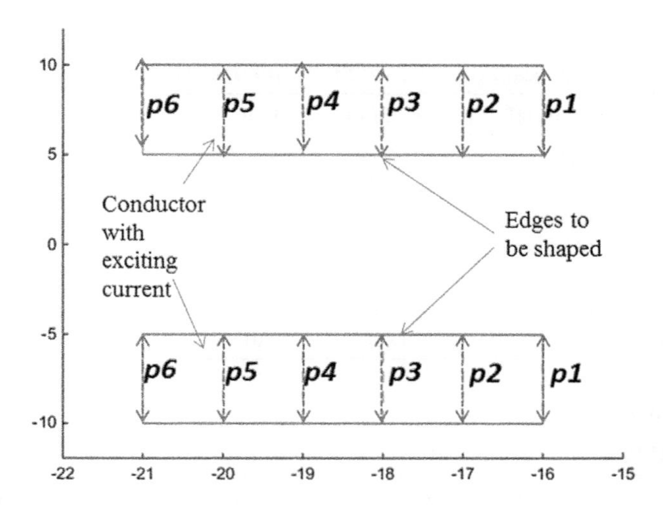

FIGURE 9.12
Design parameters.

$$F = F(\{p\}) = \frac{1}{2}\sum_{i=1}^{6}\left[T^i - T_0^i\right]^2 \tag{9.35}$$

The optimization process, using the GA, keeps adjusting $\{p\}$ until F is minimized, making T^i come as close to T_0^i as it can – as close as it can rather than exactly to T_0^i because our design goal as expressed in Equation (9.33) may not always be realistic and achievable. At that point {p} would represent our best design. In our model, since the tumor needs to be treated, the desired temperature T_0^i is set to be $T_0^i = 42.5°C$.

The computational process of this hyperthermia treatment problem is similar to the electrothermal problem which has been described in Section 9.1. But, instead of the heat equation, we have to solve the bio-heat equation, Equation (9.22).

The main purpose of this work is to shape the heating coil, so as to get the treatment temperatures at tumors without damaging the healthy cells. In analyzing with our model, the human body, tumor, and conductor are considered as homogeneous media with these typical material properties. For simplicity, we assumed a constant value for the blood perfusion rate in various biological tissues. Electrical and thermal properties are taken from Gabriel et al. (1996) and presented in Table 9.2. The physical parameters of the blood are given in Table 9.3.

The exciting current density in one conductor which is carrying current is set to 30 A/m^2 and the other conductor is set to –30 A/m^2. The exciting frequency is kept relatively low in the 60 Hz–5 kHz range to avoid a very fine mesh. The parameterized mesh for the initial shape of the numerical model is shown in Figure 9.13. The numerical model is meshed with 1625 nodes and

TABLE 9.2

Physical Properties of the Numerical Model

Material	μ_r	σ_e (S/m)	σ_t (W/(mK))	J_0 (A/m²)
Air	1	3×10^{-8}	0.0257	0
Tumor	1	0.07	0.56	0
Muscle	1	0.15	0.22	0
Skin	1	0.1	0.22	0
Bone	1	0.7	0.013	0
Coil-1	1	6×10^{7}	401	30
Coil-2	1	6×10^{7}	401	−30

TABLE 9.3

Physical Parameters of Blood

Tissue	Q_{met} (W/m³)	ρ_b (kg/m³)	C_b (J/kg/k)	T_b (K)	ω_b (1/s)
Muscle/Skin	300	1020	3640	310.15	0.0004
Tumor	480	1020	3640	310.15	0.005
bone	120	1020	3640	310.15	0.00001

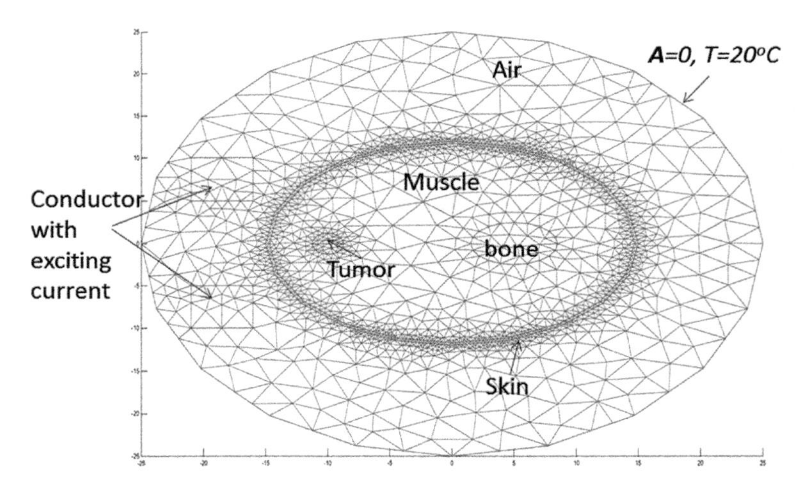

FIGURE 9.13
Mesh for the numerical model.

3195 elements. The radiation and convection effects at the boundaries are neglected, taking the boundary temperature to be the room temperature at 20°C and the magnetic vector potential $A = 0$ at boundary.

The simulation result for the numerical model is shown in Figure 9.14. This shows the equi-temperature lines for the human thigh with tumor. The temperature is measured on muscle and inside the tumor along the x axis where

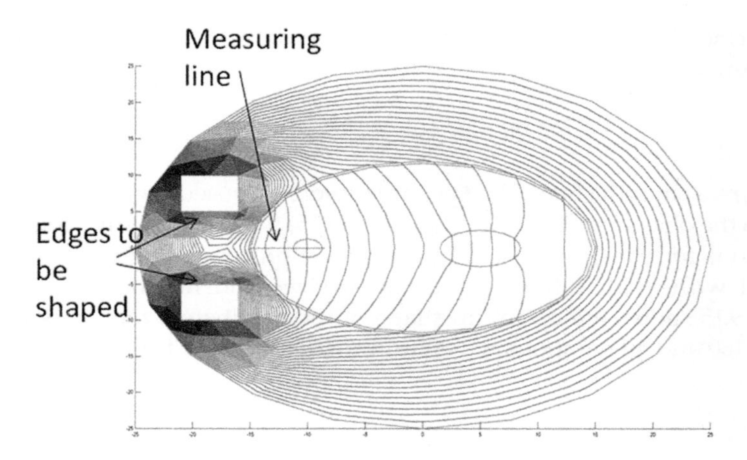

FIGURE 9.14
Equi-temperature lines for the initial shape.

the measuring line is marked (in green). The measuring line is divided into ten points where readings were taken and tabulated as in Table 9.4. For hyperthermia treatment, the temperature at a tumor should be greater than 42°C and should not be greater than 44°C. Therefore, the desired temperature is set to be 42.5°C and to achieve that the edges of the conductors are shaped as shown in Figure 9.14.

In the optimization process using the GA, multiple solutions were created and checked as to whether it is minimizing the objective function of Equation (9.35). The tolerance boundaries for each p_i were set to $1.5\text{ cm} \le p_i \le 8.5\text{ cm}$ for the conductor at the top and $-1.5\text{ cm} \le p_i \le -8.5\text{ cm}$ for the conductor at the bottom. As such we follow, Subramaniam et al. (1994) and extend

TABLE 9.4
Reading at the Measuring Line

Measuring Point	x (cm)	y (cm)	Initial Shape T (°C)	Optimum Shape T (°C)
1	−11	0	53.92	43.08
2	−10.75	0	53.37	42.99
3	−10.5	0	52.70	42.91
4	−10.25	0	51.92	42.82
5	−10	0	51.13	42.74
6	−9.75	0	50.91	42.65
7	−9.5	0	50.69	42.57
8	−9.25	0	50.48	42.49
9	−9	0	50.26	42.41
10	−8.75	0	49.92	42.33

their principle, so as to maintain a non-undulating shape by imposing the constraints

$$p_1 \leq p_2 \leq p_3 \leq p_4 \leq p_5 \leq p_6 \tag{9.36}$$

to ensure a smooth shape. The penalties were imposed by adding a penalty term to the object function F in Equation (9.35) whenever it fails to satisfy the conditions for constraints. The optimization process using GA was experimented with, with different population sizes and numbers of iterations. Figure 9.15 shows the optimum shape of the conductors for the population size of 160 and 40 iterations which minimize the object function F to a value

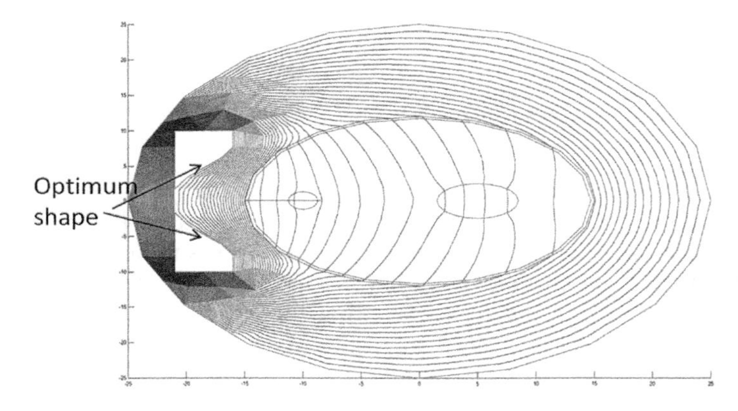

FIGURE 9.15
Optimum shape of the conductors to treat tumor.

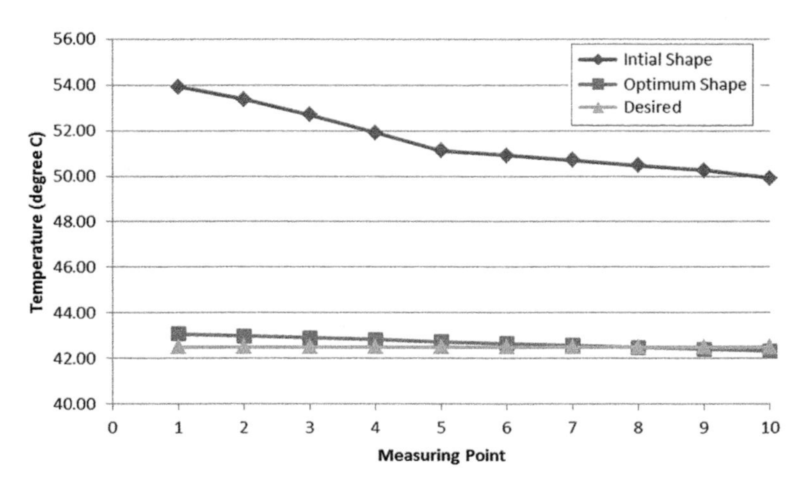

FIGURE 9.16
Initial, optimum, and desired temperatures.

of 0.97. When the conductor is at its optimum shape, the temperature at the measuring line is measured and tabulated as the last column of Table 9.4.

In conclusion, the temperature inside the tumor for the initial and optimum shapes of the numerical model is plotted with the measuring line points shown in Figure 9.16. For comparison, the desired temperature is also plotted in the same figure. The temperature for the optimum shape is in a very good range to treat the tumor as it is not beyond 44°C and not below 42°C. It is impossible to keep the temperature everywhere in a tumor the same. Our results show that we can keep it in the temperature range which would treat the tumor effectively.

References

Adeli, H., and Kumar, S., 1995, "Concurrent structural optimization on massively-parallel supercomputer," *ASCE J. Struct. Eng.*, Vol. 121, no. 11, pp. 1588–1597.

Allen, S., Kantor, G., Bassen, H., and Ruggera, P., 1988, "CDRH RF phantom for hyperthermia systems evaluations," *Int. J. Hyperthermia*, Vol. 4, no. 1, pp. 17–23, Feb.

Arora, J.S., and Hang, E.J., 1976, "Efficient optimal design of structures by generalized steepest descent programming," *Int. J. Numer. Meth. Eng.*, Vol. 10, pp. 747–766.

Atrek, E., Gallagher, R.H., Ragsdell, K.M., and Zienkiewicz, O.C., 1984, *New Directions in Optimal Structural Design*, Wiley, Chichester.

Attardo, E.A., and Borsic, A., 2012, "GPU acceleration of algebraic multigrid for low-frequency finite element methods," Antenna and Propagation Society International Symposium (APSURSI), Chicago, July, 8–14.

Battistetti, M., Di Barba, P., Dughiero, F., Farina, M., Lupi, S., and Savini, A., 2001, "Optimal design of an inductor for transverse flux heating using a combined evolutionary-simplex method," *COMPEL – Int. J. Comput. Math. Electr. Electron. Eng.*, Vol. 20, no. 2, pp. 507–522, June.

Beveridge, Gorden S.G., and Schechter, Robert S., 1970, *Optimization: Theory and Practice*, McGraw-Hill, New York.

Bogdan, O., Tudorel, A., and Mariana, D.R., 2012, "Improving the performance of the linear systems solvers using CUDA," Proceedings of Challenges of the Knowledge Society (CKS), pp. 2036–2046, Bucharest, Romania.

Brauer, H., Ziolkowski, M., Tenner, U., Haueisen, J., and Nowak, H., 2001, "Verification of extended sources reconstruction techniques using a torso phantom," *COMPEL – Int. J. Comput. Math. Electr. Electron. Eng.*, Vol. 20, no. 2, pp. 595–606, June.

Cancer Research UK, 2012, "What is Coley's toxins treatment for cancer?," 31 Aug., 2012. http://www.cancerresearchuk.org/about-cancer/cancers-in-general/cancer-questions/coleys-toxins-cancer-treatment. [online, accessed 27 June, 2015].

Carey, G.F., Barragy, E., McLay, R., and Sharma, M., 1988, "Element-by-element vector and parallel computations," *Commun. Appl. Numer. Meth.*, Vol. 4, no. 3, pp. 299–307.

Cecka, C., Lew, A.J., and Darve, E., 2011, "Assembly of finite element methods on graphics processors," *Int. J. Numer. Meth. Eng.*, Vol. 85, pp. 640–669.

CEDRAT, n.d. a, "FLUX 10 2D and 3D Applications New Features." http://www.jaewoo.com/material/magazinefolder/jwnews/0710/10_New_Features.pdf. [online , accessed 01 Aug., 2013].

CEDRAT, n.d. b, "Forge." http://forge-mage.g2elab.grenoble-inp.fr/project/got. [online, accessed 26 July, 2014].

CEDRAT, n.d. c. http://www.cedrat.com/en/software/got-it.html. [online, accessed 21 April, 2014].

Cendes, Z.J., Shenton, D. and Shahnasser, H., 1983, "Magnetic field computation using Delaunay triangulation and complementary finite element methods," *IEEE Trans. Mag.* Vol. MAG-19, no. 6, Nov.

Chang, I., 2003, "Finite element analysis of hepatic radiofrequency ablation probes using temperature-dependent electrical conductivity," *Biomed. Eng. OnLine*, Vol. 2, p. 12, May.

Chen, T., Johnson, J., and Robert, W.D., 1995, *A Vector Level Control Function for Generalized Octree Mesh Generation*, Springer, Vienna.

Chou, C.-K., 1988, "Application of electromagnetic energy in cancer treatment," *IEEE Trans. Instrum. Meas.*, Vol. 37, no. 4, pp. 547–551, Dec.

Conroy, C.J., Kubatko, E.J., and West, D.W., 2012, "ADMESH: An advanced, automatic unstructured mesh generator for shallow water models," *Ocean Dyn.*, Vol. 62, pp. 1503–1517, Nov.

Cormen, T.H., Leiserson, C.E., Rivest, R., and Stein, C., 2011, *Introduction to Algorithms*. The MIT Press.

Coulomb, J.L., Meunier, G., and Sabonnadiere, J.-C., 1982, "An original stationary method using local jacobian derivative for direct finite element computation of electromagnetic force, torque and stiffness," *J. Magn. Magn. Mater.*, Vol. 26, no. 1–3, pp. 337–339, March.

Cuthil, E., and McKee, J., 1969, "Reducing the bandwidth of sparse symmetric matrices," Proceedings of ACM National Conference, New York.

Das, S.K., Clegg, S.T., and Samulski, T.V., 1999, "Electromagnetic thermal therapy power optimization for multiple source applicators," *Int. J. Hyperthermia*, Vol. 15, no. 4, pp. 291–308, Aug.

Davis, Timothy A., 2011, *MATLAB Primer*, 8th edn., CRC Press, Boca Raton, FL.

Devon, Y., 2013, "Numerical accuracy differences in CPU and GPGPU codes," Master's Thesis, Dept. of Electrical and Computer Engineering, Northeastern Univ., Boston, MA, 2011. http://iris.lib.neu.edu/cgi/viewcontent.cgi?article=1057&context=elec_comp_theses. Downloaded 22 Nov..

Dong-Joon, S., Dong-Hyeok, C., Jang-Sung, C., Hyun-Kyo, J., and Tae-Kyoung, C., 1997, "Efficiency optimization of interior permanent magnet synchronous motor using genetic algorithms," *IEEE Trans. Magn.*, Vol. 33, no. 2, pp. 1880–1883.

Drago,G., Molfino, P., Repetto, M., Secondo, G., Lia, G.M.A., Montanari, I., and Ribani, P.L., 1992, "Design of superconducting MHD saddle magnets," *IEEE Trans. Magn.*, Vol. 28, no. 1, pp. 450–453.

Dziekonski, A., Sypek, P., Lamecki, A., and Mrozowski, M., 2012, "Finite element matrix generation on a GPU," *Prog. Electromagn. Res.*, Vol. 128, pp. 249–265.

Enomoto, H., Harada, K., Ishihara, Y., Todaka, T., and Hirata, K., 1998, "Optimal design of linear oscillatory actuator using genetic algorithm," *IEEE Trans. Magn.*, Vol. 34, no. 5, pp. 3515–3518.

Fernandez, D.M., Dehnavi, M.M., Gross, W.J., and Giannacopoulos, D., 2012, Alternate parallel processing approach for FEM. *IEEE Trans. Magn.* Vol. 48, pp. 399–402.

Fortune, S., 1987, "A sweep line algorithm for Voronoi diagrams," *Algorithmica*, Vol. 2, pp. 153–174.

Fukuda, H., and Nakatani, Y., 2012, "Recording density limitation explored by head/media co-optimization using genetic algorithm and GPU-accelerated LLG," *IEEE Trans. Magn.*, Vol. 48, no. 11, pp. 3895–3898, November.

Gabriel, S., Lau, R.W., and Gabriel, C., 1996, "The dielectric properties of biological tissues: II. Measurements in the frequency range 10 Hz to 20 GHz," *Phys. Med. Biol.*, Vol. 41, no. 11, p. 2251, Nov.

George, A., 1973, "Nested dissection of regular finite element mesh," *SIAM J. Numer. Anal.*, Vol. 10, pp. 345–363.

Geuzaine, C., and Remacle, J., 2009, "Gmsh: A 3-D finite element mesh generator with built- in pre and post-processing facilities," *Int. J. Numer. Meth. Eng.*, Vol. 79, no. 11, pp. 1–24.

Gitosusastro, S., Coulomb, J.-L., and Sobonnadiere, J.-C., 1989, "Performance derivative calculations and optimization process," *IEEE Trans. Magn.*, Vol. 25, no. 4, pp. 2834–2837.

Godel, N., Nunn, N., Warbutron, T., and Clemens, M., 2010, "Scalability of higher-order discontinuous galerkin FEM computations for solving electromagnetic wave propagation problems on GPU clusters," *IEEE Trans. Magn.*, Vol. 46, no. 8, pp. 3469–3472.

Gordon, R.G., Roemer, R.B., and Horvath, S.M., 1976, "A mathematical model of the human temperature regulatory system--transient cold exposure response," *IEEE Trans. Biomed. Eng.*, Vol. 23, no. 6, pp. 434–444, Nov.

Haupt, R.L., 1995, "Comparison between genetic and gradient-based optimization algorithms for solving electromagnetics problems," *IEEE Trans. Magn.*, Vol. 31, no. 3, pp. 1932–1935, May.

Haupt, R.L., and Haupt, S.E., 2004, *Practical Genetic Algorithms*, John Wiley & Sons, Hoboken, NJ.

Henderson, J.L., 1994, "Laminated plate design using genetic algorithms and parallel processing," *Comput. Syst. Eng.*, Vol. 5, no. 4–6, pp. 441–453, Aug.

Hestenes, Magnus R., and Stiefel, Eduard, 1952, "Methods of conjugate gradients for solving linear systems," *J. Res. Natl. Bur. Stand.*, Vol. 49, no. 6, pp. 409–436, Dec.

Hoole, S.R.H., 1988, *Computer-Aided Analysis and Design of Electromagnetic Devices*, Elsevier, New York.

Hoole, S.R.H., 1989a, *Computer Aided Analysis and Design of Electromagnetic Devices*, Elsevier, New York. Now acquired by Prentice Hall.

Hoole, S.R.H., 1989b, "Computational electromagnetism," in Chapter 8, pp. 179–216, in R.L. Coren (Ed.), *Applied Electromagnetism*, Prentice Hall, Englewood Cliffs, NJ.

Hoole, S.R.H., 1990, "Finite element electromagnetic field computation on the sequent symmetry 81 parallel computer," *IEEE Trans. Magn.*, Vol. MAG-26, no. 2, pp. 837–840, March.

Hoole, S.R.H., 1991a, "Optimal design, inverse problems and parallel computers," *IEEE Trans. Magn.*, Vol. 27, pp. 4146–4149, Sept.

Hoole, S.R.H., 1991b, "Review of parallelization in electromagnetic device simulation and possibilities for extensions," *Int. J. Appl. Electromagn. Mater.*, Vol. 2, no. 1, pp. 99–108.

Hoole, S.R.H., ed., 1995, *Finite Elements, Electromagnetics and Design*, Elsevier, Amsterdam, May.

Hoole, S.R.H., and Agarwal, K., 1997, "Parallelization of optimization methods," *IEEE Trans. Magn.*, Vol. 33, no. 2, pp. 1966–1969.

Hoole, S.R.H., Anandaraj, A., and Bhatt, J.J., 1989, "Matrix solvers in repeated finite element solution – subjective and objective considerations," Proceedings, Progress in Electromagnetics Research Symposium, p. 316, MIT, Massachusetts, July 25–26.

Hoole, S.R.H., and Carpenter, C.J., 1985, "Surface impedance models for corners and slots", *IEEE Trans. Magn.*, Vol. MAG-21, no. 5, pp. 1841–1843, Sept.

Hoole, S.R.H., Cendes, Z.J., and Hoole, P.R.P., 1986, "Preconditioning and renumbering in the conjugate gradients algorithm", in Z.J. Cendes (Ed.), *Computational Electromagnetics*, North Holland, pp. 91–99, July.

Hoole, S.R.H., and Hoole, P.R.P., 1985, "On finite element force computation from two- and three-dimensional magnetostatic fields," *J. Appl. Phys.*, Vol. 57, no. 8, pp. 3850–3852.

Hoole, S.R.H., Karthik, V.U., Sivasuthan, S., Rahunanthan, A., Thyagarajan, R.S., and Jayakumar, P., 2014, "Finite elements, design optimization, and nondestructive evaluation: a review in magnetics, and future directions in GPU-based, element-by-element coupled optimization and NDE," *Int. J. Appl. Electromagn. Mater.* doi: 10.3233/JAE-140061, Prepress online date Aug. 28.

Hoole, S.R.H., Sathiaseelan, V., and Tseng, A., 1990, "Computation of hyperthermia-SAR distributions in 3-D," *IEEE Trans. Magn.*, Vol. 26, no. 2, pp. 1011–1014, March.

Hoole, S.R.H., Sivasuthan, S., Karthik, V.U., Rahunanthan, A., Thyagarajan, R.S., and Jayakumar, P., 2014, "Electromagnetic device optimization: the forking of already parallelized threads on graphics processing units," *Appl. Comput. Electromagn. Soc. J.*, Vol. 29, no. 9, pp. 677–684.

Hoole, S.R.H., Subramaniam, S., Saldanha, R., Coulomb, J.-L., and Sabonnadiere, J.-C., 1991, "Inverse problem methodology and finite elements in the identification of inaccessible locations, sources, geometry and materials," *IEEE Trans. Magn.*, Vol. 27, no. 3, pp. 3433–3443, May.

Hoole, S.R.H. Udawalpola, Rajitha, and Wijesinghe, K.R.C, 2007,. "Development of a benchmark problem and a general optimisation package with Powell's method to develop the benchmark," *J. Mater. Process. Technol.*, Vol. 181, no. 1–3, pp. 136–141, 1 Jan.

Hoole, S.R.H., Weeber, K., and Subramaniam, S., 1991, "Fictitious minima of object functions, finite element meshes, and edge elements in electromagnetic device synthesis," *IEEE Trans. Magn.*, Vol. 27, no. 6, pp. 5214–5216.

Hsu, Yeh-Liang, 1994, "A review of structural optimization," *Comput. Ind.*, Vol. 25, no. 1, pp. 3–13, Nov.

Irons, B.M., 1970, "A frontal solution program for finite element analysis," *Int. J. Numer. Meth. Eng.*, Vol. 2, pp. 5–32.

Jennings, A., 1977, *Matrix Computation for Engineers and Scientists*, John Wiley, London.

Joe, B., 1995, "Construction of three-dimensional improved-quality triangulations using local transformations," *SIAM J. Sci. Comput.*, Vol. 16, pp. 1292–1307, Nov.

Jones, W., and March, N.H., 1985, *Theoretical Solid State Physics*, Courier Corporation, Chelmsford, MA.

Juno, K., Hong-Bae, L., Hyun-Kyo, J., Changyul, C., and Hyeongseok, K., 1996, "Optimal design technique for waveguide device," *IEEE Trans. Magn.*, Vol. 32, no. 3, pp. 1250–1253

Kakay, A., Westphal, E., and Hertel, R., 2010, "Speedup of FEM micromagnetic simulations with graphical processing units," *IEEE Trans. Magn.*, Vol. 46, no. 6, pp. 2303–2306.

Karthik, V.U., Sivasuthan, S., Rahunanthan, A., Thiyagarajan, R.S., Jeyakumar, P., Udpa, L., and Hoole, S.R.H., 2015, "Faster, more accurate, parallelized inversion for shape optimization in electroheat problems on a graphics processing unit (GPU) with the real-coded genetic algorithm," *COMPEL - Int. J. Comput. Math. Electr. Electron. Eng.*, Vol. 34, no. 1, pp. 344–356, Jan.

Keran, S., and Lavers, J.D., 1986, "A skin-depth independent finite element method for iterative solution of systems of linear equations," *J. Comput. Phys.*, Vol. 26, pp. 43–65.

Kiss, I., Gyimothy, S., Badics, Z., and Pavo, J., 2012, "Parallel realization of element-by-element FEM technique by CUDA," *IEEE Trans. Magn.*, Vol. 48, pp. 507–510.

Kershaw, D.S., 1978, "The incomplete cholesky conjugate gradients method for the iterative solution of systems of linear equations," *J. Comp. Phys.*, Vol. 26, pp. 43–65.

Krawezik, G.P., and Poole, G., 2009, "Accelerating the ANSYS direct sparse solver with GPUs," Symposium on Application Accelerators in High Performance Computing. http://saahpc.ncsa.illinois.edu/09/papers/Krawezik_paper.pdf. Downloaded 1 Nov., 2013.

Krishnakumar, S., and Hoole, S.R.H., 2005, "A common algorithm for various parametric geometric changes in finite element design sensitivity computation, *J. Mater. Process. Technol.*, Vol. 161, pp. 368–373.

Kroeze, H., van de Kamer, J.B., de Leeuw, A.A.C., Kikuchi, M., and Lagendijk, J.J.W., 2003, "Treatment planning for capacitive regional hyperthermia," *Int. J. Hyperthermia*, Vol. 19, no. 1, pp. 58–73, Feb.

Kurgan, P.G.E., 2011, "Treatment of tumors located in the human thigh using RF hyperthermia," *Prz. Elektrotech.*, Vol. 87, no. 12b, pp. 103–106.

Lawson, C.L., 1977, "Software for C1 surface interpolation," in *Mathematical Software III*. Academic Press.

Lee, D.T., and Schachter, B.J., 1980, "Two algorithms for constructing a Delaunay triangulation," *Int. J. Comput. Inf. Sci.*, Vol. 9, pp. 219–242.

Ma, X.W., Zhao, G.Q., and Sun, L., 2011, "AUTOMESH2D/3D: robust automatic mesh generator for metal forming simulation," *Mater. Res. Innov.*, Vol. 15, pp. s482–s486, Feb.

Magele, C., and Biro, O., 1990, "FEM and evolution strategies in the optimal design of electromagnetic devices," *IEEE Trans. Magn.*, Vol. 26, no. 5, pp. 2181–2183, Sept.

Mahinthakumar, G., and Hoole, S.R.H., 1990, "A parallel conjugate gradients algorithm for finite element analysis of electromagnetic fields", *J. Appl. Phys.*, Vol. 67, pp. 5818–5820.

Marchesi, Michele L., Molinari, Giorgio and Repetto, Maurizio, 1994, "A parallel simulated annealing algorithm for the design of magnetic structures," *IEEE Trans. Magn.*, Vol. 30, no. 5, pp. 3439–3442, Sept.

Marinova, I., and Mateev, V., 2012, "Inverse source problem for thermal fields," *COMPEL – Int. J. Comput. Math. Electr. Electron. Eng.*, Vol. 31, no. 3, pp. 996–1006, May.

Marrocco, A., and Pironneau, O., 1978, "Optimum design with lagrangian finite elements: design of an electromagnet," *Comput. Meth. Appl. Mech. Eng.*, Vol. 15, pp. 277–308.

Mavriplis, D.J., 1995, "An advancing front Delaunay triangulation algorithm designed for robustness," *J. Comput. Phys.*, Vol. 117, pp. 90–101, March.

Meijerink, J.A., and van der Vost, H.A., 1977, "An iterative solution method for linear systems of which the coefficient matrix is a symmetric M-matrx,". *Math. Comput.*, Vol. 31, no. 137, pp. 148–16

Merge Sort, n.d., http://webdocs.cs.ualberta.ca/~holte/T26/merge-sort.html. [Accessed 11 Feb., 2015].

Miriam, L., and Ramachandran, J., Thomas, W., and Devon, Y., 2012, "Open CL floating point software on heterogeneous architectures – portable or not?," Workshop on Numerical Software Verification (NSV).

Mohammad, K., Moslem, M., and Ali, R., 2013, Optimization of loss in orthogonal bend waveguide: genetic Algorithm simulation, *Alexandria Eng. J.*, Vol. 52, no. 3, pp. 525–530.

Muller, J., 1996, "On triangles and flow," Special Section on Software Agents in Electronics.

Nabaei, V., Mousavi, S.A., Miralikhani, K., and Mohseni, H., 2010, "Balancing current distribution in parallel windings of furnace transformers using the genetic algorithm," *IEEE Trans. Magn.*, Vol. 46, no. 2, pp. 626–629, Feb.

Nakata, T., Takahashi, N., Fujiwara, K., and Kawashima, T., 1991,"Optimal design of injector mold for plastic bonded magnet," *IEEE Trans. Magn.*, Vol. 27, no. 6, pp. 4992–4995.

Nazareth, J.L., 2009, "Conjugate gradient method", *Wiley Interdiscip. Rev. Comput. Stat.* Vol. 1, no. 3.

Niu, S.Y., Zhao, Y.S., Ho, S.L., and Fu, W.N., 2011, "A parameterized mesh technique for finite element magnetic field computation and its application to optimal designs of electromagnetic devices," *IEEE Trans. Magn.*, Vol. 47, pp. 2943–2946.

NVIDIA, n.d., "Parallel Programming and Computing Platform | CUDA | NVIDIA | NVIDIA." http://www.nvidia.com/object/cuda_home_new.html. [online, accessed 23 Feb., 2015].

Okimura, T., Sasayama, T., Takahashi, N., and Soichiro, S.I., 2013, "Parallelization of finite element analysis of nonlinear magnetic fields using GPU," *IEEE Trans. Magn.*, Vol. 49, no. 5, pp. 1557–1560.

Painless, n.d., http://www.cs.cmu.edu/~quake-papers/painless-conjugate-gradient.pdf

Pennes, H.H., 1948, "Analysis of tissue and arterial blood temperatures in the resting human forearm," *J. Appl. Physiol.*, Vol. 1, no. 2, pp. 93–122.

Pham, T.H., and Hoole, S.R.H., "Unconstrained optimization of coupled magneto-thermal problems," *IEEE Trans. Magn.*, Vol. 31, no. 3, pp. 1988–1991, May.

Pironneau, Oliver, 1984, *Optimal Design for Elliptic Systems*, Springer-Verlag, New York.

Preis, K., Magele, C, and Biro, O., 1990, "FEM and evolution strategies in the optimal design of electromagnetic devices," *IEEE Trans. Magn.*, Vol. 26, no. 5, pp. 2181–2183, Sept.

Richter, C., Schops, S., Clemens, M., 2014, "GPU acceleration of algebraic multigrid preconditioners for discrete elliptic field problems," *IEEE Trans. Magn.*, Vol. 50, no. 2, pp. 461, 464, Feb.

Robilliard, D., Marion-Poty, V., and Fonlupt, C., 2009, "Genetic programming on graphics processing units," *Genet. Program. Evolvable Mach.*, Vol. 10, no. 4, pp. 447–471, Dec.

Roemer, R.B., 1991, "Optimal power deposition in hyperthermia. I. The treatment goal: the ideal temperature distribution: the role of large blood vessels," *Int. J. Hyperthermia*, Vol. 7, no. 2, pp. 317–341, Jan.

Rothlauf, Franz, 2006, *Representations for Genetic and Evolutionary Algorithms*, Springer, Berlin.

Rupert, J., 1995, "A Delaunay refinement algorithm for quality 2-dimensional mesh generation," *J. Algorithms*, Vol. 18, pp. 548–585, May.

Russenschuck, S., 1990, "Mathematical optimization techniques for the design of permanent magnet synchronous machines based on numerical field calculation," *IEEE Trans. Magn.*, Vol. 26, no. 2, pp. 638–641.

Saludjian, L., Coulomb, J.-L., and Izabelle, A., 1998, "Genetic algorithm and Taylor development of the finite element solution for shape optimization of electromagnetic devices," *IEEE Trans. Magn.*, Vol. 34, no. 5, pp. 2841–2844, Sept.

Sathiaseelan, V., Mittal, B.B., Fenn, J., and Taflove, A., 1998, "Recent advances in external electromagnetic hyperthermia.," *Cancer Treat. Res.*, Vol. 93, pp. 213–245.

Schafer-Jotter, M., and Muller, W., 1990, "Optimization of electromechanical devices using a numerical laboratory," *IEEE Trans. Magn.*, Vol. 26, no. 2, pp. 815–818.

Schneiders, R., 2000, "Algorithms for quadrilateral and hexahedral mesh generation," in Proceedings of the VKI Lecture series on Computational Fluid Dynamic, pp. 2000–2004.

Schoberl, J., 1997, "NETGEN an advancing front 2D/3D-mesh generator based on abstract rules," *Comput. Visual. Sci.*, Vol. 1, pp. 41–52, July.

Shewchuk, J.R., 1996a, "Applied Computational Geometry towards Geometric Engineering," in Proceedings, First ACM workshop on Applied Computational Geometry.

Shewchuk, J.R., 1996b, *Triangle: Engineering a 2D Quality Mesh Generator and Delaunay Triangulator*, Vol. 1148, pp. 203–222. Springer-Verlag, Berlin, Germany.

Shewchuk, J.R., 2002, "Delaunay refinement algorithms for triangular mesh generation," *Comput. Geom.*, Vol. 22, pp. 21–74, May.

Si, H., 2006, "TetGen, a quality tetrahedral mesh generator and three-dimensional Delaunay triangulator, 2007." http://tetgen.berlios.de.

Si, H., 2015, "TetGen, a Delaunay-based quality tetrahedral mesh generator," *ACM Trans. Math. Softw.*, Vol. 41, pp. 1–36, Feb. Accessed 10 July 2015.

Siauve, N., Nicolas, L., Vollaire, C., Nicolas, A., and Vasconcelos, J.A., 2004, "Optimization of 3-D SAR distribution in local RF hyperthermia," *IEEE Trans. Magn.*, Vol. 40, no. 2, pp. 1264–1267, March.

Silvester, P., 1969, "High order triangular finite elements for potential problems," *Int. J. Eng. Sci.*, Vol. 7, pp. 849–861.

Silvester, P.P., 1970, "Symmetric quadrature formulae for simplexes," *Math. Comp.*, Vol. 24, pp. 95–100.

Silvester, P., 1978, Construction of triangular finite element universal matrices," *Int. J. Numer. Meth. Eng.*, Vol. 18, no. 12, pp. 237–244.

Silvester, P., 1982, "Permutation rules for simplex finite elements," *Int. J. Numer. Meth. Eng.*, Vol. 18, no. 7, pp. 1245–1259.

Silvester, P.P., and Ferrari, R.L., 1983, *Finite Elements for Electrical Engineers*, Cambridge University Press, Cambridge, MA.

Silvester, P.P., and Ferrari, R.L., 1996, *Finite Elements for Electrical Engineers (3rd edition)*, Cambridge University Press, Cambridge, UK.

Silvester, P., Cabayan, H.S., and Browne, B.T., 1973, "Efficient techniques for finite element analysis of electric machine," *IEEE Trans. Power App. Syst.*, Vol. PAS-92, no. 4, pp. 1274–1281, July/Aug.

Simkin, J., and Trowbridge, C.W., 1991, "Optimization problems in electromagnetics," *IEEE Trans. Magn.*, Vol. 27, no. 5, pp. 4016–4019, Sept.

Sivasuthan, S., Karthik, V.U., Rahunanthan, A., Jayakumar, P., Thyagarajan, R., Udpa, L., and Hoole, S.R.H., 2015, "A script-based, parameterized finite element mesh for design and NDE on a GPU," *IETE Tech. Rev.*, Vol. 32, pp. 94–103, March.

Sivanandan, S.N., and Deepa, S.N., 2008, *Introduction to Genetic Algorithms*, Springer, Berlin.

Starzynski, Jack, and Wincenciak, Stanislaw, 1998, "Benchmark problems for optimal shape design for 2D eddy currents," *COMPEL - Int. J. Comput. Math. Electr. Electron. Eng.*, Vol. 17, no. 4, pp. 448–459, Aug.

Stea, B., Rossman, K., Kittelson, J., Shetter, A. Hamilton, A., and Cassady, J.R., 1994, "Interstitial irradiation versus interstitial thermoradiotherapy for supratentorial malignant gliomas: a comparative survival analysis," *Int. J. Radiat. Oncol. Biol. Phys.*, Vol. 30, no. 3, pp. 591–600, Oct.

Stewart, G.W., 1973, *Introduction to Matrix Computations*, Academic Press, New York.

Su, P., and Drysdale, Scot, 1997, "A comparison of sequential Delaunay triangulation algorithms," *Comput. Geom.*, Vol. 7, pp. 361–385, April.

Subramaniam, S., Arkadan, A., and Hoole, S.R.H., 1994, "Constraints for smooth geometric contours from optimization," *IEEE Trans. Magn.*, Vol. 37, Sept.

Sullivan, D., 1990. "Three-dimensional computer simulation in deep regional hyperthermia using the finite-difference time-domain method," *IEEE Trans. Microw. Theory Tech.*, Vol. 38, no. 2, pp. 204–211, Feb.

Talischi, C., Paulino, G.H., Pereira, A., and Menezes, I.F.M., 2012, "PolyTop: a Matlab implementation of a general topology optimization framework using unstructured polygonal finite element meshes," *Struct. Multidiscip. Optim.*, Vol. 45, pp. 329–357, Jan.

The CGAL Project, 2013, "*CGAL User and Reference Manual*," CGAL Editorial Board 4.2.

Tsuda, N., Kuroda, K., and Suzuki, Y., 1996, "An inverse method to optimize heating conditions in RF-capacitive hyperthermia," *IEEE Trans. Biomed. Eng.*, Vol. 43, no. 10, pp. 1029–1037, Oct.

Uler, G.F., Mohammed, O.A., and Chang-Seop, K., 1994, "Utilizing genetic algorithms for the optimal design of electromagnetic devices," *IEEE Trans. Magn.*, Vol. 30, no. 6, pp. 4296–4298.

Vaidya, A., Yu, S.H., St. Ville, J., Nguyen, D.T., and Rajan, S.D., 2006, "Multiphysics CAD-based design optimization," *Mech. Based Des. Struct. Mach.*, Vol. 34, pp. 157–180, July.

van Laarhoven, P.J.M., and Aarts, E.H., 1987, *Simulated Annealing: Theory and Applications*, Kluwer Academic Publishers, Dordecht/London.

Vanderplaats, Garret N., 1984, *Numerical Optimization Techniques for Engineering Design*, McGraw-Hill, New York.

Venkataraman, P., 2009, *Applied Optimization with MATLAB Programming*, 2nd edn., John Wiley, Hoboken, NJ.

Vishnukanthan, K., and Markus, K., 2012, "Parallel finite element mesh generator using multiple GPUs," in 14th International Conference on Computing in Civil and Building Engineering, pp. 1–8, Publishing House ASV.

Visser, W., 1968, "A finite element method for the determination of non-stationary temperature distribution and thermal deformations," in Proceedings of the Conference on Matrix Methods in Structural Mechanics, Air Force Institute of Technology, Wright Patterson Air Bace ASEE, pp. 925–943.

Ward, M.R., 1971, *Electrical Engineering Science*, McGraw-Hill.

Weeber, K., and Hoole, S.R.H., 1988, "Comment on 'A Skin Depth-Independent Finite Element Method for Eddy Current Problems'", *IEEE Trans. on Magn.* Vol. MAG-24, NO. 6, p. 3261, Nov. 1988.

Weeber, K., and Hoole, S.R.H., 1992a, "A structural mapping technique for geometric parametrization in the synthesis of magnetic devices," *Int. J. Numer. Meth. Eng.* Vol. 33, pp. 2145–2179, July 15.

Weeber, K., and Hoole, S.R.H., 1992b, "The subregion method in magnetic field analysis and design optimization," *IEEE Trans. Magn.*, Vol. 28, no. 2, pp. 1561–1564, March.

Weeber, K., Vidyasagar, S., and Hoole, S.R.H., 1988, Linear-exponential functions for eddy current analysis, *J. Appl. Physics*, Vol. 63, no. 8, pp. 3010–3012.

Whitehead, A., and Fit-Florea, A., 2011, Precision and Performance: Floating Point and IEEE 754 Compliance for NVIDIA GPUs, nVidia technical white paper, 2011. https://developer.nvidia.com/sites/default/files/akamai/cuda/files/NVIDIA-CUDA-Floating-Point.pdf

Wong, M.L., and Wong, T.T., 2009, "Implementation of parallel genetic algorithms on graphics processing units," in M. Gen, D. Green, O. Katai, B. McKay, A. Namatame, R.A. Sarker, and B.-T. Zhang (Eds.), *Intelligent and Evolutionary Systems*, Springer, Berlin, Heidelberg, pp. 197–216.

Wu, W., and Heng, P.A., 2004, "A hybrid condensed finite element model with GPU acceleration for interactive 3D soft tissue cutting," *Comput. Anim. Virtual Worlds*, Vol. 15, no. 3–4, pp. 219–227.

Xin, J., Lei, N.,Udpa, L., and Udpa, S.S., 2011, "Nondestructive inspection using rotating magnetic field Eddy-current probe," *IEEE Trans. Magn.*, Vol. 47, pp. 1070–1073, May.

Zienkiewicz, O.C., 1977, *The Finite Element Method in Engineering Science*, 2nd ed., McGraw-Hill, London.

Zienkiewicz, O.C., and Cheung, Y.K., 1965, "Finite elements in the solution of field problems," *The Engineer*, pp. 507–510, Sept.

Zienkiewicz, O.C., and Lohner, R., 1985, "Accelerated 'relaxation' or direct solution? Future prospects for FEM," *Int. J. Numer. Meth. Eng.*, Vol. 21, pp. 1–11.

Index